P9-DHA-197

PATHS OF FIRE

PATHS OF FIRE

AN ANTHROPOLOGIST'S INQUIRY INTO WESTERN TECHNOLOGY

Robert McC. Adams

PRINCETON UNIVERSITY PRESS   PRINCETON, NEW JERSEY

Copyright © 1996 by Princeton University Press
Published by Princeton University Press, 41 William Street,
Princeton, New Jersey 08540
In the United Kingdom: Princeton University Press,
Chichester, West Sussex
All Rights Reserved

Library of Congress Cataloging-in-Publication Data

Adams, Robert McCormick, 1926–
Paths of fire : an anthropologist's inquiry into
Western technology / Robert McC. Adams.
p.   cm.
Includes bibliographical references and index.
ISBN 0-691-02634-3 (cloth : alk. paper)
1. Technology and civilization.
2. Civilization, Western—History.   I. Title.
CB478.A34   1996
303.48′3—dc20   96-12733   CIP

This book has been composed in Times Roman

Princeton University Press books are printed
on acid-free paper and meet the guidelines
for permanence and durability of the Committee
on Production Guidelines for Book Longevity
of the Council on Library Resources

Printed in the United States of America

10   9   8   7   6   5   4   3   2   1

To Ruth

Contents

Tables

Preface

IT IS A COMMONPLACE that changes linked to the advance of technology are a driving force of our times. With the close symbiosis that has developed between technology and the accelerating tide of scientific discovery, breakthroughs in the basic understanding of nature simultaneously feed upon new technological capabilities and are quickly translated into them. Novel industrial processes and products explode the field of perceived needs and preferences. Markets increasingly become global in their reach as much of the world's population finds itself caught up in overlapping webs of electronic communication. New tools of data storage and analysis provide a basis for conducting research and pursuing comparative advantage in ever-intensifying international competition. Simultaneously opening new avenues of personal expression and recreation, they re-shape the satisfactions as well as frustrations of daily life. Burdened though we are with chronic educational deficiencies, ours has ineluctably been driven toward becoming a knowledge-based society under these powerful, convergent influences.

Great as are some of the apparent incentives, our readiness to exploit the rewards of advancing technologies exacts its costs. More often than not, unforeseen areas of improved efficacy or widened application are later found that heighten these prospects further and take them in entirely new directions. But at what risk, and to whom? Advancing technology has also meant that risks may be devastating but long-delayed, beyond our capacity to detect at the outset. What *is* an acceptable role for government monitoring and regulation, which necessarily constrains, and sometimes can distort, the range of techno-economic choices?

Difference in the timing of risks and rewards is often a crucial consideration. As a society, we steeply discount the future. Thus, longer-term consequences receive little attention until they strike us in full force. Pollution and the destruction of environmental resources and amenities provide preeminent examples. Still, it must be noted that technological advances themselves, too easily identified as the active forces of change, are in reality only the inanimate means employed toward this end at some human agency's discretion. Responsibility lies instead with the readiness of human institutions to submit to forces of short-term expediency that heavily diminish any accounting for outcomes. Moreover, when the bleak future arrives, what other means are there than technological ones with which to reverse the damage?

In the short run, government intervention in the management of technological change may appear to be conducive to, even necessary for, national
success in a fiercely competitive world. But it, too, carries familiar risks of
policy inflexibility and economic inefficiency. A need to strike an appropriate balance between such risks and opportunities is implicit in public
discussion of a national "technology policy" and "industrial policy." However, these narrowly polarized alternatives only beg deeper questions: To
what extent is the generating of truly significant new technologies an autonomous, specialized, creative process, not readily subject to policy
choices of any kind? And ethically, what *should* be done, and how should
the social consensus be sought to shape advancing technology in the service of the unfolding requirements of the Good Society of the future?

Compounding such questions and thereby introducing an additional
layer of risks and opportunities is the fact that large areas of international
competition now are no longer subject to the primary constraints of national frontiers. Like corporate ownership itself, corporate strategies rest on
extensive networks of marketing and licensing agreements. The international movement of industrial facilities is thus in the main responsive to
international marketing and labor-cost considerations. In search of lowered
costs or higher productivity, locational mobility has become a key to corporate survival, but with potentially devastating local effects.

All in all, ongoing technological change presents us with a highly complex, contradictory set of challenges. Systemically linked in ways that are
often counterintuitive, these challenges include irregular, nonlinear paths
of advance that defy prediction. Enveloping and invading our lives at every
level, they call for choices in which short-run and long-run considerations
are forcibly blurred by attendant uncertainties. The field of relevant variables widens uncontrollably, to embrace much of society and culture as
well as the economy. As Leo Marx, a historian deeply conversant with the
subject, provocatively asks, "Why not start with the intuitively compelling
idea that technology may be *the* truly distinctive feature of modernity?"[1]

This book is certainly not alone in asking how all this has come about and
how we, as a society, should think about it. There is a long-established
tradition of inquiry into the history of the sciences that touches on it, and a
vigorous, young discipline directly devoted to the history of technology
itself. Economic and social historians have made major contributions from
other directions. Further perspectives are provided by corporate strategists
and makers and critics of federal policies for technology.

Although I have tried to draw judiciously upon all these sources, this
book primarily embodies the reflections of an anthropologist, an apparent interloper from a discipline prevailingly uninvolved with issues of
modern technology. However broad the hunting license of a field of inquiry

that proclaims nothing human is alien to it, how can an anthropologist's insights add to work that is already well advanced along other research frontiers?

Consistent with a general shift of anthropological attention toward aspects of contemporary civilization, I suggest that our insights may be useful in three principal respects. First is a relative openness to complexities of interaction that are not readily reducible to simple regularities or precisely measurable quantities. At the possible expense of methodological parsimony, this account will seek to avoid drawing any line around the set of social features that are viewed as systemically linked to technology and relevant to understanding its many paths of growth and change.

Second, most anthropologists tend to maintain a consistently integrative, contextual emphasis. In order to understand specific paths of change and continuity, the idea of context directs us toward an inductive, broadly eclectic approach that takes few hierarchies of structure or importance for granted. Causal relationships, where the evidence allows them to be identified at all, are rarely viewed as fully determinative and more often as merely predisposing, probabilistic, or facilitative.

And third, common to most anthropologists is a contrarian readiness to search out diverse, improbable kinds of patterning, to be skeptical of commonly accepted categories or boundaries, and to employ varying temporal and geographic scales as tools of inquiry. That accounts especially for the very long-time perspective that I have taken—one that begins with Near Eastern antiquity, at the point of origin of the urban institutions on which modern technological society is based, and extends to the present day.

Much of my earlier work was devoted to the unfolding sequence of urban life in southwestern Asia, from late prehistory to the threshold of the modern era. In shifting to a fresh and unfamiliar subject, I admittedly have had to rely heavily on secondary works and to skirt many issues of uncertainty or contention that can only be the province of specialists. The justification lies, I suppose, in the belief that an ongoing dialectic between specialists and generalists has something to offer that is not likely to emerge in any other way.

To what degree is technology, like science, a branch of knowledge in its own right? Some authorities believe that a disciplinary pedigree became attached to technology only quite recently, and at least partly as a derivative of the achievements of science. If one accepts this standpoint, at least for the eras before the onset of general literacy, are we doomed never to know whether technology was more than an unreflective understanding of "how to," a hodgepodge of recipes and chance findings embedded in magical formulas?[2]

For most of pre-modern history this pessimistic judgment may have some validity. However, some important reservations are in order. First, we should never allow ourselves to overlook the larger truth that technology, an enormously elaborated "extrasomatic means of adaptation," is a unique property of the human species. If certain technological developments account for some of our troubles, many others are also the foundation on which human civilization originated and has (if discontinuously) moved forward in giant steps across millennia. Second, few secular trends remain uniform across such a vast arena as the science-technology nexus. Hence there is always reason to hope that documentation of illuminating exceptions will have been fortuitously preserved. And third, there is a difference between what can be learned specifically about the source of a given technical process or contrivance, and larger patterns of similarity or divergence that might illuminate such discoveries. Even if the search for explanations of single contributions often proves discouraging, a survey of larger patterns may have more rewarding results.

It is just such a concern for patterning that has provided the stimulus for this study—the recognition that major technological advances have frequently been composed of waves or clusterings of sometimes seemingly unrelated inventions and innovations. What were the larger social, cultural, and economic settings of these clusterings? Which contextual features help to explain why transformative bursts of innovation occurred when and how they did and not in other, more quiescent, intervening periods? Alternatively, what impact did rapid technological progress have on the societies in which it occurred? Has the role of technology in history been primarily that of a dependent or independent variable—driven or driver? Only by large-scale comparison, beginning in the ancient world and progressing to the modern, could these questions reasonably be addressed.

My own growing interest in these questions coincided with declining opportunities to pursue them in fieldwork in the Near East, and especially in Iraq. Moving away from the detached stance of an observer of relics of behavior that were safely removed in time from any possible intervention, I turned to a more active engagement in academic and scientific problems of organization. Both at the University of Chicago and in the National Academy of Sciences–National Research Council complex, one new group of questions that naturally arose involved how directions of research are set, sustained, and changed. Another focused on the linkages between science and society—with technology first among them.

As such, it may be reasonable to think of this essay in some respects as a vast flanking movement, undertaken to envelop if not capture from the rear, as it were, the main elements of a persistent process of change begun in the remote past that our current methods and opportunities for study leave tantalizingly hidden in the mist. Speaking in similar spirit and of like circumstances, Alfred Kroeber once noted that "[e]xperience has shown that it is hopeless to storm, by a frontal attack, the great citadels of causality underlying highly complex groups of facts."[3] But the difficulty is, of course, that flanking movements seldom encounter easy going and instead have to contend with whole systems of other, equally well-defended citadels. So in the end, as has been the case here, they become no less satisfying ends in themselves.

By seeking to invoke broad consistencies in the direction and processes of technological change that span five or six millennia, I admittedly run the risk of comparing entities and their social contexts that are too fundamentally different to sustain the effort. But rewarding engagement with colleagues and programs at the Santa Fe Institute in recent years convinces me that models of thought need not be in any sense bounded by subject matter or discipline. The Institute has taken leadership in moving away from the reductionistic simplicity of general laws toward sciences of complexity that concentrate on the extreme sensitivity of most systems to initial conditions, on the chaotic effects of positive feedbacks, and consequently on indeterminate, nonlinear dynamics.

Why not, therefore, try to draw what suggestive insights we can from fields that, however different in substance, are less constrained by the narrowness of focus, ambiguity, and incompleteness that presently characterizes the archaeological and early written record of the Middle East? By concentrating on turning points and irregularities in rate across an extended period of time, there is some hope of learning how to distinguish less arbitrarily between breaks and continuities in processes of technological change.

The subject of this study began with a modest inclusion of the theme of technology in graduate courses of a more general character that I taught at the University of Chicago and as a visiting professor at Harvard University and the University of California, San Diego in the late 1970s and early 1980s. This led to more direct approaches in formally prepared lectures. One such occasion, perhaps the first on which I tried to take a more comprehensive view of the field, was as part of the Tenth Anniversary Lecture Series of the Department of Anthropology of the Johns Hopkins University in 1986, with support from the Maryland Humanities Council and the Wenner-Gren Foundation. At this initial stage Gillian Bentley's library searches in 1985–86 provided assistance in deepening my initial acquaintance with the literature on the subject.

For some years following, single lectures remained a compressed but adequate format for what I began to visualize as a lengthy, programmatic journal article. The fora for these included Hunter College of the City University of New York, an annual meeting of IBM's senior research scientists, the Israel Academy of Sciences, the Darryl Forde Lecture Series of the Department of Anthropology, University College, London, and the Bren Lectureship at the University of California, Irvine. I am indebted to participants on each of these occasions for their observations and criticisms. Suggestions and critical comments made on a draft paper prepared for the London lecture in 1988 by Grace Goodell, Robert K. Merton, Alex Roland, and Arnold Thackray were instrumental in offering encouragement and beginning to set a direction for a considerably more ambitious enterprise.

Until the fall of 1993, however, with administrative responsibilities at the Smithsonian Institution continually supervening, my efforts were largely confined to a progressively wider scanning of the literature and a preliminary accumulation of notes and other materials. But from that time forward, having informed the Institution's Board of Regents of my intention to retire a year or so later, this book began to take shape. An early draft was sufficiently far advanced by the following summer to receive valuable comments from Robert M. White, then president of the National Academy of Engineering, and from an anonymous reader for a publisher. Subsequent to my retirement in September 1994, I am much indebted to President Bruce MacLaury of the Brookings Institution for generously making available its impressive library and other facilities. Adding, amending, and editing continued through another term as a visiting professor in the Department of Anthropology at the University of California, San Diego in the spring of 1995, where some of this material could be tested in class lectures, and it concluded early in 1996 during my term as a fellow at the Wissenschaftskolleg zu Berlin.

Throughout the process of revision, I have benefited to an extent that is difficult to acknowledge adequately from an ongoing editorial dialogue with Peter J. Dougherty, publisher for social science and public affairs for Princeton University Press. With his help the main lines of the argument of the book have emerged with greater consistency and clarity—and with unquestionably greater economy as well, overlooking some personal regret over the necessary disappearance of many of my numerous excursions into anecdotal detail. He has my warmest thanks for this guidance, although, needless to say, the responsibility is mine alone for errors and inconsistencies that remain.

A final word may be in order on my personal standpoint regarding technology. My early experience in the late 1940s and early 1950s as a mill hand in the "Big Mill" in South Chicago, the old (now demolished) South Works of the U.S. Steel Corporation, was, as I reflect on it, not lengthy, but permanently defining. That was in some respects still a time of innocence. One could not help taking pride in industrial might, in being part of a giant source of jobs and wealth. Not overlooked, but somehow casting a shadow that only heightened the image of power, was the all-too-evident side effect of a surrounding, chronically depressed neighborhood that seemed to slide deeper into anonymity as great columns of acrid smoke corroded even its street signs. I decided then and still believe today that we begin and end with what we produce.

PATHS OF FIRE

1

Paths of Fire: The Idea of Technological Change

MODERNITY, while it envelops and defines us, has as its conceptual essence only an evolving ambience and many loosely related states of mind. Few of its more specific attributes are closely coupled to shared perceptions or meanings. Yet our modern world is inconceivable in the absence of a long series of technological achievements that are held in common and largely taken for granted.

As with other human products, most new technologies were invented and applied to serve immediate ends. Continuously shaped by human design and use, their introduction also modified prior contexts of design and use. Yet the process of innovating and deploying new technologies has not ordinarily been as narrowly pragmatic as this suggests. Evident for at least several centuries, first in Britain and later and more strongly in the United States, has been a consistent aim to design new technologies that would not only meet existing needs or demands but would create and satisfy previously unrecognized ones. New technologies converge with modernity in embodying a primary direction—toward indefinitely expanding the realm of individual powers of expression, understanding, autonomy, and well-being.

Technological transformations in Britain and the United States typically have been complex, dynamic, variously motivated, and largely unpredictable. Under what conditions were they initiated? Why have innovations tended to be strongly clustered in space and time? What distinguished the handful of technological "breakthroughs" from the almost endless supply of other, more modest or unsuccessful innovations? Are the principal sources of novelty to be found in spontaneous modifications of practical routines, in the pursuit of entrepreneurial opportunity, or in spinoffs from science and the disinterested pursuit of knowledge? Why do institutions, firms, and individuals—or these two nations at different periods in their history, for that matter—differ in their readiness to devote resources and creative energies to taking world technological leadership?

Only a few generalized answers to questions like these will emerge from this study. No doubt that is at least partly explained by the most salient of its conclusions—that the course of technological change is continuously subject to many heterogeneous, often conflicting influences. Characterized as "path-dependent" for this reason, it is cumulative and yet largely lacks broad as well as specific regularities of process.

A second conclusion is that the closest relationships consistently link technological change to socioeconomic forces. And it is the socioeconomic forces rather than spontaneous technological developments that tend to be primary and determinative. Apart from occasional, limited, and contingent episodes of apparent technological "momentum"—"sites, sectors, and periods in which a technology-oriented logic governs,"[1] no convincing case can be made that technology has ever consistently served as an impersonal, extrasocietal agency of change in its own right.

Relationships between science and technology constitute a third area of inquiry that will be consistently pursued, although it lends itself less well to a concisely summarized proposition that can be stated at the outset. Very briefly, it is a dynamic, evolving relationship of increasing synergy and mutual interdependency as we approach the modern period. Until quite recently, it has primarily been viewed from the more consciously documented perspective of the sciences. This has meant that the active, creative contributions of a more technological character—those of early craftspeople, laboratory technicians, machine builders and instrument makers, designers, and later those of engineers—have not been adequately credited for their importance. A renewed effort to restore the balance, adding to the richness and complexity of technoscientific interactions in all periods, is undertaken here.

Similarly underweighted, but unfortunately beyond the reach of this or any study largely based on secondary accounts, is the role of the firm. Major, long-lived corporations may have surviving archives and so escape the difficulty. But smaller undertakings, in many cases the principal seedbeds of technological innovation and experimentation, have universally fared less well. The outcome is that any account—this one included—failing to identify new sources to correct the balance tends to perpetuate a picture that is distorted in the direction of both size and success. The story that cannot yet be told is the one that adequately takes into account plans and efforts that were cut short, orphaning their technological contributions, by the ever-present likelihood of business failure.

Human beings are the active agents of technological change, but innovations are its units. They obviously vary greatly in their change-promoting, destabilizing potential, from minor, quickly superseded improvements to the few in every historic epoch that have ultimately proved to be of decisive importance. In most of these latter instances, the greatest breaks in continuity occurred at the outset of closely related clusters or series. Such breakthroughs then tended to have an accelerating, focusing effect on lesser innovations that followed. Cumulatively, however, those lesser, later innovations often added indispensably to the success of the original inventions.

It is a common error to regard inventions and innovations as more or less independent acts of origination. Although the record usually turns out on closer inspection to be marred by ambiguities, at least in principle the substance and timing of each of these achievements seems conducive to being more sharply defined than the continuous processes of change that they precipitated. That in turn indirectly lends weight to a belief in the autonomy of the technological sphere as a whole. Yet almost all of these undeniably creative acts were dependent, in the first instance, on accumulated insights and capabilities stemming from earlier discoveries. And surely no less important than this technical heritage was the role of potential users, consumers, or investors. Frequently, their suggestions or demands contributed substantially to the process of innovation by envisioning needed features and stimulating the search for them.[2]

Hence it is a reifying distortion to isolate inventions and innovations as triggering events that, at least in each major case, started a new clock ticking with a self-contained, autonomous set of propensities for change. It supports the misleading implication that each such discovery tends to contain within itself, *ab initio*, a consistent set of directions for its unfolding further development. Only by recognizing the embedded, interactive aspect of what are also originative processes are we adequately prepared to understand later deflections in the direction of technological change that respond to changing contextual circumstances rather than to any internal logic.

Technological change, this book will argue, is complex, non-linear, and in fundamental ways indeterminate. It is continuously waylaid by forces and uncertainties from many directions, many of them unexpected. A few individual inventors like Thomas A. Edison may have almost single-handedly stimulated the technology of their times and set its course for years to come. Far more commonly, however, innovative contributions have fallen within a broad, temporally unfolding front of complementary efforts involving many collaborators or competitors. Thus, innovations are better understood less as independent events that unleashed new sequences of change in their own right than as periodically emergent outcomes of wider, interactive systems.

Consistent with the same "contextual" viewpoint, technology itself should not be identified exclusively with its traditional, technique- or artifact-oriented aspects. Human society has always had to depend on what can be broadly characterized as primarily technological means. How else can we organize ourselves functionally and hierarchically, extend our mastery over the constraints and uncertainties of our environment, and improve our collective ability to set goals and move toward their attainment? Inventions and innovations are necessary but not sufficient steps in the pursuit of these objectives. Their ultimate importance depends on the extent to which they are not only adopted for the uses originally envisioned but incorporated in descending levels of further, previously unanticipated innovations.

Obviously, not all innovations are uniformly positive in their effects. Technological means are, after all, just as crucial for coercion and exploitation as for more benevolent purposes. Still, the balance is overwhelmingly positive. The densely interactive, endlessly reshaped mass of artifacts, techniques, designs, and organizational strategies, and the bodies of tacit as well as formal knowledge on which these all rest—in effect, the technological base of the social order—has grown, irregularly but cumulatively and without significant reversal, throughout the entire human career.

Furnishing raw materials for change, technological innovations have a role in some respects reminiscent of that of genetic mutations in biology. But they differ in a number of important ways. First, they are decidedly less random, more situationally responsive or context-dependent, more prone to occur in closely interrelated clusters, and likewise more transformative in the scale and scope of their consequences. No less important, they

> do not adhere strictly to their finder or creator, but are shared, at least to some extent. In many cases the sharing is intentional, in others despite efforts to keep findings privy. But in any case, that the new technology ultimately goes public means that technology advances through a "cultural" evolutionary process. The capabilities of all are advanced by the creation or discovery of one.[3]

It is thus a reductionist blind alley to concentrate on individual creativity in explaining innovations.

The domain of deeper understanding toward which this book is directed cannot be rigorously circumscribed. Receiving most consistent emphasis are what can be characterized as technologically relevant personnel, including their social backgrounds and outlook, institutional attachments, training and experience; the degree of their engagement in scientific or managerial as well as more strictly technical activities; and their evolving threshholds of technoscientific capability. More selectively and broadly, its focus at times widens to include the relationship of technology to governmental initiatives and structures; population, settlement, and resource-extraction systems; and business and economic organizations and strategies.

Broad as this listing is, any anthropologist like myself will readily acknowledge that it, too, is naturally embedded within encircling, successively larger and more diffuse cultural systems and their attendant patterns of meanings. Inherent in the subject of technology is a tendency to concretize and reify, to take for granted that the "real" world is adequately expressed in the intrinsic properties of unambiguous facts and material objects. There is indeed a place for an essentialist approach of this sort, especially when, as in this book, it is intended that a wide-ranging reconnaissance of surface topography will encourage later plowing to greater depths. But an adjoining, permanent place must also be found for another stance, even if the two can only coexist in a degree of tension, that is better able to

> capture human complexity [whose] goal is generating alternatives to traditional historiographies that construe motivation in terms of interests, causality in terms of economic, or some other, determinism, and that privilege things and events over relationships and processes.[4]

As it is considered here, in short, technology is an enormous, vaguely bounded array of material and nonmaterial, new and traditional elements. The loosely linked systems into which most of these elements are grouped can neither function nor be understood except as parts of an embracing social organism. In other words, there is no free-standing, self-enclosed framework of technological explanation. As has long been generally recognized, doctrines of "hard" technological determinism fail to provide a realistic basis for serious social or historical analysis. After all, as is observed in a valuable recent overview of the issue:

> In spite of the existence of an engineering profession, technology is not an organized institution; it has no members or stated policies, nor does it initiate actions. How can we reasonably think of this abstract, disembodied, quasi-metaphysical entity, or of one of its artifactual stand-ins (e.g., the computer), as the initiator of actions capable of controlling human destiny?[5]

A Contextual View: Limitations, Distortions, and New Directions

The study of technology has been a curious lacuna in the relatively seamless web of disciplines. Too frequently it exists in the shadow of the history of science, largely drawing its methodology and theoretical models from that source. Nothing is surprising about this. It surely reflects the more self-conscious concern in the sciences for prompt and public dissemination of new findings, as well as the long-prevailing dependence of scientific reputations on priority of discovery. If further reinforcement is needed, it is perhaps provided by the preponderant role that universities, with their devotion to basic research, have over the last century or so come to play in intellectual life. But the consequence is, in any case, that for a deeper, less derivative understanding, technology demands a distinctive approach of its own.

Science, for all of its intense specialization, in principle constitutes a corpus of comprehensive, generalized knowledge with recognized frontiers and strategies for its advancement. Its practitioners form at least a loosely bounded community with shared commitments and values. There is no community of technologists in a similar sense, and few individuals would so define themselves. Technology is a more weakly integrated sphere of activity, its parts more readily taking on the coloration of their different contexts. Virtually all members of society participate in technology to some extent, holding very divergent views of what they know of it and directing their efforts toward highly heterogeneous objectives. Yet technology is like science, and is increasingly joined with science, in contributing to our individual as well as aggregate human potential, and in being largely cumulative in character.

Until the recent past, there is rarely much hope of attaining lasting certainty in the study of the history of technology. It requires high tolerance for weakly supported, only conditionally superior hypotheses. Many of the major increments in technological change came through small, multiple, gradual, and often almost unnoticed improvements. Characteristically, these tended to be more far-reaching in their cumulative effect than was apparent at the time of their initial introduction. Given the informality of the process, records of those later improvements tend to be poor, incomplete, self-serving, or altogether lacking. In not a few cases information was purposely left unrecorded and even shrouded in mystery to assure economic or social advantage. Diffusion of new innovations largely depended on precept, example, and word of mouth, or on hands-on acquaintanceship with novel, imported artifacts.

Nor is this the end of the limitations to be overcome. Exogenous changes in social receptivity or economic demand have consistently been more determinative of the directions taken by technological trends than properties or developments inherent in technology itself. Sensitive, diachronic analysis is clearly necessary to isolate factors—some brief but decisive, others diffuse and only slowly accretive—that contributed to specific sequences of change. The field of analysis is subject to radical, continuing adjustment, from a particular craft workshop or factory floor to the competitive interactions of an entire industry, and thence to problematical connections with progenitors or derivations that sometimes may begin by seeming implausibly remote. Technical and economic considerations cannot be rigorously partitioned, although they were frequently invoked independently, by different individuals, at different points in the hierarchy of decision processes. Much of the available data on older technologies takes the form of actual productive arrangements and artifacts. Whether these still survive or are only partially recoverable from drawings, archaeological excavations, and museum specimens, they are always difficult to relate to the criteria employed in the original production decisions.

In the face of obstacles like these, further progress in understanding the development and role of technology clearly requires not only the marshaling of very disparate kinds of direct and indirect evidence but a careful and comprehensive evaluation of *context*. Context is, by definition, an unbounded and heterogeneous domain. There is almost no limit to what may, under some circumstances, need to be considered in order to understand specific paths that technology has taken: the uncharted dynamics of consumer preference; military requirements, mercantilist strategies, and other manifestations of state intervention; markers of status gradations and routes of social mobility; resource costs and availability; orientations toward tradition and novelty; tolerance for risk and uncertainty; and, of course, residual values as well as more dynamic religious callings. Separately or in combination, we must be sensitive to the parts that all such contextual factors played in creating milieux that help us to understand technological advance.

This breadth of relevant sources to be sought out and issues to be examined characterizes many forms of anthropological and especially archaeological inquiry. But let me take care not to overemphasize the connection between taking a contextual approach to the evolution of technology and the potential anthropological contribution to the subject. In their present training, little equips most anthropologists for the meticulous source criticism and textual analysis that is everyday grist for the mills among historians, or for the theoretical rigor and quantitative sophistication that is increasingly entering economic history from economics. Although other perspectives have begun to command a substantial place in recent years, traditions of participant observation among ethnographers, and of site-focused excavations among archaeologists, remain the dominant ones. Those traditions tend to take implausibly for granted that the most significant social interactions, extending even to those involving power and resource relationships, were between relatives and neighbors nearest at hand.[6]

At the same time, adding perspectives drawn from the credo of participant observation to the battery of present approaches offers more than an abiding effort to obtain a wider view. It turns attention in many new directions—among them, toward a micro view of industrial settings, where much of the real process of technological change unfolds in long, complex, usually almost undetected sequences of interaction. Similarly awaiting the promise of fuller utilization is techno-archaeology, a subspecialty or extension of ethno-archaeology that would assist in interpreting ambiguous data about vestiges of past technologies.[7] Materials research is likely to make a significant contribution as well, as a companion discipline alongside geo-archaeology, zoo-archaeology, and archaeological botany. Present experimental approaches will no doubt begin to play a part in clarifying the advantages and drawbacks of modifications of hypothetical production processes. In a word, anthropology has a role to play in bringing the testimony of artifacts, drawings, manuals, inventories, sales records, court proceedings, and contemporary shop-floor workers and laboratory analysis more fully within a single synthetic framework.

Perspectives drawn from studies of living societies also can serve as powerful, much needed supplements and correctives when dealing with historic records that have been routinely constrained by the observational and conventional filters of those who were charged with recording them. To take the most obvious example, the grossly disproportionate roles accorded to the two sexes in the overwhelming majority of technological records currently tends to be passed over in silence by most contemporary works on the history of technology. Professionally at pains to account for gender roles in their own immediate fieldwork, anthropologists are less likely to set aside as unproblematical and undeserving of closer study the huge disjunction between the myopia of the written sources and the everyday experience that they have always inadequately reflected.

This prevailing exclusion of women from all but the most consumer-based and subordinate association with the historic development of technology is a characteristic of the record that needs more than merely to be acknowledged. Sex-linked differences in qualities of skill, focus of interest, or creativity that would explain it are undemonstrated—and highly unlikely. Gender-linked differences in the roles open to men and women, on the other hand, are social constructs, and these have an astonishingly pervasive uniformity across cultures and civilizations.[8]

In this respect, the inadequacy of existing records is an important obstacle that is difficult to overcome. Patents are an instructive case in point, and much will be said later about the general deficiencies in patent records as an adequate representation of the sources of significant inventions. But for women in particular, with less disposable income and less access to technical, legal, and business training and assistance than men, the taking out of patents in their own names was until fairly recently only rarely a practical possibility. Until the passage of the series of Married Women's Property Acts toward the end of the nineteenth century (in both England and the United States), patents would automatically become the property of the husband. It is thus not surprising that women are credited with only about 1 percent of patents that were awarded during the nineteenth century. Since then it has risen, at first very slowly but now at an accelerating rate.[9]

Our subject, however, is not the general distribution of sources of invention and innovation within society at large but the emergence of proto-industrial and industrial technology. This is the part of the considerably larger technological field on which this book will concentrate—the one culminating in industrial scale, methods, organization, and ultimately the propensity to generate new inventions as a fundamental property of the system itself.

Scope and Historical Acceleration of Technology

Fernand Braudel has written that "[i]n a way, everything is technology," and again that "[t]echnology ultimately covers a field as wide as history and has, of necessity, history's slowness and ambiguities. Technology is explained by history and in turn explains history."[10] For the great French historian, the subject escaped from any ordinary set of disciplinary constraints and hence could benefit from the constructive tension between contextualizing approaches and more specialized studies. It extended without clear distinction from the level of everyday life where "routine predominates," although there are "gentle slopes along which the whole mechanism slides," upward into a realm of economic life that is supported by these routines "but implies calculation and demands vigilance."

To enter into the virtually boundless territory that Braudel designated as technological seems to require that we take into account not only the designers, makers, and users of technological products but all those involved in decisions which either deal directly with, or are merely affected by, those products. Carried to this extent, technology may become virtually a synonym for all of the instrumental means for reconfiguring as well as reproducing the social order. Braudel himself, however, was not primarily concerned with its adaptive and transformative significance. For him technology was, like the social order itself, "an enormous mass of history barely conscious of itself . . . its immense inertia, . . . its old sometimes antediluvian choices."[11]

At Braudel's slow-moving, lowermost level, attempts to understand the role of technology have tended to focus on continuities in how or what things are done or made.[12] The object is, almost by definition, in the foreground—tools or devices, the ends they serve, and the parsimony of design and use to which uncounted repetitions of uses have led. There is an inherent emphasis on—even a reverence for—traditional practices: "codified ways of deliberately manipulating the environment to achieve some material objective."[13]

Complementarities and feedbacks lending coherence and impact to larger, interacting, and multifunctional technological systems, or to technology at large as it existed within a particular cultural or historical context, receive little attention. What is given greatest salience may be little more than a hodgepodge of individually reified and routinized activities, each with a distinctive, relatively unchanging function.

I am skeptical about the utility of an artifact-oriented view of technology, as opposed to a more systems-oriented one, for the study of any period of history. For all of Braudel's enormous authority as a contextualizing historian, it has overtones of long-discarded scholastic assumptions about the existence of a wall of separation between affairs of the mind and hum-drum activities of the hand. By tending to decompose technology into dis-sociated parts, it neglects elements of choice in the selections that must be made continuously within a whole technological repertoire, although these are perhaps the principal source of flexibility and adaptive change.[14]

The element of consistent and conscious reproducibility cannot be over-looked, of course, but the objectives for which this is done must be broad and plural rather than singular. Harvey Brooks's formulation thus is an improvement: "knowledge of how to fulfill certain human purposes in a specifiable and reproducible way . . . something that can be reconstructed in principle through specifiable algorithms." But even more important is his further qualification that

> technology must be sociotechnical rather than technical, and a technology must include the managerial and social supporting systems necessary to apply it on a significant scale. . . . Today, managerial innovations are becoming an increas-ingly important aspect of technology. . . . Management, insofar as it can be de-scribed by fully specifiable rules, is thus a technology, and indeed every large bureaucratic organization can be considered an embodiment of technology just as much as a piece of machinery.[15]

As this suggests, we need to take into account an ongoing evolution in the character of technology itself. The relationship of a pre-modern farmer or craftsperson to a tool (or set of tools, for that matter) is likely to have been a highly personal and habitual one. Occasional improvements might bring gains in efficiency and yield, or reductions in cost, but wider, sys-temic effects were otherwise limited. But Brooks offers a modern contrast, whereby with computers and communications systems, "not only the soft-ware, but also the organization that goes with the system, are inseparable from the physical embodiment of the technology, and are often the most expensive and innovative parts of it."

Indeed, the pace of change today clearly has become more revolutionary than evolutionary. To ignore this emergent, transformative element, and to conceptualize technology as consisting only of fixed sequences of stan-dardized acts yielding standardized results, becomes more and more mis-leading as we approach modern times. Formal definitions, if they are to be helpful at all across a broad span of centuries that reaches into the present, need to consider the creative, purposive, systemic features not only of tech-nology in general but of most individual technologies as they are put into practice.

Above all, it is important to recognize the multiplicity of factors affect-ing technological change. Some of those factors take the form of percep-tions of technical possibility, and as such are inextricably a part of the innovation and design process:

Technologies always embody compromise. Politics, economics, theories of the strength of materials, notions about what is beautiful or worthwhile, professional preferences, prejudices and skills, design tools, available raw materials, theories about the behavior of the natural environment—all of these are thrown into the melting pot whenever an artifact is designed or built. . . . *They might have been otherwise.*[16]

Many other factors, however, are functionally unrelated and essentially external in origin. Some of these are pervasive, long-term, cultural preferences or social forces that at least may tend to induce some regularity of process. But there are also some that are short term and localized—"flashes of insight, happy accidents, vagaries of financing, chance meetings, and a host of other events of a nonsystematic character."[17] Omnipresent elements of capricious choice, uncertainty, miscalculation, and variability in the effectiveness of individual leadership also can lead to abrupt, unpredictable shifts.

As Braudel recognized, taking a more inclusive field of view comes at the cost of correspondingly less rigorous methods and more ambiguous boundaries. Attention to the broad sweep of historically specific circumstances does not easily accommodate itself to formal, generalizable mathematical models. As the field of view widens, qualitative judgments grow in importance while measurable entities and isolated, well-defined relationships recede. This tends to support a synthesizing—but correspondingly eclectic—methodological stance if we are to aim at better understanding some of the truly generative discoveries and ideas that technology has contributed, and to deal with their wider social consequences and sources.

The increasing pace of modern technological change has already been mentioned. No doubt it reflects a steeply increasing range of applications of new technologies resulting from many new paths of convergence with basic scientific research. What has been put in place is nothing less than a systematic process to generate an accelerating flow of new inventions, innovations, and their applications. But that, too, cannot be understood as an isolated, autonomous development. It is largely a reflection of the transformative powers of modern capitalism in its current, globally competitive forms.

Proprietary capitalism, to borrow the illuminating typology of William Lazonick, characterized the epoch of the Industrial Revolution and much of the nineteenth century with its "integration of asset ownership with managerial control." Reliance primarily on market coordination subsequently gave way, Lazonick argues, to planned coordination within business enterprises as a superior means of organizing productive resources. Management capitalism, which separated ownership from control, is now giving way to wider frameworks of cooperative business organization that he terms "collective capitalism." These channel but do not eliminate market competition, while at the same time closely linking interrelated productive and marketing activities under a growing degree (as especially in contemporary Japan) of state tutelage and overt support.[18]

Lazonick emphasizes the enabling power of these institutional innovations, tending to see as natural and unproblematical the corresponding growth that is required in the basic, enabling supply of technological innovations. Similarly, F. M. Scherer, invoking the earlier judgment of Joseph Schumpeter, argues that greater weight of responsibility for growth should be assigned to entrepreneurs than to inventors.[19] On the other hand, no less forcefully, Simon Kuznets maintains the contrary emphasis. While recognizing that there is and must be an interplay of the technological and institutional determinants of growth, he describes as primary the ramifying effects of technological changes on the organization of production and management, and on conditions of work and the required educational level of the work force.[20] More recently, Giovanni Dosi and his collaborators apparently attempt to straddle the issue. While acknowledging the "coupled dynamics . . . between technology and other social and economic variables," their formulation of the balance between innovation and entrepreneurship tips toward the former in seeing the " 'cumulative and self-propelling advance of technology' [as] a sort of *primus inter pares* within the menu of ingredients for economic development."[21]

I know of no way definitively to reconcile these viewpoints. In part, they reflect the different analytical standpoints of their proponents—Lazonick's centering on the creative, microeconomic strategies of firms and Kuznets's on macroeconomic aggregates and the national strategies that may enhance them. At least for the present, it must suffice to say that there *is* a continuing interplay. Whether furnishing an independent line of causation or not, a cumulatively overwhelming mass of technological innovation has been deployed across this span of more than two centuries within a substantially changing institutional context of firms competing in a market.

But we need to foreshadow the rise of capitalism, not simply to follow the main lines of its evolution. It was a central feature of Braudel's larger synthesis that a decisive shift in the balance of economic vision and effort accompanied the emergence of modern Western society. More and more heavily involving the self-directed, consciously innovative pursuit of new opportunities and enterprises, this transformation was for him the decisive force accounting for technology's accelerating movement away from its earlier condition of "immense inertia."

His view has much in common with propositions accepted by most modernization theorists two generations or so ago, and with the influential viewpoint developed at about the same time by a group of "substantivist" anthropologists, economists, and historians led by Karl Polanyi.[22]

A dichotomy between traditional and modern societies was accepted as fundamental.

At least in its original, fairly sweeping form, this position today has significantly fewer adherents. Historically attested attitudes and behaviors do not neatly divide themselves into capitalist and precapitalist categories. In many cultures and historical epochs, and for social aggregates as well as for individuals, the overwhelming empirical impression is one of immense variety.[23] Significant clusterings of market-oriented practices can be traced far back in history as well as in the premodern ethnographic record.[24] Even in sectors of modern Western society where market relations today play a dominant part, market "anomalies" and apparent failures as a result of the confluence of other institutional forces are now a subject of increasing scholarly scrutiny. From an opposite direction, price-setting markets can be identified in early modern England in which buyers and sellers nevertheless were so deeply embedded in common networks of trust and mutual indebtedness that profit-making motives and behavior virtually disappear from view within a moral milieu of community solidarity and cooperation.[25] Broad changes like these in the current outlook of the social sciences argue strongly for a moderation of the uniqueness and decisiveness of the transition to capitalism at a particular point in time on which Braudel placed so much emphasis.

Similar qualifications apply to his vision of technology as a multilayered, almost oceanic phenomenon. Questions evoked by that vision bring us closer to the central issues involved in our own subject matter. While characterizing the movements of the deepest currents as barely perceptible until the quickening pulse of capitalism, Braudel seems to have regarded those currents as wholly impersonal, unrecognized, outside of history—in effect, exogenously driven forces. Unlike the more superficial ebb and flow above them, however, the deeper currents also were for him correspondingly more decisive, cumulative, and indeed permanent in their effects.

In Braudel's view, the more significant elements of technological change were, until fairly recent times, unintended and largely unrecognized. Perhaps they should be thought of as the cumulative product of a long aggregation of small improvements in practice, supplemented by random, accidental discoveries. No routines are, after all, absolutely rigid at the hands of all practitioners. There are always personal variants, as well as special contingencies that call for ad hoc, trial-and-error modifications. Where they proved superior, examples of both kinds surely have led to improvements in even the most habitual practices.[26]

But in this view, little is left of earlier technological change but dissociated processes and devices, each undergoing marginal changes and improvements in response to isolated needs and rationales. Each becomes, in effect, a separate commodity, accepted or rejected on its individual merits. I find it difficult to believe that, at virtually any level of social complexity, there were not more integrative visions and cohesive patterns of association than Braudel supposes among what were for him essentially separate spheres of technical activities. Part of the apparent absence of evidence for them may be a reflection of the limitations of our earlier sources. But also needing to be overcome may be the reluctance of modern scholars to recognize patterns of motivation different from the economically "rational," individually acquisitive ones that capitalism has made so central.

To define the emergence of capitalism as the crossing of a kind of behavioral and cognitive gulf implies an a priori denial that consistent, longer-term patterns of technological advance can be identified that spanned that gulf. Yet a generalized pattern of technological advance can be identified and deserves our attention as one of the cumulative, irreversible accomplishments of humanity. Even without considering such "second-order" consequences as transformations in life expectancy and quality of life, it includes such basic and yet universally valued elements as "the substitution of machines for human efforts and skills, the substitution of artificially produced energy for animal and human energy, and the exploitation of new raw materials, including, increasingly, man-made ones."[27]

Still another set of long-term regularities is likely with respect to how technological advance occurs. Brian Arthur has informally offered an alternative characterization that contains, and seems to require, no reference to capitalism as a necessary institutional setting for its operation: "technology isn't really like a commodity at all. It is much more like an evolving ecosystem. In particular, innovations rarely happen in a vacuum. They are usually made possible by other innovations being already in place. . . . In short, technologies form a highly interconnected web . . . a network."[28]

It is also important to recognize how closely we have come to associate technological advance with enhancements in purely personal gratification and unrestrained freedom of individual action. If these were universal aspirations before the advent of capitalism, we might expect that we could reliably trace the presence or absence of technological improvements from their secondary impacts on growing inventories of personal property. The immense devotion of energy to cathedral-building during the Middle Ages is merely one of many examples demonstrating that this has not always been so. It is no less plausible to assume that substantial improvements in technical equipment or productive efficiency sometimes were devoted to freeing greater time for religious observance, family or community rites of passage, or merely leisure.

Population growth has no doubt been another alternative outcome of improved productive efficiency. Especially under conditions of uncertainty, larger family units heightened security not only for families as a whole but for their individual members. Another widespread human aspiration involves the opportunity to participate in larger-scale, stimulatingly more diverse social interactions. Yet to the extent that these and similar motivations have had significant parts to play in history, they qualify if not contradict the pursuit of personal advantage as the consistently dominant outgrowth of improved technological performance.

As already noted, there are increasing difficulties if we allow the field of meaning of contemporary technology to be identified primarily with Braudel's slowly moving, deepest substratum. Characteristic of the technologies that now have begun to dominate our lives is an entirely different set of features. To begin with, change is omnipresent. Rapid product obsolescence is driven by intense competition for market share and economies of scale and process improvement. Highly organized, essentially permanent research and development (hereafter R & D) efforts were limited to only a handful of firms and industries a few decades ago. Now they have become common as the need to maintain a continuing flow of new innovations is more widely recognized.[29]

No less obvious is another characteristic of modern technology—its far greater, and growing, complexity. As social units increase in scale and interdependence, most technologies no longer can serve singular, self-sufficient ends at the hands of individuals. Exemplified by automated office devices, their effectiveness depends instead on being used in combinations that may have originated with, but now cross-cut, the specializations within society itself. At the same time, they must incorporate provisions for flexibility to adapt to unanticipated, turbulently interacting shifts in objectives and operating environments. Adding a further dimension of complexity is the fact that, within these larger technological ensembles, the experimentation, growth, saturation, and senescence phases of different subcomponents seldom coincide and have become increasingly difficult to predict.[30] At both a global and a national scale, we are encountering an increasingly unified science-and-technology enterprise.

The recent, worldwide recession has taken its toll on major, internationally active corporations as well as smaller firms, and on the scale of employment in leading industrial sectors. But a braking in the pace of technological advance is not in evidence. In fact, technological innovations have been substantially responsible for continuing gains in productivity that threaten to make increasing numbers of unskilled and semiskilled workers permanently redundant.

Those directing and working within the science-and-technology enterprise, on the other hand, are at least relatively less affected. They continue to pursue goals that they have had a significant hand in choosing, exercising relatively greater freedom of approach and maintaining higher security of employment than most of their counterparts in other sectors with comparable training and professional attainments. Providing a degree of protection against cyclical economic downturns have been several generations of striking economic successes with new, high-tech products and processes. Mutually supportive corporate linkages also help to insulate their work from politically inspired brakes or shifts of direction.

This science-and-technology nexus is at the heart of powerful trends toward global interdependency. The greater sense of identification with place of an earlier generation is eroded by a kaleidoscopic succession of near and remote interactions, their respective influences over peoples' lives no longer conforming with laws predicting a decline with increasing distance. At the core of the shift are increasingly massive flows of electronic information that constitute the vital coordinating mechanisms linking the world's institutions and economies. But dependent on these flows is a wider set of transnational phenomena.

The increasing dominance of international corporations is one. They are at the same time vigorously competitive and continuously maneuvering for cooperative marketing and cross-licensing agreements and potential buyouts and mergers. A necessary condition for this complex interaction has been the growth of worldwide, almost instantaneous, virtually unregulatable capital flows, facilitated by electronic transactions through offshore financial markets that largely evade control by national economic authorities. And with these developments has come "the accelerated homogenisation of consumer preferences, product standards and production methods on a global scale"[31]

In short, modern technology has become an integral part—the essential part—of a worldwide complex of economic and institutional forces that is at once deeply destabilizing and highly creative.

Innovation Clusters, Series, and Trajectories

It would be strange indeed if the rate of technological advance were smoothly continuous. Innovations are, in effect, the elementary building blocks of technological change, and the manner in which innovations are generated and put to use militates strongly against their even distribution in time. To begin with, as already noted, innovations differ greatly in importance. It is only natural that clusterings of subsidiary improvements and applications quickly follow the more fruitful and fundamental breakthroughs as their potential is recognized and widely exploited. At the same time, the heterogeneity of innovations must be a source of irregularity in the rates at which they occur. Merely to identify a few broad categories, they alternately destroy or modify older processes or products, create new ones, employ new combinations of resources, and develop new linkages with markets. Such differences among innovations seem likely to be a source of different, unrelated periodicities.[32]

However, of even greater importance as a source of the punctuated, episodic character of technological advance are factors of human vision and motivated leadership. Much of what is known about this we owe to Joseph Schumpeter (1883–1950), the outstanding figure in the study of industrial entrepreneurship. Noting that compressed surges or waves of innovation generally lay at the root of major economic advances and tended to be focused on narrow sectors of industrial activity, he readily accepted, even welcomed, the disruptiveness of their impacts.

Describing them as "more like a series of explosions than a gentle, though incessant, transformation," Schumpeter interpreted these shocks as the main engine of capitalist growth. By creating new opportunities for profit, they encouraged an inflow of entrepreneurial talent and capital investment and a consequent growth of new industries. This generally led to intensified competition, he believed, and for a time secondary innovations would sustain the momentum. Presently, however, an exhaustion of even the derivative technological potential of these subsequent initiatives culminates in a depression which terminates only after a later shock.[33]

As implied in this generalized sequence, account needs to be taken of an important terminological distinction between inventions and innovations. Largely as a result of Schumpeter's stimulus, it is common today to differentiate between the two. The term invention is reserved for genuinely original acts of discovery, occurring under circumstances that are largely outside of and immune from economic processes. Innovations, on the other hand, are integral to ongoing capitalist economic development. They involve the combination, modification, and application of themes drawn from an existing pool of knowledge, but the agents primarily responsible for introducing them are not independent scientists or engineers but entrepreneurs and teams working under their immediate directions.

Unlike inventions, which tend to focus on a central discovery, Schumpeter recognized that innovations may consist of long, unrelated series of individually minor, if often cumulatively decisive, improvements. Under this distinction, both invention and innovation play a part in technological advance. Basic, essentially unpredictable discoveries usually precede refinements and improvements in applications but sometimes also grow out of them. Science is probably closer to the category of discoveries or inventions, and engineering closer to innovations and accompanying applications. Separating the two, however, has inevitably involved drawing a line

of division across a continuum. Hence there has been an essentially pragmatic tendency to reserve the term inventions for patentable ideas and to
apply innovation much more loosely to newly introduced improvements.[34]

There is seldom a clear, let alone predictable, line of derivation leading
from an invention to the innovations later associated with it. Inventions are
usually generated in a controlled environment, responsive to their originator's untested perception of a need or opportunity. Later they are introduced into a larger technological and marketing matrix that may call for
improvements as well as for their fundamental reorientation or redefinition.
As merely one of many examples, nylon emerged from a search for new
polymers to strengthen existing product lines. It only became DuPont's
most commercially important discovery after being successively fitted into
new niches first for stockings and later for tire cords, structural plastics,
and carpeting.[35]

All of this is consistent with Schumpeter's basic view of capitalism itself. He saw it, in short, as an evolutionary process, its continuously dis-
equilibrating changes largely a product of waves of innovation. An unmatched capacity to generate growth, not stability, is the measure of the
success of this "perennial gale of creative destruction."

Herein lies the essence of the entrepreneurial function—a role blending
rational calculation with charismatic leadership. While not ordinarily directly generating new technical innovations, successful entrepreneurs must
intuitively envision their potential and seek out or devise for them new and
more profitable production functions, marketing niches, and sources of investment capital. The entrepreneur may also be a capitalist, of course, but
the assumption of risk is not essential; "qua entrepreneur he loses other
people's money."[36]

Far from accepting consumer preferences as a given, in the manner of
neoclassical economics, Schumpeter boldly asserted that the entrepreneur
could and should set out to create or re-direct these tastes as necessary.
Tolerant of oligopoly as an encouragement to research-based innovation,
he attached little importance to mechanisms of innovation diffusion or to
demand as a significant contributor to the business cycle. But these clearly
more controversial views do not detract from the fundamental character of
his contribution on entrepreneurship and innovation.

Where a number of economic sectors are affected more or less simultaneously by innovation clusters, related changes are likely to occur on a
pan-societal scale. The Industrial Revolution in England might be considered the prototypical example of this, and the U.S. era is clearly another.
Beyond mere simultaneity, new technologies that cluster in time have
not infrequently enhanced the range of applicability as well as the market
demand for one another. Such was the case in this country during the nineteenth century, first with railroads and steelmaking and then with the steam

turbine, electric power generation, and electric lights. A similar, somewhat later example of convergence is provided by petroleum technology, the internal combustion engine, plastics, and polymers. In our own time, microelectronics extends its massively creative, supportive influences to many other centers of innovation at once.[37]

As already noted, some clustering of innovations is to be expected even in the absence of entrepreneurship. Irregularities of rate, interspersing relatively long quiescent periods with shorter impulses of rapid change and growing complexity, are not a feature of technological progress alone. Economist Brian Arthur has plausibly suggested that they will commonly characterize unstable hierarchies of dependence or "food webs" where the "individuals or entities or species or organisms co-exist together in an interacting population, with some forming substrates or niches that allow the existence of others." Positive feedbacks make the behavior of these systems nonlinear, with even modest increases having a greatly magnified effect through the creation of attractive new niches. "Hence we would expect that such systems might lie dormant in long periods of relative quiescence but burst occasionally into periods of rapid increase in complexity. That is, we would expect them to experience punctuated equilibrium."[38]

Quite similarly, the factor of cross-fertilization and mutual reinforcement can intensify clustering in the case of technology. Arthur illustrates this "growth in coevolutionary diversity" with the example of multiplying specialized products and processes in the computer industry. Modern microprocessors created niches for memory systems, screen monitors, modems, and other hardware, which in turn begat further niches for new operating system software and programming languages, and beyond that for desktop publishing, computer-aided design, electronic mail, and so on. All of the latter innovations fell quickly into place once the breakthrough innovation of the microprocessor had been achieved.

Again, Arthur describes the "structural deepening" of technological diversity as an irregular process with an inherently multiplicative dynamic of its own. Early jet engines, for example, were quite simple in concept. But once their superiority led to widespread acceptance, new needs quickly were recognized—for higher performance, to respond to more extreme stresses and temperatures, and to extend operating range and reliability. Multiple compressors, air intake control systems, cooling systems, afterburner assemblies, fire-detection and de-icing assemblies, engine-starting systems, and much more all were soon added as engine power climbed to levels thirty to fifty times higher than had been possible originally. "But all these required further subsystems, to monitor and control their performance when *they* ran into limitations."[39]

A number of attempts have been made to provide a quantitative expression of the clustering of innovations as a function of time. As conveniently summarized in a recent work by Louis Girifalco, all have successfully re-produced convincing, impressively peaked clustering patterns alternating with troughs of pronouncedly low innovative activity. There is little agree-ment, however, either on the lists of advances selected as important in these studies, or on the dates individually assigned to them. Nor are there good grounds for optimism that either the subjective elements in the process or the obscurities in the basic data can be readily overcome. Suffice it merely to say, then, that "the clustering of technological innovation will be taken as a fact."[40]

Longer-run regularities are strongly argued by some, and there is consid-erable, if debatable, evidence in support of them. More debatable still is the proposition that they exhibit a regular periodicity of about a half-century, and that, as Schumpeter originally held, this can be closely tied to business cycles. Advocates of so-called long waves identify them, in particular, with lengthy, fairly regular "Kondrarieff" cycles of rising and falling prices, aggregate output, and employment.

Brian Berry, a recent and vigorous advocate of long waves, accurately identifies one key problem as "evidentiary." Can such waves in fact be identified?[41] "Brute-force empiricism," however, is not enough to establish their existence. Skeptics continue to argue against them also on first princi-ples. Doubts are raised, for example, as to whether a plausible nexus of causality has yet been sketched that can account for a regular and repeated relationship over time between recurrent swarms of new products and pro-cesses on the one hand, and the general level of profitability and business expectations on the other. It is no less plausible to maintain as an alterna-tive that "the recurrence of innovation clusters has been more in the nature of *an historical accident* than endogenously generated fluctuations in the rate of innovations."[42]

Quite apart from the regularity of the cyclical aspect of long waves, therefore, is the question of whether wider socioeconomic changes are con-sistently triggered by some deeper regularity inherent in technology itself, or instead by a wide and unpredictable variety of other contextual variables and even accidents. Berry, defending the former alternative, has outlined a sequence of phased developments that he and others describe as techno-economic trajectories or paradigms. Each, he suggests, is

a combination of interrelated innovations that embodies a quantum leap in poten-tial productivity, opens up an unusually wide range of investment and profit op-portunities, and produces major structural crises of adjustment in which social and economic changes are necessary to bring about a better match between the new technology and the system of social management.[43]

An early example of such a determinate chain or trajectory, according to Berry, began with the telegraph (1837–39). Consequent upon its invention were important innovations in many aspects of communications, including a national grid and the Atlantic undersea cable. A host of subsidiary innovations followed, leading to great improvements in the technical standardization, coordination, efficiency, and safety of the railroad and shipping industries. Much more effective and quickly responsive financial institutions, credit arrangements, and securities markets became possible. Secondary effects rippled outward, stimulating, for example, the flow of capital investment and hence the pace of industrialization.[44]

A considerable degree of linkage between innovative technologies and changing structural features of the society and economy certainly emerges as plausible from this account. But nothing yet, and no compelling avenue of further investigation yet visible, establishes the direction of flight of the causal arrow connecting the two. The most economical explanation is that the linkage is a highly interactive one, with arrows flying continuously in both directions at once.

Systems: Artifacts of Human Design and Leadership

At least insofar as they are agencies of change and in change, technological systems are jointly composed of inanimate and human elements. Their substance and modes of operations are better understood, as noted earlier, if they are thought of as *socio-technical systems*. Enormously variable as they are, the material components need no further elaboration. But it is people, not machines, whose crucial function, "besides their obvious role in inventing, designing, and developing systems, is to complete the feedback loop between system performance and system goal and in so doing to correct errors in system performance."[45]

What underlies and sustains technological systems is partly institutional and partly technical, partly rooted in material capabilities and possibilities and partly in human associations, values, and goals. Centralized control or a motivating vision is not essential for a system, for the common feature is merely one of "interconnectedness—i.e., a change in one component impacts on the other components of the system."[46]

Technological systems have several obvious characteristics, and some that are less obvious. They are complex, hence usually decomposable into hierarchies of subsystems that may be independently subject to intervention and modification—or breakdown. They persist in time, following a life cycle of varying duration from birth to maturity, and in most cases ultimately on to slow senescence and disappearance. They must have boundaries of some sort, as implied by the criterion of "interconnectedness," although these may need to be left loosely defined and somewhat dependent on the perspective of the viewer.

How does the somewhat abstract and expansive notion of technological system articulate with technology as practice? At one level, as Edward Constant has cogently suggested, we may think of communities of technological practitioners contributing different skills and disciplines to their daily work on a common class of undertakings in one or more organizational settings. Problem solving, hence change, goes on at this level primarily in the form of continuing efforts to reduce uncertainties and improve performance under more variable or stressful conditions. "Solutions to such problems are normally sought within the received tradition through incremental permutation of the conventional system." At another, higher level, however, technology-as-function cannot maintain an independent existence but must be embedded in larger systems and their organizational superstructures.[47]

In short, technological systems should be thought of as engaged with external constraints, challenges, and opportunities as well as with internal problems and the articulation of their sub-units. At these and intervening gradations or levels, common usage of the term is sufficiently open-ended to include products of conscious leadership and design at one end of its spectrum and a kind of slow, multicellular, evolutionary growth and adaptation at the other. This looseness and flexibility, it can be argued, bows to the real-world heterogeneity of systems phenomena. But it is also an indication, at the very least, of the continuing "need to think through in what sense the DuPont corporation, the auto or the leather industry, and Silicon Valley were or are systems in anything like the same sense."[48]

Even in the remote past—and certainly to a far greater extent today—technological systems tend to be associated with teleological overtones. A goal toward which these inanimate entities seem to have been directed is often taken for granted. Largely responsible for this must be the frequent personification of such goal orientations by outstanding, creatively risk-taking corporate leaders who have successfully developed new processes or products through consciously focused R & D.[49]

Classically exemplifying this entrepreneurial aspect of technological change and innovation, Thomas Hughes depicts Thomas A. Edison and Samuel Insull as the initial system builders in the electric power industry. Their forceful and eloquent pronouncements clearly sought to reinforce the impression of the inevitability of the technology to which they were committed. Calling for great breadth of vision as well as persuasiveness, it is a principal function of the system builder simultaneously to impose a coherent pattern on its originally independent elements and to cast doubt upon any alternatives.[50]

Yet "the precise place to be accorded individual economic action in the evolution of modern technological systems" is controversial ground on which to venture:

When do specific actions, taken purposively and implemented by identifiable agents (say, entrepreneurs) have the power to alter significantly the course of technological history? . . . Rather than being "made" by any*one* at any particular time, does not technology simply accumulate from the incremental, almost imperceptible changes wrought by the jostling interactions among the herd of small decision makers . . . all of whose separate motions have somehow been coordinated and channeled by markets under the system of capitalism? Should we not then insist that the nature of technological progress is best conveyed by conceptualizing it as a continuous flow, and foreswear narratives that dwell on a succession of unique actions and discrete historical "events"?[51]

However, a difficulty with following this suggestion of Paul David's is that his question probably has as many answers as there are records of individual cases on which to base them. Furthermore, the potential extent to which an individual as a purposive agent looms larger than impersonal processes of technological change is surely dependent on the narrowness of the sector of change under scrutiny. David deals at length with Edison's skillful, aggressively conducted retreat from the direct-current generating systems he had promoted, as alternating-current systems gradually carried the day. But the choice of actions taken by an entrepreneur, insofar as it overcomes the external pressures that largely induced the choice, has the simultaneous effect of sharply limiting our knowledge of alternatives that might reasonably have been chosen instead.

A recent critique that focuses on Edison in considerable detail enlarges upon these themes. The authors cast doubt on whether Edison's position was the only technologically practical alternative, or even the most economically advantageous one. They challenge, in short, "the prevailing sense of . . . individual omniscience and self-conscious historical missions," and "argue instead that . . . social networks were vital in providing the resources and opportunities that facilitated success." This argument has a "family resemblance," as they note, to one on which David himself has collaborated with Brian Arthur on the way in which inefficient technologies are sometimes "locked-in" by accidents of priority or other external considerations.[52] What has become known as "path-dependence" unquestionably is a major, recurrent feature of the historical record. Initial advantages, whether predisposing to an optimal resolution or an inferior one, have a very good chance of establishing the course that is subsequently followed.

Whatever balance must be struck in evaluating Edison's actions, innovator-entrepreneurs typically confront an imperfectly understood, highly irregular front of challenges: competing firms and systems, financing and marketing uncertainties, and a host of technical problems of varying difficulty. It may be little more than a matter of chance whether attention is initially focused on strengthening the weakest link in the chain or on some other, temporarily more urgent problem. Even if purely technological considerations predominate, these may be perceived and measured in other than purely objective terms. Thomas Hughes has persuasively argued for the widespread importance, for example, of the military metaphor of the "reverse salient" that

> appears in an expanding system when a component of the system does not march along harmoniously with other components. . . . As a result of the reverse salient, growth of the entire enterprise is hampered, or thwarted, and thus remedial action is required. . . . The reverse salient will not be seen, however, unless inventors, engineers, and others view the technology as a goal-seeking system.[53]

According to Hughes, "innumerable (probably most) inventions and technological developments result from efforts to correct reverse salients." At least in contemporary economic life, this form of goal orientation has come to play a crucial part. It encompasses attempts to address diverse, related issues of marketability, profitability, and public acceptance, as well as technical puzzle solving in a narrower sense.[54]

Having stressed the frequent association of technological systems and system building with a goal orientation, we are at risk of implicitly setting upper limits on the utility of the term *system* itself. No matter what its scale or complexity, a single class of commodities seen in the context of its design, production, distribution, and use (e.g., the Boeing 747s) unambiguously constitutes a system. Although independently managed, the textile mills that sprang up in Lowell, Massachusetts in the early nineteenth century were consistent enough in design and mode of operations to come under the rubric.[55] There was a comprehensive plan behind our construction of a network of strategic highways, so that it, too, usefully qualifies as a system within this goal-oriented framework. But the term may not be equally appropriate for our entire network of primary and secondary roads, most of which simply evolved in response to local priorities.

As we explore these upper limits of generality, we must recognize that consciousness and goal orientation are by no means necessary criteria for the existence of *all* systems. Perhaps it should be conceded that to speak comprehensively of a modern American industrial system would conflate so many differently oriented and even contradictory elements as to be meaningless. Possibilities for comprehensive planning, and for realization of those plans as they were originally visualized, are clearly circumscribed in any democratic, or for that matter merely pluralistic, society. Yet there still can be striking aspects of systematicity—imposed, in particular, by the consistent application of market principles—in the outcome.

Highly uniform regularities, for example, have been persuasively illustrated in the successive rates of growth and replacement of canal systems, railroads, highways, and finally the airline share of total intercity route mileage in the United States. Typically these follow almost identical logistic curves or straight-line log-linear transforms, and similar patterns can be found in lengths of wire in telegraph systems, the wood-coal-oil-natural gas succession in primary energy substitution, and other phenomena.[56] Singularity of goal may make systems unambiguous, in other words, but more generally, finding technological systematicity only depends on what questions are asked and where it is looked for.

Technological Diffusion and Adoption

Originality, as in inventions, may seem in principle unambiguous. But even apart from factual uncertainties, it seldom proves to be an all-or-nothing proposition. Examples of parallel, sometimes almost simultaneous, discoveries or inventions have been found to be surprisingly numerous.[57] Similarly, there are many gradations of originality in the borrowing process. The copying of a simple, self-explanatory artifact or process may require no originality at all. But where things are more complicated there is much experience to show that personal demonstration and instruction may be an absolute requirement.[58]

Taking a diffusion perspective on the transmission of technology, the initiative is assumed to be in the hands of the diffusing agency. Its outcome is seen as resulting from an autonomous decision of the lender, rather than as a transaction in which both parties have parts to play. An adoption perspective has the opposite emphasis, placing primary emphasis on the perceptions and calculations made independently by the borrower. In most—perhaps all—real-world instances there probably is a place for both perspectives in fluid interaction. One may in fact legitimately ask, as anthropologist Julian Steward did long ago, "whether each time a society accepts diffused culture, it is not an independent recurrence of cause and effect."[59]

Perhaps the simplest way to conceptualize diffusion introduces a so-called "epidemic" model. Diseases are transmitted from infected to healthy individuals, usually moving most readily along major routes of trade and communication. With many potential recipients of the disease, rates of transmission are initially very high. The rate falls, however, as the potential number of contacts declines with the declining proportion of the population that is uninfected. Ultimately, near-stability is reached.

The difficulty with the epidemic model is that, in key respects, it invokes only the diffusion perspective. Adoption is seen as unproblematic, an involuntary act of relatively undifferentiated recipients (granting that a few may have differing degrees of resistance or immunity). Little or no allowance is made for an element of the borrower's choice based on differing access to information or resources. Other eventualities are also overlooked.

There may be a concerted move toward acceptance on the part of many borrowers, for example, after the advantages of a new product or process have been convincingly demonstrated. Or it is equally possible that non-adopters will make that choice quite deliberately, perceiving some advantageous niche left open by the affirmative decision of most of their competitors.

There are still further difficulties. One is that few inventions or innovations come into existence in their final, fully developed form. We will presently take note of numerous modifications and improvements that have accompanied most of the great technological breakthroughs, cumulatively adding more to the success of the product or process than the original breakthrough itself. Recognizing that adopters also have choices to make, potential adopters may find it advantageous to delay their decision in anticipation of the further improvements that are likely to occur. As all these alternative possibilities of divergent but entirely rational borrower behavior suggest, the epidemic model has serious limitations.[60]

Still a further difficulty is that the larger competitive environment is no more stationary than the diffused product or process. Rather than writing off large, unrecovered investments, seemingly obsolescent technologies sometimes may not be abandoned but instead vigorously revived with new innovations and capital improvements. The greatly enhanced efficiency of water wheels resulting from John Smeaton's technical studies in the eighteenth century, for example, no doubt substantially slowed their replacement by steam engines as a source of motive power for nineteenth-century factories.

These cautionary considerations suggest that technological transfer and adoption tends to be a transaction involving two or more unequal partners. Rationality on the part of either partner may be hard to recognize or define under these circumstances. A recent international symposium of specialists gave considerable attention to this question, against the claim of a surprisingly voluminous database of some 3,000 "well-documented and quantified" examples of such diffusion. Yet even with an empirical footing of this impressive size, the rapporteur's summary of the discussion can only feed the fires of doubt about the extent of rationality in transactional scenes:

> Prices convey certain kinds of information and send certain signals, but their roles and causal functions from a diffusion perspective would seem to be less important, especially in the early phases of diffusion, than mainstream economic theory would indicate. What emerges is the importance of specific history, . . . captured in the phrase . . . path dependence.[61]

With diffusers and adopters jockeying for position and weighing their respective advantages in a climate of considerable uncertainty, negotiations over technology transfer frequently tend to be complex, volatile, and protracted. The breeding of hybrid corn, which has been the subject of classic work on the rates at which innovations diffuse, is only one of many examples. While the idea originated no later than 1918, its implementation in commercial quantities in any state was delayed for nearly two decades— and in some states for more than three decades.[62]

In such instances, a number of alternative explanations typically are possible. Under the adoption perspective that focuses on the distinctive attributes of the individual recipient, the primary factor might be "innovativeness and personal interaction or communications effects." The economist Zvi Griliches, who conducted the original study of hybrid corn, sought to explain the lags largely by differences in calculations of profitability, while also noting that different hybrids needed to be independently bred for each area.[63]

For diffusion as an obviously variable historical process, it is clearly important to recognize that technologies generally overcome their initial flaws through gradual, if not necessarily continuous, evolutionary development. Feedback information from users is one of many sources of improvements.

> The existence of these sources of positive feedback brought about by the irreversible, dynamic, decreasing cost effects of the diffusion of a new technology implies that small initial advantages or disadvantages . . . can cumulate readily into large advantages or disadvantages in comparison with alternative technologies. A particular product design, process technology, or organizational system thus can become "locked in," while rival technologies are "locked out" through the workings of decentralized competitive market processes.[64]

Contingent elements, path dependency, and variants of adoptive behavior all converge virtually to refute the possibility of order and regularity in technological diffusion. Yet it is evident that superior technologies—those that promise enhanced well-being, less costly substitutes for human effort, improved ways of marshaling natural and human resources and organizing work, more efficient modes of transport, or simply extensions in the accessible ranges of experience and enjoyment—have always spread irresistibly and continue to do so. Disturbances and impediments of various kinds may deflect or delay, but do not permanently block, technological advance.

The Embeddedness and Indeterminacy of Technological Change

Whether technology itself ultimately covers a field as wide as history, as Braudel maintained, may be debatable. Beyond question, however, is the fact that it is a vast, ill-defined, hugely variable domain. The many subsystems of which it is composed are only loosely linked together, and their ties with the larger social order are also highly differentiated and seldom move in unison. But for the ties that are most dynamic and transformative we are led to look in two principal directions.

Foremost through most of history have been the forces of economic growth, increasingly identified with markets. More generally, however, a much wider range of activity is implied—directing the paths of capital investment, managing systems of production and distribution, assessing and assuming risks, and providing the entrepreneurial leadership necessary for new technologies to be successfully deployed.

The second direction of close and interactive ties that play a part in accounting for technological change is the one we identify today with basic science and the disinterested pursuit of knowledge. Over the course of the last century or so it has come to assume fundamental importance. Today, in fact, it will presently be shown that there is an increasing convergence between at least parts of the basic research complex and market-driven economic institutions and activities. That the rate of convergence is still accelerating, with no end to the process yet in sight, reaffirms the dynamism of this direction of technology's interdependent linkages as well as its economic ones. But what makes this possible is the cumulative growth of science's practical relevance and powers, its capabilities to contribute directly to the fulfillment of economic goals on the basis of new understandings of fundamental principles. Those capabilities must be understood not as a constant attribute of the work of scientists and their antecedents but as a gradual historical emergent over the last two centuries or so.

Consistent with this general position, and also expressed at the outset of this book, is the conviction that "hard" technological determinism does not provide a useful vantage point from which to understand technological change. My own interest in technology may have originally stemmed from the issue that was at one time heatedly debated among archaeologists, on the role of technology as a putative "prime mover" in society. To think of technology in this way implies a philosophical predisposition to suppose that the physical means of human adaptation to the social as well as natural environment can exist to some extent outside the social order creating them.

Unqualified forms of this doctrine find few if any advocates today. Indeed, if we were to accept Braudel's insistence that technology covers a field as broad as history, labeling the undifferentiated mass of everything technological as a prime mover would contribute little to deepening historical understanding. But there are several, not necessarily mutually exclusive, alternatives that are more commonly encountered.

The first is the more restricted assignment of a degree of causal responsibility for particular institutions or behaviors to specific technological features. This is a position associated, for example, with the work of medieval historian Lynn White, Jr. What is clearly called for is neither a blanket acceptance nor denunciation of technological primacy when it is argued in these terms, but rather a careful specification of proposed, specific causal linkages that will permit them to be subjected to critical scrutiny.[65]

The second alternative, proposed by Nathan Rosenberg and strongly favored here, is a kind of

> "soft determinism," in which one historical event did not rigidly prescribe certain subsequent technological developments, but at least made sequences of technological improvements in one direction easier—and hence both cheaper and more probable—than improvements in other directions. Technological nature is by nature cumulative: major innovations constitute new building blocks which provide a basis for subsequent technologies, but do so selectively and not randomly.[66]

Mention should also be made of a third instance, a widely accepted proposition in economics that has overtones not unsympathetic with regard to technological determinism or prime movers. Technological advance has long been regarded as a key component in the otherwise unaccounted for "residual" responsible for positive rates of economic growth.[67] To be sure, this choice of variables is rather strictly circumscribed. The focus is only on economic growth within a modern industrial, not universalistic, framework. Also, the concern is for incremental, quantitative, rather than qualitative or structural, change. And there is an implicit atmosphere of suspended disbelief or conjecture about the concept itself: "The usual routine, in the absence of anything better, is to treat technological progress as the ultimate residual. One identifies as many of the components of economic growth as one can, and what is left provides at least an upper limit to the contribution of technological change."[68]

The larger issue, surely applying not merely to modest innovations but to at least some inventions of major importance, is the extent to which they can only be understood as having emerged from, and remaining firmly embedded in, a fuller context of social interactions. "Demand pull"—in this context, market demand for discovery of specific products or processes—is one broad category of such interactions. Relevant here is Jacob Schmookler's finding that time series for heightened investment tend to coincide with, but to slightly precede, heightened rates of patent applications. This seems to be more easily explained on the assumption that upswings in investment were responses to accelerations of business activity after the troughs of economic downturns, but that upswings in patenting were responsive in turn to this form of economic demand.[69]

Suggestive as this analytical approach seems to be, the significance of "demand pull" remains controversial. In an interesting exchange on the subject, Harvey Brooks maintained that "the market really does not demand anything that does not exist," while Alfred Chandler offered in rebuttal a series of historical examples (the substitution of kerosene for whale oil, the locomotive, the telegraph, and the intersection of the invention of the cotton gin with innovations in British textile machinery) in which it seems to have done so.[70]

Presumably the effectiveness of demand as a stimulus to inventiveness varies in different historical, economic, and scientific settings, and is quite likely to be different for each identifiable case. Can the autonomy of inventive acts that extend beyond the frontiers of present knowledge be termed a technological "supply push," corresponding to "demand pull" from the opposite direction? Difficulties are apparent in rigorously specifying either "supply" or "demand" for this purpose. In a way, therefore, the entire distinction between the two may be heuristic rather than operational.[71]

Here the issue of the contributions of science forces its way into any "contextual" consideration of the sources of technological change. Once again, of course, the answer to this question cannot be static. Modern manifestations of both science and technology are radically different than their predecessors of only a generation ago, which in turn bore little resemblance to the antecedents of both at the time of the Industrial Revolution.

Science and technology are complementary, reciprocal, interdependent, and symbiotic—and are becoming progressively more so. Although strains in the relationship exist, they arise primarily from economic pressures and the institutional setting and not from differences in substantive findings, theories, methodologies, or daily activities. The two have come to constitute overlapping groupings along a broad continuum, although outmoded stereotypes persist toward the more distant, opposing ends of the continuum.

The relative recency of this development must be re-emphasized. An earlier interpretation of the relationship, enshrined in Vannevar Bush's report to President Roosevelt, *Science, the Endless Frontier*, envisioned an essentially one-way relationship in which new ideas and discoveries were generated in basic scientific research and moved progressively outward into technology and applications. That interpretation, even if already somewhat tendentious in its own time, was enormously influential only two generations ago. Certainly it does not stand today without very important qualifications.

As we proceed backward into the nineteenth century and earlier, relations between science and technology become progressively more tenuous and questionable. As a way of thought of increasing experimental and analytical rigor, earlier science's influence on the outlook of inventors and practicing technologists is likely to have been great—but is difficult to document persuasively or in detail. What is often called the scientific revolution of the seventeenth century and the Industrial Revolution of the late eighteenth and early nineteenth centuries were too distinct in content, and too widely separated in time, to demonstrate any pattern of direct dependency of the latter on the former. Moreover, most accounts of great scientific contributions have systematically underrated the dependence of science itself on instruments and instrument makers whose place is unequivocally in the technological realm.

There is no difficulty in demonstrating the extent to which progress in science has been linked to new instrumental windows of observation like the telescope and the microscope. Increasingly in recent years, instrumental advances have made it possible to detect, measure, and even manipulate entities ranging from subatomic particles to the radiative signals from almost inconceivably remote galaxies, and to do so across progressively shorter and more accurate intervals of time.

But we must also reject, as Nathan Rosenberg has recently insisted, the "rather crude sort of technological determinism" that this may seem to imply, and to recognize instead that "the relationship between technology and science is far more interactive (and dialectical)." Instrumental advance, after all, is seldom if ever an entirely autonomous process. Frequently it is dictated by anticipations of scientific need or opportunity. And at the same time, the breadth and depth of the impacts of particular instruments defy prediction and differ enormously:

> [I]mprovements in observational capabilities were, by themselves, of limited significance until concepts were developed and hypotheses formulated that imparted potential meaning to the observations offered by new instruments. The microscope had existed for over 200 years and many generations of curious observers had called attention to strange homunculi under the lens before Pasteur finally formulated a bacterial theory of disease, and the modern science of bacteriology, in the last third of the nineteenth century. The expansion of observational capability had to be complemented, in turn, by the formulation of meaningful interpretive concepts and testable hypotheses before the microscope could make large contributions to scientific progress. Thus, path dependence was critical in the sense that a particular body of science, bacteriology, had to await progress in an observational technology. But such progress was, of course, not sufficient for the scientific advance, only necessary.[72]

What questions can we ask of technological change with some hope of finding answers? *How* it occurred? Yes, in a generalized way and within limits: There are many examples of cascades of related changes, creating a kind of "systems" logic or technological momentum that would carry on for at least a limited period until its inner potential to go forward had been exhausted. That can be balanced against "demand-pull" factors, including financial, production, and marketing considerations as well as the rapidity and effectiveness of diffusion processes.

Why is more difficult, and perhaps unanswerable except as a matter of purely subjective satisfaction. A materialist approach perhaps would satisfy itself with "historically specific configurations of resources, techniques, methods of production and factor endowments."[73] But in human affairs the whole is almost always more than the sum of its inanimate parts. The contributive role of an emergent scientific outlook to otherwise unrelated developments in technology is one example of this. The rise of Japan, in spite of (or is it partly because of?) its extraordinary resource constraints is another. Not many will be satisfied, therefore, by explanations that are so narrowly limited to purely materialistic factors. Causality really resides only in the whole of an embracing, ongoing, interacting system.

As noted earlier, the linearity with which systems seemingly unfold is usually only apparent retrospectively. Participants, even well-informed, centrally placed participants, typically are unable to form an accurate sense of the potential of a new technology for which they have considerable responsibility. This is partly because of the importance of path dependency that has already been mentioned; unexpected convergences of forces not originally deemed consequential or even relevant, and even accidents, can have a decisive influence on the later directions that new technologies actually follow. But it is also because new technologies are almost inevitably seen as supplements to existing technological infrastructures and placed within existing systems contexts.

Nathan Rosenberg has drawn together some marvelously telling examples of this inability to perceive the consequences of technological change. They include the failure of Western Union to perceive the potential of the telephone, as reflected in its refusal to purchase Bell's telephone patent for a mere $100,000 in 1877, and the firm's readiness to withdraw from the telephone field in exchange for Bell's agreement not to enter the telegraph business. Similarly, Marconi only visualized radio as permitting point-to-point communication in settings (as between ships at sea) that could not be served by permanently wired connections. Again with computers, IBM's senior executive foresaw no commercial demand and believed that the single computer already in operation in the New York headquarters "could solve all the important scientific problems in the world involving scientific calculations." Finally, Bell Laboratory was initially unwilling to pursue a patent for the laser, on the grounds that electromagnetic transmissions on optical frequencies were of no importance for the Bell System's interests in communications.[74]

There is no significant evidence here for the short-sightedness of any of the decision makers who were responsible for such myopia at the birth of these technologies. What is demonstrated instead is the inherent unpredict-ability of outcomes as seen only from the vantage point of their initial discovery, and the difficulties of visualizing new systems different from those that have already become familiar. Technology emerges not only as chains of discovery that cascade outward in many unrelated directions and to unpredictable distances, but as the progenitor of systems that are, prior to the inadvertent entrance into them, essentially unknown worlds.

2

The Useful Arts in Western Antiquity

AN ACCOUNT of technological change with any pretense of comprehensiveness probably should begin far back in the early hominid past. This book makes no such all-embracing claim. The end of this chapter will carry us almost to the threshold of an industrial age in the early eighteenth century, while permanently narrowing the focus of the more concentrated chapters that follow to England and the United States. But a long backward look is more than merely congenial to an author who for several decades made his living as an archaeologist. It introduces, if only in more impressionistic outline, a strikingly similar set of themes and processes of change to those we will subsequently take up in greater detail.

For our first million years or so as an identifiable, hominid presence, stone tools shaped and refined for increasingly specialized uses provide the fullest record of our antecedents. Less is known of the uses to which many more perishable materials were put, but where local conditions are favorable for preservation there are sometimes tantalizing suggestions of how much is otherwise missing. There is also increasingly reliable testimony from bones and plant remains, as well as from soils and geological formations. From these diverse sources it is often possible to reconstruct long-term environmental variations, niches in which hominid occupation tended to concentrate, and even seasonal variations in the subsistence quest.

Technology is all these traces of adaptive activity, but it is more. Provisions for shelter against predators and inclement weather can be well documented, although with an unfortunately heavy bias on more easily locatable caves rather than open sites. Various lines of indirect evidence bear on the scale and composition of local groups as predatory, extractive, and cohabiting communities committed to their own survival and reproduction. Knowledge of the multiple uses of fire was a crucial technology that was added along the way, to be forever and inextricably associated with our species. Spoken language was more than a technology, of course; for example, it provided the means of recognition of a group's distinctiveness and solidarity. But as an enormously flexible, rapid, and efficient means of communicating differentiated roles in common actions and weighing common responses to needs or challenges that could not be met with ordinary routines, it was, in fact, the supreme human technology.

Omitting many details that testify to the increasing complexity and effectiveness of social organization and the food quest, and also of growing regional variation, this brings us down to the end of the Pleistocene epoch ten or twelve thousand years ago. Dating uncertainties and enormous gaps in the record make it difficult to discuss the sequencing of all these earlier changes. Quite possibly they (or their constituent elements) occurred in irregular spurts, matching the later patterning to which this book is devoted. But no clear answer is possible, and to provide an accurately qualified one calls for a paleolithic specialist—which I am not.

Taking the form of waves or pulsations that were at least relatively rapid, the subsequent beginnings of agriculture are closer to the theme of this study. Varied complexes of plant and animal domesticates were bred and selected from wild, locally encountered progenitors. A new seasonal round of cultivation and husbandry was introduced, presently requiring newly specialized toolkits. Human groups found themselves encountering new vulnerabilities and uncertainties in spite of the prevailing advantage of settling in larger, more sedentary social units. Qualitative increases in community size and durability led to altered responsibilities and complementarities of gender and life cycle, and produced a cascade of innovations in forms and construction techniques for shelters that were now wholly artificial.

Certainly in comparison with the long spans of barely perceptible change of the pre-agricultural past, this was a remarkably rapid and far-reaching transition. It seems to have occurred first in the Near East. But far more significant than any regional priority is the fact that it happened several times and in both hemispheres, within a span of no more than a few thousand years after a million years of human evolution. The immense distances separating the nuclear areas where this occurred, and the different arrays of domesticates and accompanying technologies, argue strongly that these were essentially parallel, independent processes. It is difficult to imagine a more suggestive demonstration of the basic dependence of remarkable, seemingly free-standing pulses of technological advance on the prior crossing of some ill-defined, clearly very general level of cultural adventurousness and competence.

The Rare Pulsations of Antiquity

The next great pulsation even more clearly attests to the context-dependent path of technological change and its many lines of intersection with wider cultural milieus. On present evidence the beginnings of civilization occurred first in Mesopotamia during the middle or latter part of the fourth millennium B.C. That seminal period was very quickly followed by similar developments along the Nile, and not long afterward in the Indus valley. There soon appeared—always with variations around a common core of greater hierarchy and complexity—similar patterns in China and the New World. One may reasonably postulate a greater degree of interconnectedness among these nuclear areas than with the beginnings of agriculture, but once again diffusion does not appear to have been a decisive factor.

Particularly the original, Mesopotamian, manifestation has been not unreasonably characterized as an urban revolution, and likened to the Industrial Revolution.[1] Like the latter, the former epoch was comparatively brief, spanning at most a few centuries and possibly much less. Alongside major, concurrent innovations in public institutions were others in the realm of technology. What we know of them is obviously still limited to relatively imperishable categories of material. However, it is noteworthy that they have in common qualities of technical precocity and high artistic creativity.

Stone cylinder seals, for example, appearing first at this time, then remained a hallmark of Mesopotamian culture for millennia. This almost iconic class of artifacts was endowed, almost from its first appearance, with a "creative power . . . such that we meet among its astonishingly varied products anticipations of every school of glyptic art which subsequently flourished in Mesopotamia."[2] Their distinctive designs, when rolled across lumps of clay, could seal containers or even storerooms against illicit entry, and identify the civic or religious authority responsible for them. Thus they were an integral part of the technology of administration. Fashioned (and initially also partly decorated) with the bow-drill, they were also exemplars of the new technology of the wheel that is first manifested at roughly the same time in wheeled vehicles and wheel-made pottery.

Similarly, an examination of jewelry and other metal objects from the so-called Royal Tombs of Ur, a few centuries later, has revealed "knowledge of virtually every type of metallurgical phenomenon except the hardening of steel that was exploited by technologists up to the end of the 19th century A.D."[3] There were comparably far-reaching developments, reflecting most of the technical and stylistic patterns that prevailed for centuries or even millennia afterward, in monumental architecture, the mass production of ceramics, woven woolen textiles, and in all likelihood woodworking.

Concurrent and more seminal still was the Mesopotamian invention of writing. While clearly an administrative advance of unmatched power and versatility, there may be some hesitancy in describing it also as a technological achievement. There is nothing to suggest, for example, that early writing was directly instrumental in the refinement or dissemination of other craft techniques, or that any specialists other than scribes and some priests and administrators were themselves literate.

On the other hand, writing clearly consisted of a cluster of innovations that played a crucial role in allowing craft specialization of all kinds to go forward under the protection and support of temple and state institutions. It almost immediately became the essential technology, in particular, for managing the food reserves needed to maintain a dependable ration system with wide if not universal entitlements. Apart from sustaining cadres of craft, administrative, and military specialists on a part- or full-time basis, that system made possible an unprecedented marshalling of male and female labor for great public works like city walls and canal systems. Writing also made an important contribution to the amassing and allocation of scarce raw materials by the great palace and temple organizations that may have forced the initial construction of cities and certainly dominated them.

Most of these raw materials, such as copper ingots, timber that was superior to the soft, quick-growing woods of the Mesopotamian lowlands, and many kinds of stone, could only be obtained from great distances. The military or trading expeditions that procured them likely were undertaken at the initiative of the great organizations. Not a few of these technical achievements, particularly those affecting armaments and transport vehicles, both contributed to and depended upon the strength and administrative effectiveness of these urban institutions. In other words, there is a nexus of functional interdependencies closely linking writing to the florescence of the crafts and urban societies nourishing the demand for their products. Writing is thus a central aspect of Mesopotamian technology, even if most of the individual specialists we more narrowly associate with Mesopotamian craft technologies had little or no share in directly producing it.

The sequence of steps by which writing and literacy became perhaps the most salient characteristic of Mesopotamian civilization has been greatly clarified in recent years. Writing no longer appears as a great and sudden inspiration or invention, profound in its consequences but inexplicable in the individual genius required for its origins. Instead, as Hans Nissen and his colleagues observe in a recent work of great importance, "a whole series of precursors" have been identified, all of them "primitive ways of mastering problems of information storage."

Only after this, late in the fourth millennium B.C., did the small but most decisive step take place—shifting from clay tokens or models to their pictographic representations on clay surfaces. Clay, it should be noted, had already become familiar as an information-carrying medium through its use for sealings. Once this further step was taken, the greater versatility of pictographs must have been immediately evident to all. The conventionalizing of the pictographic symbols that made writing intelligible accordingly would have begun almost immediately, as would its diffusion and acceptance throughout the cultural and social sphere in which it had arisen.

The earliest uses to which writing was put were overwhelmingly as records of administrative transactions. Apart from the few examples of lists that may have been mnemonic aids in the conventionalization process, no records even remotely suggest a wider set of motivations for writing itself.[4] Moreover, as Nissen writes elsewhere, the administrative processes that can be identified in the archaic texts

> differ from those known from later periods only in detail. This includes the use of specific forms for specific transactions, the arrangements of the documents in a way that the "reader" was able to recognize the headline or the total on short glimpse; thus the chief administrator was able to keep control over a large number of actions with the least effort and time spent. . . .[5]

Later development of cuneiform writing systems need be mentioned only briefly. They proceeded more slowly, continuing through the third millennium and even later. Nissen and his colleagues note that "writing assumed new functions without losing older ones." In one direction, pictographs became more conventionalized cuneiform signs. Progressively reduced in number, they were improved by many small modifications. Gradually they became a system that could communicate with at least some of the richness and precision of a verbal message, rather than merely recording some form of association between persons, objects, and quantities by making closely adjoining impressions of the appropriate symbols on a clay tablet.

In another direction, accounting procedures, supplemented now with clarifying verbal specifications, became more effective and formal. Balanced accounts appeared toward the end of the third millennium, calculating theoretical debits with the real production of individual fields or labor teams. Metrological systems and numerical notations were simplified, reducing (although not eliminating) errors in arithmetical calculations.[6]

This sequence of developments, limited as it is to writing as only one among many contemporaneously emerging technologies in early Mesopotamian cities, furnishes a prototype for the differently based findings of the present study. It reaffirms the relatively accelerated rate, as well as the impulsive irregularity, with which new solutions to pre-existing problems can develop in naturally unfolding, not inexplicable ways. It further suggests that the problems initially addressed and solved need not exhaust the full potential of the original solution. And as there can be cascades of subsequent, previously unforeseen refinements and applications, so also a sequence like this can convey a sense of technological momentum and a transformative power of its own.

But with the available evidence for the ancient Near East as limited as it is, it will always be difficult to draw all of the enigmatic details together into an inclusive and coherent picture. What we are are primarily left with, in short, is a compelling impression of a brief creative burst. It culminated, but also was followed by, lengthier periods in which new needs, relationships, and institutions took shape only more gradually. The burst itself is known to have coincided with the development of cities, the intensification of social stratification, and the formalization of state institutions and their attendant ceremonies and monumental architecture. On the other hand, it apparently was not accompanied by substantial subsistence shifts, an overall rise in population pressure, or other infrastructural changes. All that can be said is that extraordinary technological advances seem to have been linked to—and perhaps were at the heart of—a still more extensive social and cultural transformation.

Unfortunately, Mesopotamian technology was not a focus of scribal or administrative concern. With the passage of time, much fuller, better understood documentation becomes available for the crafts, but the limitations remain essentially the same. Particularly for the Third Dynasty of Ur, at around the end of the third millennium B.C., cuneiform texts permit us to recognize handicraft production with a complex division of labor. Especially in the case of textiles, this could take place at an industrial scale, involving thousands of female workers in varying conditions of slavery and dependency. On their recruitment and training, the organization and supervision of their work, the facilities and methods they were able to employ, or the channeling of their output to meet diverse demands, the texts are virtually silent. What is recorded has merely to do with their provisioning with rations and carefully allocated raw materials, the latter sometimes imported by specialized traders from great distances.[7]

To a limited extent, analyses of archaeologically recovered craft objects that must have been produced in this way can supplement the textual picture. Such analyses are especially helpful with regard to techniques of fabrication and utilization and the sources of exotic materials that were employed. Regrettably, textiles are not preserved; their special importance is that, for resource-poor Mesopotamia, they were the principal export. Still a further difficulty is that archaeological attention has tended to be skewed in the same direction as that of the authorities who directed the scribes: away from common use and toward elite activities and standards of consumption that are disproportionately reflected in excavated luxuries, monumental buildings, and tombs.

From a technological (if no other) viewpoint, a lengthy, largely quiescent epoch ensued after the explosive changes in the late fourth millennium that accompanied the birth of Near Eastern cities and civilizations. Later antiquity witnessed only a few conjunctures of innovation, and these seem to have had relatively narrower and much slower social impacts or concordant changes.

The first, in the late second and early first millennia B.C., centered on the introduction of iron metallurgy. Smelted iron had long been known and used in very small quantities, so that the shift away from bronze cannot be attributed to the mere discovery of a new material with valuable inherent properties. At levels of metallurgical skill that had been attained at the beginnings of what is characterized as the Iron Age, iron's relative hardness was not yet a reliably attainable characteristic. As it became so, new or improved tools—picks, shovels, axes, shears, scythes, chisels, saws, dies, lathes, and levers—cumulatively transformed craft capabilities in fields like transportation and building.

But these developments follow, and hence cannot account for, the onset of the Iron Age. Progressive deforestation of much of the ancient Near East, as a consequence of pastoral overgrazing and wood-cutting for copper ore reduction as well as cooking, may have been a more significant factor. Contemporary iron ore reduction processes are not well understood, but were probably several times as fuel-efficient as that of copper. Also helping to explain the shift may have been the wider dispersal of iron ore deposits.[8]

The widespread substitution of iron for copper thus may have been economically driven. But it should be kept in mind that the production processes involved were of a pyrotechnological character. This may help to explain why much concurrent experimentation and innovation in other crafts, leading to improvements in glazes, glasses, and frits, and possibly also in metallic plating, had a pyrotechnological basis.[9]

The next succeeding punctuation of the technological lull occurred later in the first millennium B.C. Two contributory streams can be distinguished, although by Hellenistic times they had largely coalesced. Originating in the Near East were coinage and alphabetization, both stimulants to administrative advances and the long-distance movement of ideas in conjunction with commerce. As such, they constituted a kind of information infrastructure for the diffusion and utilization of proto-scientific as well as technological understanding.

A second contributory stream was centered in the Aegean. Although the Greek world is much more widely heralded as the birthplace of a formal discipline of natural philosophy, it can be credited with some technological contributions as well. If we take into account not merely the number of innovations but the extent to which they found practical application, however, the overall record has to be characterized as a conspicuously limited one:

[T]he Greeks and Romans built a high civilization, full of power and intellect and beauty, but they transmitted to their successors few new inventions. The gear and the screw, the rotary mill and the water-mill, the direct screw-press, glass-blowing and concrete, hollow bronze-casting, the dioptra for surveying, the torsion catapult, the water-clock and water-organ, automata (mechanical toys) driven by water and wind and steam—this short list is fairly exhaustive, and it adds up to not very much for a great civilization over fifteen hundred years.[10]

Only one of these devices deserves special mention for its economic significance: water wheels for the milling of grain. These too are generally believed to have been very limited in number until a much later time. In all of classical antiquity, there are fewer than a dozen known literary references to any use of water power at all.[11] Still, anyone with an archaeological background must harbor a small seed of doubt as to the representativeness—let alone completeness—of our information on their use. The lack of interest in not merely technical but also economic matters on the part of the great majority of classical writers does not inspire confidence in the significance of the paucity of textual references to labor-saving devices of any kind. That is paralleled, as we shall see shortly, by the insulated attitudes of medieval scholastics toward major technological advances going on all around them.

The dismissal by most classical scholars of the importance of water wheels for grain-milling, relying on the literary evidence, is a conclusion on which Örjan Wikander and others have recently mounted a frontal attack. Criticizing the lack of effort "to establish the fundamental premise of the discussion: the actual technological standard of Roman society," he holds that the orthodox impression of the extreme rarity of the water mill "has only been produced by the systematic disregard of the evidence for its existence in Antiquity." Archaeologically attested examples of their use, he is able to show, slowly grew in number over time. The paucity of literary references to water-driven mills, in comparison with medieval ones, can be plausibly questioned on a variety of grounds. But even for the late classical period, the most that Wikander has been able to document for the third, fourth, and fifth centuries A.D. are from eight to eleven occurrences per century across the entire span of the empire. While further findings may well deepen the impact of this criticism, it has not yet displaced the prevailing tendency to identify the Roman agricultural regime with technological stagnation.[12]

The existing textual sources that are most informative take the form of compendia of ingenious devices. Evident in the works of authors like Vitruvius (died 25 B.C.) and Hero of Alexandria (who flourished during the late first century A.D.) are great versatility, critical acuity, and engineering competence, but they provide few clues to the context or frequency of use—or even the practicality—of what they describe. Hero of Alexandria's descriptions of pneumatic and steam-operated toys, for example, are given without reference to the wider potential utilities of the principles involved in their operations.[13] Also of little or no concern to them was the issue of priority of discovery, so that the degree to which technical discoveries or innovations fell into clusters, in space as well as time, is almost impossible to ascertain.[14]

Greek and Hellenistic natural philosophy, for all of the seminal importance with which it is often invested as the source of a unilinear tradition leading to modern science, is of only marginal relevance to the history of technology. This is in no sense to deny its conceptual advances on its Babylonian and Egyptian antecedents. Essentially new was the comprehensive effort to categorize material phenomena and to give rational, lawful explanations for them. "The capricious world of divine intervention was being pushed aside, making room for order and regularity. . . . A distinction between the natural and the supernatural was emerging, and there was wide agreement that causes (if they are to be dealt with philosophically) are to be sought only in the nature of things."[15] But as Finley affirms with a commonplace, "the ancient world was characterized by a clear, almost total, divorce between science and practice. The aim of ancient science, it has been said, was to know, not to do; to understand nature, not to tame her."[16]

As a prototypical applied mathematician, Archimedes of Syracuse (circa 287–212 B.C.) may stand as an exception. Brilliant mechanical innovations like the Archimedean screw for raising water are traditionally associated with his name. But it is suggestive of the isolation in which this was conceived that the principle of the screw apparently found no other applications for more than two centuries—until Hero of Alexandria's description of the double-screw press and the female screw-cutting machine.[17] A noteworthy feature of Archimedes's reputed contributions is that so many of them were sophisticated ballistic devices that seem to have been developed only under the extraordinary stimulus of a Roman siege of his native city.[18] Euclid's (circa 300 B.C.) geometry, in the same vein, ultimately laid the foundations for cadastral surveys. However, he himself eschewed all practical applications. And while Galen (A.D. 129–210) in medicine and physiology and Ptolmey (circa A.D. 150) in cartography are rightly regarded as immensely influential, this largely reflects how they came to be viewed during the Renaissance rather than by their contemporaries.

We should recognize that even technologically concerned and innovative figures like Archimedes and Euclid were embedded in an aristocratic and literary culture that was profoundly anti-utilitarian. Plutarch tells us that he regarded "mechanical occupations and every art that ministers to needs as ignoble and vulgar . . . [and] directed his own ambition solely to those studies the beauty and subtlety of which are unadulterated by necessity."[19] That attitude widely prevailed in Rome as well, where many parallels can be found for Cicero's view: "Illiberal, too, and mean are the employments of all who work for wages, whom we pay for their labour and not for their art, for in their case their very wages are the warrant of their slavery. . . . And all craftsmen are engaged in mean trades, for no workshop can have any quality appropriate to a free man."[20]

Literate Greek and Roman circles, it is plain, regarded not merely manual labor and personal service but most forms of remunerated employment and even entrepreneurial activity with profound disdain. This was a slave-owning society, dependent for its leisure and wealth on unremunerated labor and the ownership of great agricultural estates. "Political administration was one thing, management something else again, and management throughout the classical period, Greek as well as Roman, urban as well as rural, was the preserve of slaves and freedmen, at least in all larger establishments, those in which the owner himself did not normally take an active part."[21]

Consistent with the limited record of achievement in technological innovation, Finley sees classical society at large as essentially agrarian and tribute-based, not oriented toward craft production as a basis for substantial interregional exchange. "The contribution of manufactures was negligible; it is only a false model that drives historians in search of them where they are unattested, and did not exist." But at least in its apodictic, unequivocal character, this generalization needs to be examined more critically. Precisely because the dominant ideology reflected in the textual sources and articles for luxury consumption is so exclusionary in its interests and attitudes, it provides a poor basis for evaluating the part played in the economy at large by the activities of slaves and freedmen.

To begin with, while the objective had little to do with interregional exchange, it should not be forgotten that the state itself had an overriding interest in certain classes of industrial or quasi-industrial production— especially military. For the Late Empire (around A.D. 400) the *Notitia Dignitatum* provides a listing of more than seventy state factories, presumably operated largely by slaves and primarily devoted to producing weapons and armor, although minting, dyes, and woolen cloth for uniforms are also mentioned. All these facilities were strategically located in a wide belt adjoining the imperial frontier extending from Britain to Syria.[22]

But there were private workshops, if perhaps for the most part not "factories," as well. Drawing upon admittedly scattered references, R. J. Hopper has called attention to the existence of diverse facilities for the production of military, ritual, and consumer goods, mostly of small to moderate scale. The largest workshop of a group engaged in the production of weapons, furniture, and bronze articles is reported to have employed 120 slaves, while a few others fall in the range of one to a few dozen. Apparently more common were still smaller enterprises, often located in an Oriental bazaar-like setting with only a skilled slave or two as employees engaged in producing consumer goods for what may have been a strictly local market.

Xenophon, in his *Cyropaedia*, attests that the division of labor in such establishments tended to be more fully developed in the larger cities.[23] For cities of the Roman empire, relying in the main on inscriptional evidence, Joyce Reynolds lends further support to this reconstruction of a less economically passive urban role than Finley supposes:

> [A] great range of occupations is attested—from scribe or scrivener to the porter in the docks and marketplaces. Prominent are men in the service trades (bakers, cooks, shopkeepers, for instance) and producers of luxuries (goldsmiths, silversmiths, purple dyers, mosaicists). But a considerable number were producing goods which were useful if not very glamorous—masons, woodworkers, leatherworkers, cloth and clothing workers, potters and the like. Most of their work was directed no doubt to supplying the local market. . . . In some cities there was production for export. Such production tends only to show itself when a local product acquired an empire-wide reputation, as did the linen of Tarsus, the wool of Tarentum or Mutina and the glass of Alexandria.[24]

How was it possible, we may well ask, for attitudes of aristocratic repugnance to persist so uniformly for centuries in the face of the apparently significant potential for economic activities of this kind? Several factors were probably at work: the difficulties of supervising servile laborers in large numbers; the absence of any tradition of active management; difficulties in acquiring capital due to a preference of banks and private funders for less publicly visible investments in mercantile ventures that offered greater returns; and the higher status accorded to investments in land. But beyond this, a very strong tradition obviously remained in place in which, if surpluses were accumulated in this manner, they were to be "returned to the civic collectivity in the form of liturgies" and to underwrite state expenditures on civic and religious festivals, military needs, and public construction.[25] And while Roman engineering accomplishments were on an unprecedented physical and geographical scale, what is more important for present purposes is that they seem to have been, fairly consistently, applications of previously existing knowledge rather than steps in some new technological direction. Merely taking note of them, we move on.

The Middle Ages: Technical Florescence and Scholastic Withdrawal

Of greater significance was a broad advance in agricultural technology that seems to have had its inception almost immediately after the final collapse of the Roman empire, especially in northern Europe. The early Middle Ages are popularly identified primarily with social fragmentation and economic paralysis, but there occurred with little notice between the sixth and tenth or eleventh centuries A.D. a series of important advances in agricultural methods.[26] Among the significant new devices and techniques introduced at the time were improved harnesses, nailed horseshoes, and field rotation systems. Although first appearing earlier, the more widespread use at this time of the heavy moldboard plow was especially critical for Britain. Only with the turning action of this implement could the heavy clay soils of the fertile river valleys and lowlands be opened for cultivation by channeling drainage between the furrows.

The early medieval cluster of agricultural improvements (some of them no doubt having originated earlier) coincided with substantially reduced population levels throughout western Europe. In the later Middle Ages, by contrast, potentially important further improvements in agricultural technology apparently became known but were employed only on a very limited basis.[27] Progress was made in bringing new lands under cultivation, in greatly extending the reach of markets for agricultural commodities through seaborne trade, and probably in animal husbandry. The significance of these developments was held in check, however, by the relative stagnancy of agricultural regimes as a whole. This coincided with—and probably also contributed to—growing population pressure, declining living standards, and increasing mortality that culminated in the Black Death in the mid-fourteenth century.[28]

In different ways, both the early and late medieval instances tend to controvert the thesis propounded by Ester Boserup that technological innovation is driven forward primarily by the need to intensify land use in order to feed growing populations.[29] An instance of significant change seems to have coincided with a population nadir, rather than with population growth to levels that the existing resource base could not support. The contrast surely reinforces the view that there are seldom any "prime movers" or exogenous, infrastructural forces that either impose a directional thrust upon otherwise quiescent technology or emerge from within technology itself.

In late antiquity and at the outset of the Middle Ages, we need to look instead to profound changes that were underway in land management. That was a time of the dissolution of great, passively (if not negligently) managed slave latifundia into smaller tenures more immediately under the control of subsistence producers. Epitomizing the mentality of earlier Roman landowners is Columella's observation that "they avoided annually recurring expenses and considered that the best and most certain form of income was to make no investments," or the elder Pliny's advice (also in the first century A.D.) that "nothing can be less profitable than to cultivate your land to the fullest extent."[30] In the stagnancy of late medieval agriculture, too, underinvestment is among the explanations continuing to find favor in periodic recrudescences of the lively debate on the European transition from feudalism to capitalism.

Sharply contrasting with this was an enormous thrust of other forms of technical advance in the later Middle Ages. It has even been characterized as nothing less than a "medieval industrial revolution."[31] While the disparity of our sources makes comparison difficult, a case can be made that these changes "were greater in scale—by a very large factor—and more radical in kind than any since the start of civilisation."[32]

> Not only were the conditions of agricultural and industrial work transformed by the more efficient uses of power and the invention of new techniques and devices, such as tidal mills, water-driven bellows, iron-casting and the compound crank, but people's lives were affected by the introduction and spread of mechanical clocks, eye-glasses, chimneys and other innovations. In the phrase of a historian of twelfth-century theology, medieval man began to live in a "mechanism-minded world."[33]

The effects of these changes were heavily concentrated in the towns and on the larger, better managed manorial estates. Playing a special role in the diffusion and elaboration of the new technology were monastic orders like the Cistercians, whose drive for economic self-sufficiency led to a far-reaching employment of waterpower in many aspects of industrial production. But at least the water mill for grinding grain had an extraordinarily pervasive effect on the countryside at large. Listed in William the Conqueror's late eleventh-century census are 5,624 of them at more than 3,000 locations, thus providing one for about every fifty registered households.

Innovations that were dependent first on water and later, in a more limited way, on wind power, were ingenious and diverse. Over time, they contributed to virtually every productive technology. Cam-actuated trip-hammers driven by water were used in the hemp and fulling industries as early as the eleventh century. Shortly following this and dependent on the same mechanism were water-driven hammers for extracting slag from blooms in iron smelters, as well as cam-operated bellows for increasing furnace temperatures.

By the thirteenth century water-powered sawmills were in use, and the rotary motion of waterwheels had found a place in cutlery mills for the grinding and polishing of metals. Further improvements in bellows in the fourteenth century permitted a semicontinuous runoff of liquified iron, substantially reducing labor requirements and therefore cost; wood-lathes, pipe-borers, and rollers for producing metal sheets followed not long afterward. Full-rigged ships, enormously more capable of oceanic travel under adverse wind conditions, were developed during the fifteenth century. By the sixteenth century, as coal began to come into use on a commercial scale, waterpower was employed for mine-hoists, pumps, and fan ventilation. Paper mills powered by water were an early seventeenth-century innovation.[34]

To be sure, the context in which innovations like these multiplied and found wide use may not have been one that directly encouraged mundane or pragmatic responses to new insights or opportunities. "Virtually every medieval commentator . . . described manual labor in idealized, often ritualistic rather than productive terms, and—especially beginning in the twelfth century—subscribed to the idea that particular classes of Christians are naturally suited to particular types of labor."[35] But formal theological and scholastic attitudes aside, the growing differentiation, scale, and success of economic enterprise provided a highly favorable environment for technological advance.

> There would seem to have been in the West a greater degree of openness to innovation than was to be found in the more complex and perhaps less turbulent societies of Byzantium, Islam, India, and China of that age . . . indeed, the rapidity with which novelties spread indicates a popular receptivity to anything useful . . . we are obviously dealing with an age eagerly exploiting mechanical power.[36]

Once again, moreover, we must look beyond the narrow realm of technology-as-routinized-technique to a wider, implementing framework that also comprises information, calculation, and communication infrastructures. Arabic numerals, introduced in Italy in the early thirteenth century, enormously facilitated calculations of all kinds. It was virtually a precondition, for example, for the vital, later introduction to the worlds of commerce and finance of double-entry bookkeeping.

The accelerating pace of economic life in the towns naturally led to a new stress on linkages between them. Most economical and hence most important was river transport, increasingly supplemented and ultimately largely superseded by canalization. Canals themselves were of course much older. Northwest European examples intended largely for drainage purposes were dug in Roman times if not earlier and played an especially important part in the reclamation of low-lying areas in Holland throughout the Middle Ages. But commercial use required the introduction of effective locks, which had occurred in the same region by the end of the fourteenth century. Further improvements followed, sharply limiting the losses of water by reducing basin volume to accommodate vessels of modular size. These culminated a century later in Leonardo's miter gate, the basic principle of which still remains in use today.

Britain entered into this new era of economical transport along lock-assisted navigable waterways only after the mid-sixteenth century. Altogether lacking was the stimulus of centralized, state-directed policies of planning and investment. In fact, throughout the Industrial Revolution, the development of canals relied on local capital and entrepreneurship for their construction and operation.[37]

As the identification of canal improvements with commerce suggests, there is a wider context within which this openness to innovations of practical utility becomes intelligible as something more than an isolated attitudinal feature. Many of the same predisposing factors must be involved as those which played a part in the precocious, perhaps unique, development within late medieval western Europe of the component institutions and entrepreneurial spirit that undergird capitalism itself. It cannot be an accident, for example, that among the important innovations of the time were some, like the introduction of the keel and rudder and the general adoption of internal framing prior to external hull construction, that played an important part in development of maritime trade on a more reliable basis.

This is not the place for a detailed discussion of the historical and social context giving rise to capitalism, but neither can we entirely exclude a brief listing of some of the key factors as entirely irrelevant. It may be enough to suggest that there would probably be substantial consensus on the importance of the following:

> The early parcelization of sovereignty; the extraordinary legal authority of European towns, estates, and later, of merchant corporations; the separation, and frequent opposition, between lay and religious jurisdictions; embedded distinctions between private and patrimonial property, and the system of heritable aristocratic land tenures rather than revocable, royally controlled prebends; the tradition by which states procured extractive rights in return for formal guarantees of noble and urban privelege were all factors unique to Europe (particularly Western Europe) and, on the surface at least, had significant implications for the historic development of autonomous, market-dependent enterprise.[38]

The practical as well as metaphoric role of the introduction of clockwork is a good illustration of the complex, far-reaching effects of new technological innovations. The measurement of time had long drawn the attention of medieval scholars, so that this was in any case a domain in which the interests of scholars and craftsmen probably converged to an unusual degree. With the onset of natural philosophy the need of astronomers and navigators for accurate time-keeping made the principles and the construction of clockwork a central concern. Scattered references speak of the heterogeneous origins of clockmakers as a group, but also testify to their relatively high degree of literacy and to the serious involvement of individuals with scholarly backgrounds.[39]

Weight-driven mechanical clocks had been introduced shortly before the end of the thirteenth century. Their origin is obscure, deriving either from an adaptation of long-familiar devices to simulate the motions of heavenly bodies or from discoveries more directly linked to a mundane concern for accurate daily time-keeping,[40] but the prominence they quickly achieved in the massive towers around which town life centered is unquestionable. Spring-driven mechanisms followed by the mid-fifteenth century, leading to great reductions in size, radical improvements in technical and artistic quality, and a corresponding explosion of private demand. The introduction of the pendulum in 1658 made possible a major advance in accuracy even for household clocks. Not surprisingly, the rapid growth of the industry corresponding to these developments was accompanied by its full-time professionalization.[41]

Little need be said of the obvious relationship of clocks to the rationalization of administrative activity and the more efficient scheduling of economic life. Beyond this, major Renaissance figures were closely involved, sometimes obsessed, with clockwork. Dante Alighieri, who completed the *Divine Comedy* only a few decades after the mechanical clock's invention, compared the motion of its gears to the dances of the blessed spirits. Francois Rabelais saw in the movements of a well-trained army the "mutual concord of the wheels of a clock." Johannes Kepler expressed it as his aim "to show that the celestial machine is not to be likened to a divine organism but rather to a clockwork." Descartes translated the clock from a literary metaphor to a scientific analogy, and thence into a deterministic, mechanical interpretation of physiology. And Robert Boyle, directly identifying the universe to "a great piece of clockwork," regarded the clock as an exemplar in his writings on the "new mechanical philosophy."[42]

A widening metaphorical impact of clockwork is abundantly evident in these and many other contemporary references. They convey an attitude of unmixed admiration for the technical intricacy and regularity of clocks, together with a growing awareness of the social significance of the new dimensions of precision in time-keeping that clocks were at least symbolically associated with imposing. At a different level, it is apparent that clocks also figured in the portentous shift of imagery by which human artifices became a kind of Gestalt for the universe itself and the act of its creation.

As part of the same process, the functioning of clockwork developed into a conceptual model applicable to everything from a living body to a well-ordered state. However imperfectly at first, clocks concretely illustrated the potentially flawless working together of complex combinations of parts that gave rise to a newly emergent abstraction, that of a *system*.[43] As technological devices, in short, their influence on both realms of thought and the conduct of social life was at once subtle and profound.

There is little to show, however, that any of the new devices and processes exemplified by clockwork either resulted from or had a substantial impact on the largely self-enclosed, backward-facing world of late medieval scholarship:

> Our literary records of the Middle Ages were in large part compiled by scholars: the paucity in them of technological documentation—concerning not merely the use of tools like the carpenter's brace and lathe, but major industries such as paper-making and iron-working—is very conspicuous. The historian of medieval technology is notably better served by the artist than by the scribe. This could hardly have happened, had more than a very few scholars been impressed by the empiricism which brought in the windmill, the magnetic compass, the mechanical clock, and so on.[44]

Some historians of science, taking an "internalist" approach to their subject matter, have argued that the decisive conceptual shifts whose lineal descendants are the modern physical sciences principally involved novel theoretical approaches to problems that were already traditional in scholastic circles. Thus Rupert Hall, among others, has maintained that the theoretically rigorous core of what he viewed as the scientific revolution had been incubating among late medieval schoolmen for several centuries.[45] But notably absent throughout that long antecedent phase were the qualities of openness to observation and experiment that so strongly characterized much of seventeenth-century natural philosophy.

Instead, the scholasticism of the later Middle Ages was on the whole qualitative and unconcerned with the confirmation of its speculative ideas. Burgeoning technological stimuli that would tend to disturb this frame of thought were the product and property only of individuals associated with the crafts. While this helps to explain why little or no formal cognizance was taken of such stimuli, it enlarges our sense of the disjunction that would presently come under the stimulus of new leadership.

It is equally important to remember, however, that there were surely other, even if almost wholly unattested, channels of communication. We must avoid at all costs an exaggerated dichotomy in which

> full-grown scientists face simple craftsmen. . . . Dr. Hall maintains that the spectacular achievements of applied technology did not "arrest the attention of scholarly scientists," but all that had been accomplished with gunpowder, the magnetic compass, and the press most certainly influenced the environment in which the young men who were the scientists to be grew up and were formed. If people such as Francesco Di Giorgio and Leonardo, who were above all and essentially artists, delighted in drawing mills, gears and machines instead of flowers, fishes and butterflies as their contemporary Chinese fellow painters did, the explanation lies in those anonymous but overpowering influences that the environment exerts. . . .[46]

It thus seems reasonable to assume that technological achievements not only had a profound effect on popular modes of perception and thought but also must have penetrated to some extent into the more insulated world of scholars and their patrons. Indicative of a widely prevailing receptivity within the populace at large is the frequent portrayal of God Himself as "the archetypal craftsman, the master artisan who shaped the raw material of the universe into the harmonious and infinitely diverse contents of the physical world."[47] Still, since it would have had to operate across formidable social and intellectual barriers, the extent of this impact also should not be overdrawn:

> Technical knowledge is a gift of God and a part of human wisdom; the mechanical arts are debased. Such ambivalent attitudes make every attempted generalization provisional. . . . The fact that the theoreticians of medieval culture could maintain diverse and often contradictory attitudes toward those human activities we associate with technological progress suggests that their speculations were in good measure reactive rather than generative; that is, they wrote in response to changes in the "structures of everyday life" that were being created by others.[48]

It is hardly surprising that boundaries—between communities of craftspeople and scholars, or the corresponding ranges of their experience and outlooks—are almost impossible to draw definitively. Exemplifying the prevailing fluidity, studded with internal contradictions, is the complex and enigmatic figure of Paracelsus (circa 1493–1541). Later generations perceived him as at an extreme in fusing the literal and metaphorical, in failing to understand abstractions, and in confusing symbols with reality.[49] Even in his own time he was widely denounced as a spiritualist, religious metaphysician, and astrologer. However, he also was an effective and uncompromising voice in opposition to much of traditional science and medicine. His credulousness appears to have been balanced by an unprejudiced and empirical attitude, and the rough unconventionality of his behavior by long hours of observation in the laboratory and sickroom.

Walter Pagel, who credited Paracelsus with fundamentally new and improved concepts of disease and modes of therapy, concludes that he was "a great doctor and an able chemist," and that "it is difficult to overrate" the effect of his achievement on the development of both of these fields. Yet all of his contributions are difficult to extract from the matrix of violently opinionated natural magic in which they were embedded.[50]

The Baconian Rapprochement of Scholar and Craftsman

Francis Bacon (1561–1626), although his call to arms energized many of the leading natural philosophers of the later seventeenth century, also must be characterized as holding a range of transitional doctrines whose affinities and contradictions cannot be fully disentangled. Fiercely denouncing horoscope-based astrology as an exact predictor of events, he nonetheless emphasized that the human spirit was particularly subject to celestial influences and

> still believed in the traditional doctrine of the magical power of imagination fortified by credulity ... he dismisses the excessive claims of the Paracelsans and their "miracle-working faith," but accepts as "nearer to probability" "transmissions and operations from spirit to spirit without mediation of the senses," aided by "the force of confidence." ... Magic, then, is wrong because it makes experiments unnecessary, and Bacon liked doing and planning experiments.[51]

In the realm of ideas, Bacon's intellectual parentage of the new and sweeping empiricism associated with natural philosophy was widely acknowledged by contemporaries. So, too, was the example of his unswerving dedication to experimentation and to painstaking observation of the natural world. "Nature to be commanded must be obeyed," as he proclaimed in *Novum Organum*. Yet these seemingly clear and forthright messages appear in a larger body of occasionally inconsistent and often naive doctrines. "The effecting of all things possible—the goal of Baconian utilitarianism, as we examine it in Bacon's own words, partakes more of the alchemist's dream than of humdrum engineering. It was not wholly by accident that the example of operative knowledge which came most readily to his mind was the making of gold."[52]

What appear as puzzling ambiguities to us, however, need not have left the same, self-defeating impression upon his contemporaries and early successors. No less epochal a figure than Isaac Newton (1642–1727) maintained a prolonged, intense involvement with alchemy. So, too, did John Locke (1632–1704) and Robert Boyle (1627–1691). "[P]ossibly the last three men from Restoration England whom one would have expected, only a generation ago, to find so engaged," are now known to have exchanged alchemical secrets and pledged one another to silence.[53] Under the accepted rubric of natural philosophy there was a striking acceleration of discoveries and a profound and general deepening of attention devoted to the natural world. But it fell far short of attaining a scientific outlook in the modern sense.

The rise of natural philosophy has commonly been characterized as the scientific revolution. As the temporal clustering and epochal scale of seventeenth-century discoveries under this rubric attest, something revolutionary by any standards was indeed going on. To describe it specifically as a "scientific revolution" is, however, decidedly more problematical. I find myself persuaded by an alternative view, recently advanced by Andrew Cunningham and Perry Williams, that such a usage has during the last decade or two become progressively more difficult to sustain:

> The problem with the "scientific revolution," we maintain, is not that insufficient research has been done, or that an up-to-date historiography is needed, or

more "external factors," or more sociology of knowledge, or discourse analysis, or whatever the current intellectual fashion happens to be. The problem is that historians are ceasing to believe in a single scientific method which makes all knowledge like the physical sciences, or that science is synonymous with free intellectual inquiry and material prosperity, or that science is what all humans throughout time and space have been doing as competently as they were able whenever they looked at or discussed nature; so small wonder that a concept developed specifically to instantiate these assumptions ceases to satisfy. . . . It is only recently that a few historians of science have regularly begun to leave the term "natural philosophy" as it is. This has opened up the possibility of recognizing that while natural philosophy was itself indeed an investigation of the natural world which was sometimes empirical and sometimes even experimental, yet it was nevertheless one which was radically different from "science" in the modern sense.

For the whole point of natural philosophy was to look at nature and the world *as created by God*, and as thus capable of being understood as embodying God's powers and purposes and of being used to say something about them.[54]

Their proposed alternative is compelling, if largely on "contextualist" grounds that are likely not to satisfy historians of science of "internalist" persuasion. Setting aside altogether the association of a revolutionary character with any scheme of periodization, they simply place the origins of science itself in the Age of Revolutions of the late eighteenth and early nineteenth centuries. That was, after all, the period "which saw the origin of pretty well every feature which is regarded as essential and definitional of the enterprise of science: its name, its aim (secular as distinct from godly knowledge of the natural world), its values (the 'liberal' values of free enquiry, meritocratic expert government and material progress), and its history" (p. 427).

Our more direct interest, of course, is not in the historic development of the outlook of modern science but in the evolving relationship between scientists and technicians (or in a pre-modern setting, their precursors). Across long periods of time, this has varied enormously—in the respective social standings of these two bodies of thought and practice, in the complementarity or dominant directionality of the dialogue and cooperation between them, and in their orientation toward the production of useful knowledge. One need only think of the disparity between the emphasis on abstract philosophical principles pertaining to nature in ancient Greece as compared, for example, with the massive engineering achievements of Roman imperial policy that, along with any kind of science, were barely noticed by its literate classes. Or again, while the sixteenth and seventeenth centuries saw extraordinary advances in astronomy, optics, mechanics, and related fields, they were at the same time "somewhat backward in technological change." Thus the practical consequences of that earlier stream of thought, once widely regarded as perhaps the major source of the Industrial Revolution, now are more commonly regarded as furnishing no more than an "extremely diffuse" link.[55]

Francis Bacon must be credited with having percipiently noted what remains even today perhaps the key difference between processes of scientific and technological change: " 'The overmuch credit that hath been given unto authors in sciences, in making them dictators,' to be followed and annotated, means that 'the first author goeth furthest, and time lesseth and corrupteth.' But in the 'arts mechanical, the first deviser comes shortest, and time addeth and correcteth'."[56]

Long before industrialism reinforced the lesson with endless examples, Bacon foresaw that technological transitions would be retarded by having to be adapted by trial and error to larger production, marketing, and social requirements. Whether or not he also anticipated this, supporting technical skills often have to be formalized and diffused before inventions can be put to use. Pre-existing technical complementarities and mounting capital requirements, as well as competitive refinements in older technologies, also consistently delay the replacement of older processes.

Scientific "revolutions," therefore, are more likely than technological ones to be abruptly ushered in by great discoveries. The major phases of technological advance, while exhibiting a similarly punctuated character, at a detailed level are more often many-faceted composites. The more deliberate, multi-authored character of technological advance becomes a source of difficulty when models of change are based primarily on better-understood examples from the history of science. To represent processes of technological innovation only by their initiating discoveries tends to exaggerate the discontinuities that are involved. By importing a model that is appropriate only for the sciences, such an approach fails to recognize that the importance of those first discoveries often depends very heavily on later contributions to their effectiveness.[57]

There are, however, dangers in the other direction as well. The attention currently being devoted to revolutionary turning points in the history of science, by strengthening the contrast with the slower evolution of technology, can lead to an exaggeration of the basic differences between the two. Thomas Kuhn, for example, perceiving that science and technology have often followed independent historical courses, characterizes technology as basically acephalous and chance-driven rather than consciously focused or directed:

> Once discovered, a process in which chance may have played an overwhelming role, technological innovations were embodied in artifact and local practice, preserved and transmitted by precept and example. . . . Though I freely admit exceptions, I think nothing but mythology prevents our realizing quite how little the

development of the intellect need have had to do with that of technology during all but the most recent stage of human history.[58]

This sweeping separation between science as an affair of the intellect and anything connected with practical understanding and applications depends too largely, it seems to me, on "internalist" criteria as to what constitutes a science—or for that matter, a system of ideas. Technological advances may indeed be generally slower, more anonymous, and less driven forward than are those in science by powerful new theories or syntheses. But that is not a basis for denying a thought component to technology—for holding that only scientists generate new knowledge and technologists merely apply it.[59] New knowledge is involved in the objects and processes technology creates, and in the contextual obstacles it must overcome in order to do so.

This leads to a further, complementary aspect of the science-technology relationship. Granting that one important component of science is a "flow of cerebral events," another major component is "the craft of experimental science which is in part the history of technology."[60] And while the importance of an internalist orientation in studies of the history of *science* must be acknowledged, it does not follow that we must deny the possibility of a different, socially constituted, idea content for *technology*. For technological discoveries to be largely chance-driven, they presumably would have to lack intellectual coherence and systematicity. But do they?

Intensively pursued improvements on relatively narrow fronts of application, like those at the heart of the Industrial Revolution, plainly were seldom, if ever, outcomes of chance. No doubt some individual contributions may have arisen from haphazard discoveries or experiments rather than from a search for deeper regularities. Much more common, however, was a distinctively technological search for optimal solutions to complex problems that relied on experiments designed to find reliable approximations even in the absence of a useful quantitative theory.[61] Typically, the goals toward which such experiments were directed were for the most part supplied by practical, externally defined needs or incentives rather than by a concern to solve abstract, internal puzzles. In that sense, technology advanced through its attachment to social needs and demands rather than in isolation.

Attention is most commonly paid to scientific discoveries that have led to technological advances. With new devices or processes as the usual intermediary step, there are indeed discoveries that have had profound, largely unanticipated social and economic consequences. But it is no less clear that there were diverse and important technological as well as social influences on the birth of modern science. At least some stimulus to the strong practical or utilitarian orientation that developed in seventeenth-century natural philosophy, for example, was provided by contemporary developments in commerce and industry. One such case, which occurred toward the end of the sixteenth century, involved the discovery of logarithms by John Napier, an important advance in calculations of compound interest. William Oughtred's invention of the slide rule a few decades later is another case of immediate scientific as well as practical utility, as is the roughly contemporary introduction of trigonometric calculations for distance determinations and map-making. No less relevant is Robert Merton's demonstration of a close interconnection between many themes of seemingly "pure" mathematical and scientific interest at the time and the increasing concern for the accurate determination of longitude and other improved navigational aids in suppport of maritime commerce.[62]

It is difficult, under the circumstances, to follow the internalists in seeing the birth of modern science as little more than the outcome of a re-appraisal by medieval schoolmen of traditionally acknowledged puzzles or sets of natural phenomena. The specific role of craft lore as a source for the enhancement of scientific understanding is, in addition, sometimes directly attested. Galileo's acknowledgment in his "Dialogue" of the assistance rendered by artisans in the Venetian arsenal is an example, as is the early assumption by the Royal Society of responsibility for compiling a history of artisan trades and techniques.[63]

I would not argue that a more profoundly ordered view of nature lay ready for the taking in the lore or rules-of-thumb of craft specialists. We simply do not know how far they either knew or cared why only certain traditional procedures were successful. There is little to suggest that the greatly increased readiness of the new breed of natural philosophers to make use of craft information and practice was matched (with the clock and instrument makers as the most likely exceptions) by an interest on the part of great numbers of artisans in borrowing in the opposite direction. While the debt to technology accordingly was substantial, much of "the debt was itself created by natural philosophers and other men of learning."[64]

However, a newly kindled, outward projection of interest into technological domains on the part of the scientifically inclined also can be exaggerated. While interest in drawing craft technology into the sciences unquestionably mounted, the nascent scientific community was neither undivided on the issue nor capable of making a complete break with the past. The deepest bifurcation was between the classical and Baconian scientists. Characteristic of the "new mechanical philosophy" of the Baconians was the more fundamental commitment to observation and experiment, and with it the impulse toward association with the crafts. The self-proclaimed terms "mechanic" or "mechanical" betray a further, programmatic or social concern that was distinctively Baconian. Signaling a break with traditional scholastic philosophy and metaphysical speculation, they also publicly embraced—in the English of Shakespeare's time, through and beyond that of Samuel Johnson—a kind of plainspoken, even vulgar, practicality. Yet theirs, too, was a reduced concern for philosophical or mathematical rigor.[65]

Even on the part of the Baconians, it appears that there was a wide gulf between the programmatic stance of direct engagement with the crafts and the actual conduct of a scientific enterprise. Thomas Sprat might propound it as a "Fundamental Law" of the Royal Society that "whenever they could possibly get to handle the subject, the Experiment was still perform'd by some of the Members themselves."[66] But the members were, after all, gentlemen, still deeply embedded in an older system of social relations that left little place for give-and-take with ordinary artisans who in this respect fell into the additionally subordinate category of employed servants. As Steven Shapin perceptively notes, "the order of early modern society as a whole was arrayed about the distinction between those who worked and those who thought or fought. The seventeenth-century activity called natural philosopy participated in that distinction and displayed its consequences in the relations between philosophers and technicians."[67]

Robert Boyle provides an outstanding, well-documented example. As Shapin shows, Boyle's voluminous papers make it plain that his laboratory was "a densely populated workplace." Yet they contain the names of no more than a handful of the individuals who are indirectly referred to as having operated his experimental apparatus and made the observations and judgments that were requisite. Technicians were, in effect, invisible. Their presence was alluded to mainly when experimental results or observations failed to accord with expectation or theory, and these instances constitute the major source of information on the role of technicians in Boyle's laboratory. Significantly, there was a convention in drawings accompanying seventeenth-century publications in natural philosophy to have the agents operating scientific instruments shown not in human form but as *putti*, cherubs without individuality.

Instrument Makers, Inventors, and New Economic Stimuli

An underlying theme in much of the foregoing discussion is the huge asymmetry of our earlier sources. This makes it virtually unavoidable that we see technology largely through a scholastic lens (and as we move toward modern times, with gradually diminishing distortion, still through a scientific lens). Knowledge of the history of technology is particularly skewed and limited by the sparseness of artisans' records of any kind, and especially of more self-conscious accounts of their beliefs, motivations, and practices. Evidence attesting to stimuli to science that might have stemmed from technological innovations, rather than vice versa, is rarely to be found. But that does not mean that such stimuli were not an important factor.

Can we imagine the development of Newtonian mechanics without prior advances in accurately measuring short time intervals, or of optics and biology without the contribution of achromatic lenses first to the improvement of telescopes and then to the introduction of the microscope?[68] Even today, while scientific papers quickly and fully display the emergence of new ideas from basic research, the evolving capabilities of the equipment involved in that research receive less attention. With regard to the role of technology, the available records thus are likely to be unconsciously self-serving. They lead to a neglect of the contribution to scientific-technological progress as a whole that stems from advancing instrumental capabilities.

The depth and importance of communication between scientists and artisans or technicians has increased with time, although it seems likely that there has always been some role in science for the makers of instruments as well as for the devices they furnished. Proof may always elude us as a result of the paucity of written records, but early instrument makers surely must have contributed to the verification as well as dissemination of scientific discoveries. Their main contribution, to be sure, lay in their products. Indeed, it can plausibly be argued that the "new sorts of windows they provided for scientists to look from" were a more common cause than "great leaps of human intellect" for the periodic shifts in basic scientific paradigms.[69] As they became a technical elite in the crafts, instrument makers assumed a pivotal position from which to provide not merely tools for isolating and measuring new properties but contact with a wide world of empirical observation not easily accessible to contemporary natural philosophers.

> Naturally it was the scientific instrument-makers who were in closest touch with the virtuosi, and from the 1650s onwards their dialogue became ever closer. The first important studies made with the microscope date from this period; so the first attempts to obtain vacua by pumps for experimental purposes, and in the second half of the seventeenth century a number of new instruments were introduced, such as the pendulum clock, telescopic sights, the bubble-level, and the screw-micrometer.[70]

Centering on the gradual emergence of a specialized class of scientific instruments, Albert Van Helden has traced a convergent process that gradually drew scholars and craft specialists toward one another. At its beginning, although there were exceptions, skill levels were prevailingly low in relevant fields like clock and spectacle making. Held close to traditional norms by guild regulations, craft techniques and products were at first of little assistance to the emergent community of natural philosophers:

> The spyglass, for instance, came from the craft tradition. To produce it spectacle makers put two lenses, existing products, in a tube. But the improvement of this gadget into the research telescope was quite a different matter. For this purpose very high quality glass and lenses with strengths well outside the range used for ordinary spectacles were required. The glassmaking craft was very slow to respond, and the spectacle-making craft never adjusted to the new demands. Slowly, a specialty craft of telescope and microscope makers developed in the course of the seventeenth century. . . .[71]

As this development went forward, however, it is equally important to take note of its reciprocal effect on the nature of science itself. Growing instrumental capabilities not only inspired growing reliance on those capabilities on the part of scientists but introduced an instrument-mediated view of reality that was an inducement to experimental manipulation. Craft specialists engaged in those improvements had to collaborate more closely with scientists as they did so, surely creating the grounds for more balanced interchange and interaction. And as "by 1700 instrumentation had become an accepted dimension of all of science," rapid, perceptible advances in the quality of instruments became a central pillar of support for the idea of continuing progress in science itself:

> By the end of the century, through the work of Descartes, Huygens, Newton, and Molyneux, the telescope had been put on a firm theoretical foundation. In the case of the pendulum, its isochronous nature was an erroneous article of faith with Galileo. Not until Huygens was the instrument supplied with an adequate theoretical basis. Huygens's treatment was so elegant that it established the pendulum clock as the theory-based instrument par excellence of the seventeenth century. No instrument inspired greater faith in the validity of instruments.[72]

Formal cognizance of its instruments was taken on behalf of the Royal Society in a 1681 catalog of its holdings that included a category "Of Instruments relating to Natural Philosophy." This was distinguished from a separate listing of "Things relating to the Mathematicks," which according to a commentator at that time "was scarce looked upon as Academical Studies, but rather mechanical—as the business of traders, merchants, seamen, carpenters, surveyors of land, and the like." The gradual erasure of the distinction apparently did not begin until the later eighteenth century. Quite possibly it accompanied a broadening of involvement in studies of the natural world that weakened the former dominance of the gentility. But other factors may have been the increasing sophistication of some of the

more practical instruments, with the addition of vernier scales and tele-
scopic sights, and the gradual appearance of commercial establishments
advertising products that fell into both categories.[73]

Contributing to the opportunities for direct involvement of instrument
makers and other superior craft specialists in the enterprise of the natural
philosophers was what has been called the "printing revolution." Indeed, it
is not unlikely that the ensuing spread of literacy and the introduction of
many institutional avenues to at least some elements of a more formal education led to the recruitment of some of these individuals into the ranks of
the natural philosophers. Self-education in mathematics was a desideratum
for many practitioners in fields like gunnery and surveying, and by the
early eighteenth century many cheap books of instruction were becoming
available.[74]

For natural philosophy, certainly by the seventeenth century and in fact
already earlier, the available record is sufficient in many instances to ex-
pose internal fissures and differences in orientation. There is no reason to
doubt that it was distinguished from the crafts in being relatively more
cohesive, self-directed, and autonomous as a community and form of activity. That cohesiveness was, in the view of some historians of science, fully
reflected in the emergence by the seventeenth century of a single hierarchy
of scientific aspiration if not status:

> The academic and above all the mathematical sciences were not only those that
> advanced fastest, but they were already regarded as the models for the structure
> of other sciences, when these should have reached a sufficiently mature stage. In
> an ascending scale of sophistication, it was regarded as desirable to render all
> physical science of the same pattern as mechanics and astronomy, and to interpret
> all other phenomena in terms of the basic physical laws.[75]

More recently, however, a significantly more qualified, and even somewhat contradictory picture of the boundedness and cohesiveness of the scientific community has begun to emerge. Thomas Kuhn cites new fields like
the study of magnetism, heat, and electricity as typically Baconian in outlook rather than subservient to the older academic models. Often, in fact,
their roots lay in the crafts, and their development tended to be more
closely linked with advances in instrument making than with conceptual
standpoints borrowed from other fields of science.

> Chemistry presents a case of a different and far more complex sort. Many of its
> main instruments, reagents, and techniques had been developed long before the
> Scientific Revolution. But until the late sixteenth century they were primarily the
> property of craftsmen, pharmacists, and alchemists. Only after a reevaluation
> of the crafts were they regularly deployed in the experimental search for new
> knowledge.[76]

In a different vein but with comparably disaggregative effect upon the supposed unity of the seventeenth-century scientific enterprise, Charles Webster's survey of scientific activities at mid-century leads him to conclude that

> it is inadmissible to categorize individuals involved in the investigation of the natural world into such self-contained groups as "utilitarians," "technicians" and "scientists." Although there was a degree of specialization of scientific activity, the enthusiasts tended to be involved in science at many different levels, being commonly concerned on the one hand with the theological implications of natural philosophy and on the other with agricultural improvement or mechanical innovations.[77]

Webster, in short, summarily rejects the older view that a genuinely independent sector of seventeenth-century scientific activity can be identified. He sees this as no more than a fabrication of what "it is felt ought to have existed on the basis of present-day opinion about the methodology of science."

The elusiveness of the boundary between technology and natural philosophy, and yet also the very real disjunctions in status and communications accompanying it, has been an important theme in this chapter. Much of the problem revolves around intangible extent of mutual understanding and interaction between individuals in different walks of life. But the depth of our interpretive difficulties perhaps can be epitomized by considering a single, pivotal individual who straddled both fields. Robert Hooke (1635–1703) was a commanding figure in each of them, without confirming a significant degree of synergism between them even in his own work.

While not quite in the category of Kepler, Galileo, and Newton as a scientist, Hooke was nonetheless a figure of outstanding originality and productivity. His contributions to Boyle's Law were substantial, and he generalized it to the law of elasticity that bears his own name. Other important contributions concerned the mechanics of circular motion and planetary dynamics, the color phenomena of thin plates, pioneering investigations of the importance of air to combustion and animal life, and the identification of fossils as remains of once-living creatures. Above all, his *Micrographia* was the first great publication of microscopic investigations.

With this as merely a beginning, the breadth of Hooke's other interests is nothing short of astonishing. Apart from his scientific contributions, he was an inventor of fully comparable if not greater importance, described by John Aubrey as "certainly the greatest mechanick this day in the world." He made significant additions to every important scientific instrument that appeared during the seventeenth century: the universal joint, the cross-hair sight, the iris diaphragm, worm gears as precision devices to divide small units of arc, the air pump, and the wheel barometer. His contribution to the invention of the spring-driven watch is indeterminate, but may have been decisive.[78] And apart from these personal achievements, in his correspondence he is to be found " 'discoursing' . . . with many of those who were in the van of manufacturing and commercial progress . . . about hammering and rolling mills . . . 'cast iron pillars for bridges, hard[ening] iron into steel quite through, pressing of cloth . . . roller printing . . . steel wire work . . . 'engines' and mills of all kinds, glass-making, steel-making, coal-mining, cloth-fulling, metal-working, 'chymical' operations, etc."[79]

By any more modern standard, this range of creative curiosity defies the imagination. But it remains a question whether his technological discoveries were in some real sense subordinate to (his own or other) contemporary discoveries in natural philosophy. Indeed, it is perhaps even a question of whether there was a close linkage and interdependency without the element of subordination. On closer inspection of the analytical reasoning behind several of Hooke's major technological contributions, Richard Westfall has sharply questioned the importance of the derivation from the scientific knowledge of the time that has sometimes been claimed for them:

> Should we not see in Hooke's mechanical inventions . . . evidence not of science at work but of a fertile imagination fed by the world of practical mechanics? . . . Although he pretended to call on scientific principles to support his technology . . . we should seek the foundation of much of Hooke's contribution to technology in the practical tradition of European mechanics.
>
> In one particular area, however, it can be said that the seventeenth century did witness the appearance of a truly scientific technology, and Robert Hooke was among the men most prominently associated with it: I refer to the instruments that science itself used. . . . In fact much of his catalogue of inventions was devoted to scientific instruments. Even the universal joint was conceived as part of an instrument. . . .[80]

Having considered an individual as an illustration of the character of the science-technology interface in the seventeenth century, let us turn to an illustration that offers a different but complementary perspective. The development of the steam engine once again illuminates the complex character of advances that have both scientific and technological dimensions. But no less importantly, it forces us to see technological change as a response to strong and growing economic forces.

A means to convert heat to power through the use of steam had been an official objective at least from the time that Medici engineers first attempted unsuccessfully to lift water from a depth of fifteen meters or so by means of a suction pump. These technological efforts were followed by other, equally futile ones employing steam alone, as well as by slow and periodic but fundamental advances in scientific understanding. Drebbel demonstrated in 1604 that standing water would be drawn or forced up into a cooled retort previously occupied by steam, and there was other, equally fundamental work on the nature of the atmosphere itself.[81] The existence of atmospheric pressure, never understood by Galileo, was established by Torricelli in 1643. The celebrated Magdeburg experiment by von Guericke brought to wide public notice the immense force that the atmosphere could exert.

After these preparatory developments, matters once again took a more practically oriented course. By around 1680 the Dutch physicist Christian Huygens had sought to use atmospheric force to compress a piston into a cylinder evacuated by an explosive charge, presently followed by Denis Papin's substitution of steam for gunpowder and, a decade or so later, by his successful development of a small-scale, working model for a steam engine. This led in turn to Thomas Savery's demonstration of a marginally practical steam-driven pump in 1698. Calling for steam pressures that were beyond existing levels of mechanical or metallurgical skill, however, it was limited in lifting power and never found widespread use.

The first truly practical machine was the outgrowth of a decade-long effort by Thomas Newcomen, a Devonshire ironmonger. This effectively (if still very inefficiently) succeeded in employing steam at atmospheric pressure. Newcomen's design had some similarity to Papin's. But it also included a number of critical and mutually interdependent design features that were new, and it has never been established that he was in touch with any of the Royal Society savants who might have informed him of the earlier tradition of scientific investigation. Clearly, he was "an empirical genius of awesome proportions," and it is plausible (though controversial) to conclude, "in the present state of the evidence, that the mastery of steam power was a purely technological feat, not influenced by Galilean science."[82]

Beyond Newcomen's initial invention, however, there is much more to be said about succeeding steps leading to vast improvements in, and the widening application of, steam power. And not all of this lends itself to the exclusively technological assignment of credit just given. Much of the story relates more closely to the employment of steam engines in the Industrial Revolution, and so will be taken up in the following chapter. But the beginnings may be related here.

Newcomen engines had the disadvantage of accelerating and slowing markedly with each stroke. Being thus very irregular in motion, they were not easily made suitable for supplying smooth rotative motion. For most of the eighteenth century they were used simply as pumping engines, for draining mines and supplying waterworks. Where rotative motion was required in order to drive machinery, they were ordinarily employed again as pumps, raising water for waterwheels that then became the rotative source.[83] Perhaps for this reason, systematic study of Newcomen engines became intimately connected with the studies of waterwheel design that brought together scientific theory and practical experience.

Newcomen himself saw the need for some form of roughly standardized guidance as to the capabilities of his engines; "he was obviously not a simple empiricist, not mathematically ignorant, and in endeavoring to establish rational rules for calculating engine power was following a rudimentary scientific method." But the major contribution to a systematic understanding of the performance of Newcomen engines came from Henry Beighton (1686–1743), a fellow of the Royal Society as well as one of the outstanding land surveyors, mathematicians, and engineers of his day.

> He was particularly notable for the self-acting valve gear (1718) . . . and also for the table which he produced on the performance of these engines. . . . This table, as eventually developed, provided clear directions as to the quantities of water that could be pumped per stroke, from various depths, according to the diameter of the engine cylinder, strokes per minute, and bore of pump.[84]

If not in the initial invention, then, a fruitful union of scientific theory, mathematical competence, and engineering practice at least made an early appearance in connection with the fuller, more efficient, and more effective application of steam power. Even though the picture is still blurred by many uncertainties, this illustrates that scientific-technological relationships prior to the mid-eighteenth century were growing progressively more intimate. Clearly, the potential importance of steam power was becoming better recognized across the entire science-technology spectrum. All aspects of coal's exploitation as a substitute for wood and charcoal accordingly were becoming a high technological priority. By slow and sometimes difficult steps, they were successively mastered:

> In some changes from wood fuel to coal the alterations needed were small, the use of a grate, or the introduction or improvement of a fairly simple chimney to make coal smoke tolerable. Where the thing to be heated could be put in a boiler or some other container without being in direct contact with the fuel, the change to coal might not be too difficult and this meant that dyers, brewers, salt boilers used coal early, though often to the annoyance of their neighbours. Similarly coal could supply the heat for the working of metals once they had been smelted and refined with wood fuels. . . . However, there were many industrial processes which were far more obdurate to the introduction of coal as fuel. The reason was nearly always the need in such processes for the fuel and the material which was being heated to touch each other, or be in close proximity, in a situation where the product might be spoiled by any of the many potentially pollutant chemicals generally present in coal. . . .
>
> The fact that so much technological innovation was concentrated on enabling industries to use coal had a powerful focussing effect. . . . What cannot have been

very clearly realized during the early stages of a general shift of fuel use was the enormous potential length of the path of technical development to which the search for a coal fuel technology lighted the way.[85]

It is at once evident that these advances were overwhelmingly of a technological, not scientific, character. This casts the Newcomen engine itself into a new light, forcing us to see it less as a curious offshoot from a prior line of scientific derivation than as a product of the intense economic importance that coal mining had in the meantime assumed. The engine's primary utility, after all, was for pumping water out of mines in order for them to be worked more efficiently and at deeper levels. Extremely inefficient in its fuel requirements, it had virtually no economic justification except in immediate proximity to coal deposits.

The growing commercial role of coal as a vital resource is one of the central economic trends of the seventeenth century. Without assured, relatively cheap access to this new fuel, London, growing by almost 45 percent in the last half of the seventeenth century, almost certainly could not have become the largest city of Europe. During that period, a growing demand for coal for industrial use was simultaneously added to its already well established reliance on coal for domestic consumption. As a result, by 1700 British production had climbed to the level of 2.5–3 million tons annually, estimated to have been "five times as large as the output of the whole of the rest of the world." Half of the entire tonnage of the British merchant fleet was by then engaged in the coal trade.[86]

Huge forces were at work here, even though the linear succession of their effects is far more anonymous and difficult to document in any way than the line of scientific advances that laid some of the groundwork for— but stopped short of—the Newcomen engine. It is certainly in this economic context, I submit, that we must look for the stimulus to the steam engine as an epoch-making invention. With almost equal certainty, we must assume the existence of a wide, diffuse, relatively unknown circle of entrepreneurs, experimenters, skilled mechanics, and would-be inventors. This, and not the more elevated circle of gentlemen in the Royal Society, was Newcomen's proper setting and source of creative sustenance.

Economic pressures of a new scale and intensity had clearly begun to play a major part in focusing inventive as well as entrepreneurial attention. The rapid growth of a new industry devoted to the extraction, transport, and commercial as well as domestic utilization of coal, and secondarily to the development of a new source of rotative power that it virtually required, is but one example. Of comparable importance was the beginning of a transformation in the production of textiles.

Raw wool had of course long been a staple English product and the country's principal export. The putting-out system was an organizational innovation that anticipated its processing into intermediate or finished goods in homes and small rural workshops. Originating already in the later Middle Ages, it slowly but surely diversified economic opportunities in the countryside as well as the balance of exports.[87] Therein lay the foundations for the conversion of a handicraft into another great domestic industry. But therein also lay an escape from town-based, guild-imposed restrictions on the mode of production, effectively reducing costs as well as widening demand by recruiting women and children into the production process. While regarding neither supply nor demand as primary but both as an inter-active pair, David Landes even sees this as the supply-side key to the entire industrial transformation that presently followed.[88] The technological em-bodiments that would trigger the new industry's enormous growth were to be delayed, however, until a half-century or more after the Newcomen en-gine, having had to await the massive supplementation of wool with im-ported American cotton.

Sectoral developments like these must be understood as embedded not only within a burgeoning domestic economy but within a larger setting of international commercial activity that was maintaining an even steeper rate of growth. One essential element of this was the so-called re-export trade, consisting primarily of tropical commodities that Britain had secured from its own trading efforts or dependencies in the East and West Indies. The comparatively high and elastic demand for them greatly facilitated efforts to obtain essential imports from the European continent. Total exports, of which at the end of the seventeenth century re-exports still amounted to less than a third, had achieved something on the order of a sixfold growth during that century. It is in the parallel development of all of the ancillary institutions required for this growth—banking, insurance, warehousing, networks of distribution and retailing, and the like—that the demand-side structures of vision as well as expertise gradually took shape that would undergird and help to direct the technological surges soon to come.

3

Technology and the New European Society

THE INDUSTRIAL REVOLUTION was the gigantic wave of change that initiated the modern epoch. Getting underway slowly during the last third of the eighteenth century, it permanently transformed economic and social life beyond all recognition. Britain will be forever identified not only as its hearth but for much of the following century as the major propulsive force behind the widening ripples of industrialization that moved outward across both Europe and the North Atlantic. Yet natural and almost inevitable as this now seems, it was surely not anticipated at the time and can be understood only in retrospect.

> As late as the beginning of the eighteenth century Daniel Defoe could admit, complacently, that the English were no good at making original inventions, although they were competent enough at copying other peoples' ideas and turning them to profitable account; a surprising inversion of the supremacy usually ascribed to British inventiveness by British commentators today.[1]

Following a span of only seven or eight decades from that acknowledgment of a creative deficiency in at least manufacturing terms, Britain somehow found a way to take a commanding lead in technological innovation and then to assume for the better part of a century the unchallenged status of the workshop of the world. Clearly there were new, previously unrecognized forces at work, either singly or in combination. By plotting the paths of the emergence and dynamic interaction of some of the most salient of them we can prepare the ground for a more intelligible account of the progress of the Industrial Revolution itself.

Ironically, a factor that may at first seem likely to have had a negative impact is the lack of direct state involvement in the extraordinary changes that were underway. From the accession of William of Orange in 1689 through the Napoleonic period, the attention of the monarchy was instead focused almost exclusively on military matters. It has been estimated, for example, that some 83 percent of the expenditures of all state revenues collected during this very long period were military in origin and purpose. By contrast, it is implied by an appraisal of a contemporary that only 0.5 percent of all the revenues collected during the long reign of George III were devoted to what would today be regarded as measures to encourage

social development or economic growth. Left to the initiatives and resources of private individuals was not only the building up of the country's
physical infrastructure, but the support of education and the dissemination
of useful knowledge:

> Monarchs, ministers and their parliaments . . . persisted, however, with Tudor
> and earlier traditions of encouraging foreigners to bring novel products and tech-
> nologies into the realm while actively prohibiting the emigration of skilled arti-
> sans and the export of machinery. They also continued to rely on that other
> "cheap" but rather "ineffective" method of encouraging technological progress—
> the Elizabethan patent system—as codified in the Statute of Monopolies of
> 1624.[2]

There was, in short, what appears today as a curious disproportion between state expenditures in support of the domestic economy and an
overwhelming preoccupation with military preparations and the costs of
fighting an almost continuous series of eighteenth-century wars. Just as
noteworthy, however, is the positive support frequently expressed for this
ordering of priorities by all sectors of the British landed and commercial
elite. Successful wars, it may be argued, were an enormously popular stimulant to national loyalties. But in a deeper sense, a strategy of aggressive
projection of seaborne power was a necessary foundation not only for commercial success but for the transformation of the industrial base that was to
come.

> Britain's growing extent of the market depended on its national power, exercised
> both militarily and diplomatically, to impose and enforce the Navigation Laws.
> These laws, which endured well into the nineteenth century, secured Britain's
> position as the entrepot of the world and effectively protected British manufac-
> tures from foreign competition in the home market.[3]

To be sure, there is another, quite different respect in which the absence
of state policies of domestic intervention, and even of the financial stimuli
that might have accompanied them, may have been a positive rather than
deterrent factor. Certainly the cumulative—and, until recently, unrivaled—
development of technology in the West has had its primary ideological
articulation in doctrines stressing the importance of private property, free
markets, and the ancillary institutions of a capitalist economy. The recent
historical record, at least, leaves little doubt that political command over
the operations of the economic sector has often been a serious impediment
to technology-led progress. But that only leads to a deeper question. Would
it be correct to infer from this body of chastening experience that techno-
economic advances were in this case autonomously self-generated, essentially independent of a wider social stimulus?

On the contrary, this book constitutes a preponderantly negative reply to that as a general question. But even in the more limited, eighteenth-century English context, techno-economic autonomy was more apparent than real. Adherence to the principles of laissez faire never precluded more indirect governmental support of laws and institutions whose effect was to protect and promote industry and trade.

> In practical terms, this has meant only that it was comparatively out of fashion to regulate trade, tax it appreciably, control prices or wages, or seek to iron out the spectacular differences in individual incomes. The working assumption was that industry and commerce served the general welfare, so that it was the business of government to support and encourage them.[4]

In any case, we are speaking here of *techno-economic activities*, which can exist only as they are deeply embedded in customs, laws, and institutions. It would surely be unrealistic to assign technology in a narrower sense a distinct and independent role, as an inanimate, extrasocietal force somehow detached from the same enveloping and supportive framework. Even at the micro level, as Anthony Wallace has observed in one of the rare anthropological studies directed to this subject, the effectuation of technological change only can be understood in a profoundly facilitating social context: "the necessary institutional setting for the major innovations of the Industrial Revolution was a stable transgenerational organization that provided continuity in both plant and personnel so that a technological idea could become paradigmatic in a single community of mechanicians."[5]

On a macro level as well, there were many features of the broader, eighteenth-century British social context that together clearly encouraged the rapid promotion of technological innovations. Characteristic of it, after all, was

> a porous social structure that permitted lower- and middle-class inventive artisans and innovative entrepreneurs, including religious non-conformists and recent immigrants as well as natives, to acquire wealth, property, and social position, and that also allowed upper-class gentlemen and nobles to associate with these members of the lower orders in a common concern with the solution of technological problems.[6]

That there was substantial upward social mobility in early modern England, while long taken for granted, admittedly has come under some more recent challenge. Yet the almost total absence of legally defined and maintained privileges separating merchants and industrial entrepreneurs from the aristocracy must play some part in explaining the effectiveness and freedom of action of the former as innovators. Of comparable importance was the downward mobility of younger sons of the nobility, bringing them more closely into association with these new social elements. In both of these respects Britain was sharply distinguished from continental Europe.[7]

Probably no less significant as a predisposing factor for technological advance was the association of deeply held articles of religious faith and world view with an economically progressive social outlook. The central element here, of course, is the broadly protestant "complex of scarcely disguised utilitarianism, of intramundane interests; methodical, unremitting action; thorough-going empiricism; the right and even the duty of *libre examen*; of anti-traditionalism" that Merton long ago discerned as a progenitor of seventeenth-century science.[8] Confirmatory information now allows his early discovery not only to be extended through the eighteenth century, but to apply more directly to the proponents of technology: "Only 36 per cent of 680 scientists and 18 per cent of 240 engineers were at any time connected with either Oxford or Cambridge, and throughout the century the tendency was downward."

Of the "notable applied scientists and engineers," a mere 8 percent had been educated at either of the two universities of the Establishment, less than half the number whose education at Edinburgh and the other Scottish universities suggests a background in the dissenting religious sects. Meanwhile, the relatively more humble background of more than 70 percent of this group is indicated by the fact that they apparently did not receive a university education at all.[9]

A general study of this kind is not the place to continue such an analysis in greater detail. The decisive importance of social context, and the impossibility of demarcating technology as an independent realm with laws of forward motion of its own, seem to be clearly established. Given the ambiguous nature of much of the evidence, of course, conflicting interpretations naturally persist on many subsidiary points. Suggestive of the difficulty of the issue is the continuing ambivalence within Marxist circles between tendencies toward "economism" and "politicism." Marx himself, in an early expression of one end of a considerable range of views, is responsible for the famous aphorism that "the hand-mill gives you society with the feudal lord, the steam-mill society with the industrial capitalist." More commonly, however, he cautioned against just such a deterministic interpretation.[10] We need not be embarrassed that some equivocation still continues today over exactly where to place the emphasis.

Eighteenth-Century Population, Urbanism, and Transport

Ranking high on any list of predisposing factors for the central, initiating role that England later played in the onset of the Industrial Revolution was the extraordinary urban character of London as its capital. The population of England as a whole had increased only modestly, from about five million to slightly more than six million or so, during the long interval between an earlier rise that ended early in the seventeenth century and the advent of the Industrial Revolution in the latter third of the eighteenth century. The contrast of this relatively static level with the trend in London is striking. "In round numbers," E. A. Wrigley estimates that London "appears to have grown from about 200,000 in 1600 to perhaps 400,000 in 1650, 575,000 by the end of the century, 675,000 in 1750 and 900,000 in 1800." Across the span between the mid-seventeenth and mid-eighteenth centuries, in other words, London had risen from housing about 7 percent to an unprecedented 11 percent of England's total population.[11]

Looking across the English Channel, a further contrast is no less striking. London and Paris, roughly equivalent in 1650, already outclassed all European rivals. But Paris and its sister cities on the mainland "could do little more than maintain their population size" across the century that followed. Since France was about four times as large as England, the proportion of the French population living in Paris remained static at about 2.5 percent. London, in short, not only was a substantially different kind of urban center in relation to its hinterlands but rapidly diverged further on its own unique course.

Lacking effective provisions for sanitation or for the prevention and treatment of many endemic and epidemic diseases, London of course had a significantly higher mortality rate than the country at large. The growth of London thus cannot be seen as a "natural," that is, internally generated, population increase. Wrigley estimates that immigrants attracted into the city must have constituted half of the natural increase within the country as a whole, and further concludes that at least a sixth of the total population had some direct experience of life there. Clearly this was "a powerful solvent of the customs, prejudices and modes of action of traditional, rural England" to an extent absolutely unmatched in France or elsewhere.

As a context favoring the Industrial Revolution, these population trends and contrasts are only the beginning of the story. More directly influential was the character of economic and cultural activity within London, and how the challenges of meeting the city's complex and increasing needs for food, fuel, labor, and other inputs were met. For the new demands were not merely unprecedented in scale; they called for qualitative innovations which, as Wrigley points out, "form the most promising base for a continuing beneficent spiral of economic activity."[12]

Consider first the requirement for foodstuffs, rising in direct proportion to the population. The demands of London's food market did not merely widen the radius within which increasingly specialized farms were devoted primarily to market gardening that would supply its needs, but also encouraged other forms of specialization devoted to transport, wholesaling, and the improvement of productive facilities. Trends in these directions were, in fact, already apparent in the preceding century.[13] Daniel Defoe, commenting with some hyperbole as early as 1724, spoke of the "general dependence of the whole country upon the city of London . . . for consumption of its produce." This can presumably be taken as reflecting not only the high proportionality of its needs even on a population basis alone, but also its differentially higher demands resulting from higher living standards.

Improved road transport was obviously essential for the provisioning of London, as well as for the growing volume of personal intercommunication and commercial movement for which London was again the principal stimulus. London was quickly destined to become the hub of a radiating network. Its development depended virtually entirely on privately financed ventures, although these turnpikes were in turn dependent on grants of toll-collecting authority enacted by Parliament. Road construction, having had a modest beginning somewhat earlier, became a boom in the 1760s. With hundreds of such acts passed, all major towns and cities had been interconnected within little more than a decade.

Apart from London, the growth of manufacturing activity in other provincial cities also involved increasing needs for the provisioning of the working population no longer engaged in agriculture. In timing, these needs coincided precisely with, and furnish a major explanation for, the rapid acceleration of the enclosure movement after 1760. The outcome of the latter, apart from having created an expulsive force at least partly responsible for the ensuing, major population transition into an industrial labor force, was a redirection of agricultural activity as well. The pasturing of sheep for wool production had to give way to intensified production of foods that were needed in the cities.[14]

While unquestionably it was largely an entrepreneurial response to perceived new demands, we should not neglect the secondary effects of road network improvement. Facilitating and speeding travel, better roads reduced the uncertainty, trauma, and time of travel and thus intensified the demand for it. William Jackman has estimated an average doubling every decade in the rate of coach movement between 1750 and 1830.[15] One coach run a week had been scheduled between London and Birmingham in 1740, for example. By the close of the American Revolution that figure had grown to thirty, and by 1829 to thirty-four a *day*.[16]

Barge movements along the growing network of canals grew in volume at a comparable pace but were, of course, much slower. The demand for animal traction was a small fraction of that for drawing wagons of cargo, and costs were correspondingly reduced by a factor of at least three or four.[17]

Turning from food to fuel, the demands of the great city had already outstripped available supplies of wood and charcoal well before the century in question. As a bulk cargo of relatively low value per unit of weight or volume, coal was so unsuitable for land transport that, before the advent of the railroad, its price doubled within five miles of the pithead.[18] Its economical utilization depended upon an increasingly substantial and efficient coastal shipping effort. At a later stage, beginning in earnest in about 1760 this was to be importantly supplemented by the development of a network of canals that would allow the closer approach of barge traffic to the major mining areas. No less an authority than Adam Smith was at pains to calculate the economic advantages of waterborne commerce:

> A broad-wheeled wagon, attended by two men, and drawn by eight horses, in about six weeks time carries and brings back between London and Edinburgh [608 km] near four ton weight of goods. In about the same time a ship navigated by six or eight men, and sailing between the ports of London and Leith, frequently carries and brings back two hundred ton weight of goods.[19]

Even the basic demographic trends leading up to the threshold of the Industrial Revolution thus had substantial technological as well as economic consequences. Of fundamental importance, to begin with, was the relatively static level of England's total population during the preceding decades. Had this number instead risen sharply, the dampening effect on the economic breakthrough that was to follow might have been quite serious. As Wrigley explains:

> The interplay between fertility, mortality and nuptiality must be such that the population does not expand too rapidly and this must hold true for some time after real incomes per head have begun to trend upwards. If this is not so, the cycle of events which is often termed Malthusian can hardly be avoided; there is a great danger that real incomes will be depressed and economic growth will peter out. This happened often enough before the industrial revolution.[20]

No less important than the relative stability of England's total population, however, was the contrast with London's simultaneously increasing size. Even before taking into consideration the disproportionate economic influence of the urban population, this could only be an engine of redirected, growing demand. Simply sustaining the great city came to require new initiatives in specialized, market-oriented agriculture, improved, more economical transport of consumables, and the rapid replacement of wood-cutting by a new, technologically more demanding extractive industry devoted to exploiting coal as the major fuel. As a solvent of local ties and magnet for dispossessed or upwardly mobile immigrants from all over England, London also surely was a major source of growing receptivity to further changes in production, consumption, and modes of employment. Reorganization of socio-economic pursuits thus was dictated, with effects that were felt far beyond London's immediate hinterlands.

Agricultural Diversification and Intensification

As we have seen, eighteenth-century England became a setting of potential subsistence stress. This was related, in the first instance, to the extraordinary urban growth of London. As the Industrial Revolution got underway, the stress intensified late in the century with a sharp upward turning in the rate of population increase for the realm as a whole. Fortunately, changes of comparable importance, sometimes characterized as nothing less than revolutionary in their own right, went on contemporaneously in agriculture that tended to restore the balance between population and food supply. Agricultural labor productivity had been rising from at least the sixteenth century, creating the conditions for a rural exodus that presently would permit the growth of an industrial work force. But the growth in total output that the rising demographic trend made necessary was matched by rising land productivity only during the eighteenth century.[21]

Central to these advances was the introduction of a more intensive rotation system. Alternating food grains and feed crops, it provided for a closer integration of cultivation with animal husbandry and no longer left one-third of the arable land fallow. No less essential to the newly emphasized complementarity were new forage crops, including clovers, ryegrass, and turnips (both for green forage and winter fodder). Seed drills, first introduced at around the beginning of the eighteenth century, gradually became a fairly common component of the new techniques of constant tillage.

As supplies of fodder grew during the course of the eighteenth century, there was a corresponding growth in the number of domesticated animals—and hence the source of organic enrichment represented by manure. Apart from enhanced beef production for growing urban markets, included within this was a rise of more than one-fourth in the horse power available per farmer. That in turn was associated with the replacement of traditional forms of plows by an improved, triangular one that permitted two horses and a single driver to replace teams of two farmers attending four slow-moving oxen. Partly because the maintenance of horses was less familiar to British working farmers as well as more demanding, some ninety years had to elapse between the patenting of this Rotherham plow and its widely recognized success.

Apart from increasing crop yields, these developments "enabled cultivation to be extended on to relatively infertile soils on the chalk, limestone and sandlands which had previously been sheepdown. More manure also secured heavier yields of the forage crops which meant in turn that still more fatting beasts could be kept."[22] But the rate of change, at least with regard to new technologies, was skewed as well as slowed by a deepening social division. On one side were the landowners, increasingly supported by professional stewards and advisers. Their relatively voluminous testimony "conveys the impression of intense managerial and technically innovative activity." On the other side were tenant farmers and hired laborers.[23]

> The fact is that most of the tenant farmers, who worked perhaps 80 percent or more of the cultivated acreage of the country, had neither the incentive nor the capital to experiment, and even the richest and most efficient large landowners hesitated, for political and social reasons, to introduce labour-saving machinery into rural areas depressed by chronic under-employment. In any case, the pressure on labour came at harvest time and there was little to be gained by economising in labour at other times.

Closely bound up with the evident resistance to change, in other words, was the issue of tenure. Improvements in technique generally coincided with land that had been enclosed or was in the process of enclosure. About half of the arable land was still cultivated on the open-field system in 1700, and the main effect of parliamentary acts in bringing this system to an end only fell in the half-century or so after 1760.[24] Yet in other respects, as in an extensive draining of wetlands and other programs for long-term improvement in pasturage resources, there had already been markedly increasing investments of capital as well as labor that, on better-managed operations, permitted increases of as much as 50 percent in the marketable value of crops and livestock. And the increase of net returns to labor, while less dramatic, was no less important in the sense that a growing army of industrial workers, many of them formerly employed in agriculture, could be fed with only a much more moderate increase in the agricultural labor force. Although some of the crops involved in the new system (turnips in particular) were more rather than less labor-intensive, the requirements of the system as a whole were reduced by being more evenly distributed throughout the year. More important, however, were gains associated with rising average size of farms as a result of the enclosure movement.[25]

Demand factors, mediated by the constraints of the transport system, added to these processes as a source of emerging interregional differences. That system was itself a target of massive improvements, particularly in the form of a doubling of the length of canalized waterways between 1760 and 1800.[26] London was the first and largest market for agricultural products of all kinds, stimulating the growth of a corresponding network of suppliers with shipping access to it. Similar demands were generated later by the new industrial centers like Manchester, the needs of whose rapidly rising populations could not have been met without the concurrent rise in agricultural productivity. More generally, cultivation began to be extended into regions of light soil that formerly had been devoted to sheep pastures, while claylands that were more difficult to work and drain gradually were withdrawn from the cultivation cycle and converted into permanent pasturage.

Hampering any such reconstruction as this is the "almost total absence of aggregate data," coupled with some uncertainty as to the representativeness of the voluminous anecdotal evidence. Basing calculations on many indirect sources, however, it has been estimated that total agricultural output advanced by 61 percent in quantity during the eighteenth century, and probably improved qualitatively as well. This was approximately twice the increase that had occurred in each of the preceding two centuries. It entailed a rise of some 8.5 percent in the agricultural labor force, as compared with an 81 percent rise in the total population of England and Wales. A significant margin of foodstock importation clearly had become by necessary by 1800, particularly in the form of beef, pork, butter, and wheat from Ireland.[27]

Whether it can properly be described as a "revolution" or not, what is important about the eighteenth-century changes in British agriculture was their scale, their diversification, and their timing. In scale, they were sufficiently massive to provide a subsistence base on which a large industrial work force, mostly withdrawn from earlier small-town and rural residence and labor, could be supported. In diversification, they were able to supply the new, luxury- and specialty-oriented demands of urban markets. And in timing, perhaps most important of all, most of their essential components were already in place as the Industrial Revolution began. As they were, the dislocations to come were large and disruptive enough. It was a matter of extraordinary good fortune that the needed, fundamental changes in agriculture had been so well anticipated.

Innovation, Production, and "Consumerism"

London continues to provide an illuminating point of departure in seeking to identify aggregative trends in production and consumption. As the political, commercial, and financial capital, it inevitably was also the arbiter of taste and style. This influence was not confined to the elite but substantially affected urban society at large.

London laborers, for example, were at the top of the scale in wages, and this gap reportedly was an increasing one. With comparable or greater advantages in discretionary income for other groups higher on the social scale, London also led the way in its demands for growing imports of expensive articles of mass consumption that included sugar, tea, coffee, chocolate, and tobacco. Average annual sugar consumption, for example, seems to have increased fourfold (to about 20 pounds per capita) during the first seven decades of the eighteenth century. Presumably luxuries at first, then "decencies," and finally coming to be reckoned necessities by many, such commodities satisfied new tastes and were to only a very modest extent direct substitutions for earlier, domestically supplied consumables. As the taste for them naturally spread into other areas, it intensified upward pressure on wages, and on the augmentation of money income at the expense of rural self-sufficiency, in order to satisfy the new ordering of preferences.[28]

Josiah Wedgwood, to whose activities we shall shortly turn in greater detail, provides a good illustration of the effect of London on prices and quality of goods as well as wages. Quick to realize the advantages of a market in London that was accustomed to "fine prices" for his high-quality lines of chinaware, he was able to utilize the acceptance of his products by the London elite as an encouragement in the much wider marketing of his less expensive lines.

The critical importance of a growing supply of coal for London has been mentioned in the previous chapter as well as this one. A demanding fuel in its extraction as well as in its use, coal was slow to replace charcoal in many specialized applications. Pig iron made with coal or coke, for example, was long regarded as inferior and priced accordingly. Craftsmen were compelled to evolve new means to prevent contamination of their products with coal fumes and residues. Yet it was, at the same time, fundamental in the establishment of the brick industry, which then quickly became "the prime building material of the new age," and of substantial importance in the production of nonferrous metals, pottery, glass, brewing, paper, and textiles.[29]

London's voracious demand for coal, and the development of the shipping industry that transported it, calls attention to the wider range of influence of the great seaport that was its heart. It was, for example, a nursery for the reception and mastery of technical innovations from overseas in the cotton industry, "until they were ready for transplanting to the provinces, where there was less competition for land, labour and capital."[30] It was also, of course, the source of much of the wealth that moved outward from commerce to stimulate investment in industry, agriculture, and infrastructures of all kinds. One-quarter of the the city's population is said to have been directly employed in port trades, with a multiplier effect through suppliers of secondary supplies and services that makes the importance of the port decisive.

Overseas trade, much of it routed through London, provides the best available means of coming to at least a provisional understanding of trends in production and consumption that apply to the country at large. Enhancing the importance of trade figures, reported throughout the eighteenth century largely at fixed official rates, is the wide range of goods and services for which statistical series are available. This evidence, when matched to contemporary demographic trends, provides a revealing picture. Until

1740, imports grew by approximately 40 percent and exports by 60 percent, while population growth was minimal.

> Thus, unless overseas trade expanded at the expense of internal trade (and there is no reason to suppose it did) we may deduce that there had been some advance in English real incomes per head over this period. But it could have been very small indeed. . . . We need not be surprised therefore to find that contemporary analysts who argued in terms of national income in the second quarter of the century were content to adopt, without adjustment, the estimates compiled by Petty or King fifty to seventy years previously. Nor that an attempt to combine all the available statistical indicators (trade returns, production and excise data, etc.) into a single indicator of national product yields an estimated overall rate of growth for this period of not more than 0.3 per cent per annum.
>
> The expansion that began in the 1740's was more marked. Between about 1741 and 1771, when English population grew by nearly a fifth, the volume of imports and of domestic exports increased by roughly two thirds. This was a solid achievement though it is still doubtful whether it brought with it much increase in the rate of growth of incomes per head. An estimate of the overall trend in national product suggests a trebling of the rate of growth of national product in the absolute . . . but little or no increase in the rate of growth in real product per head by comparison with the 0.3 per cent per annum calculated for the period 1701–41.[31]

There is convincing statistical evidence, in short, of a very modest rate of increase in English imports during the first two-thirds or so of the eighteenth century. The acceleration after 1740, since it coincided with the beginning of an upward turn in population growth after a lengthy static period, implies little if any increase in rate on a per capita basis. This low rate need not mirror the rate of growth of domestic GNP per capita, of course. But this was a time when some imported consumables were in increasingly widespread use. According to 1760 figures, nearly half of the families in England had already used some sugar, tea, and similar products. Hence, a rate of growth of expenditures for imports as low as this (bearing in mind that amounts rose as costs continued to decline) strongly suggests a comparably low rate of growth for English per capita disposable income.

This conclusion, it must be noted, is sharply at variance with a conclusion that some scholars have reached on the basis of other bodies of evidence, that there was a veritable "consumer revolution" in the eighteenth century. The first interpretation diminishes the possibility that the "pull" of rising consumer demand could have contributed significantly to generating the technological breakthroughs of the Industrial Revolution itself. The second, while it does not directly establish the existence of any relationship at all, at least opens the possibility wider. Hence the case for such a revolution needs to be briefly considered.

An important, if largely impressionistic and nonquantitative, body of evidence can be drawn from the generalizing observations of commentators and visitors to the country. With some frequency and few exceptions, they reportedly took positive, somewhat surprised note of what they observed as widespread prosperity:

> [C]ontemporaries were eloquent in their descriptions and explanations of what Arthur Young in 1771 called this "UNIVERSAL" luxury, and what Dibden in 1801 described as the prevailing "opulence" of all classes. . . . The Göttingen professor Lichtenberg said of England in the 1770s that the luxury and extravagance of the lower and middling classes had "risen to such a pitch as never before seen in the world"; the Russian writer Karamzin said of England in the 1780s "Everything presented an aspect of . . . plenty. Not one object from Dover to London reminded me of poverty."[32]

What has been claimed to be of particular significance here is the impression of augmented purchasing power that permitted access to "not only necessities, but decencies, and even luxuries" on the part of a very considerable proportion of the population. This is variously attributed to rising wage levels in the latter part of the century, to an earlier, prolonged drop in food prices as a result of good harvests, and to heightened family earnings as women and children began to enter the industrial labor force. But the consequence, in any case, is said to have been a substantial rise in income extending to many social strata well below the level of the aristocracy and urban elite:

> [I]t is not unreasonable to suggest that there was an agriculturally-induced increase in home demand before 1750; that between 1750 and 1780 the proportion of the population with family incomes in the £50 to £400 per annum range increased from something like 15 per cent to something approaching 25 per cent; and that these extra households made a major contribution to the market for the mass consumer products of the early Industrial Revolution.[33]

Particularly in the case of visitors, however, there are grounds for considerable skepticism about the basis for their judgments. Were they maintaining an implicit standard of comparison that tended to favor England, not only with other parts of Europe with which they were more familiar but for the readership with whom they wanted to communicate? Were they primarily familiar with the more prosperous residential districts and highroads? How well did literate travelers of the time, presumably of comfortable means, even "see" poverty—let alone go looking for it?

There is a further difficulty in attributing to a rise of consumer demand, as yet largely unexplained in its own origins, a generative force behind the Industrial Revolution. A near-coincidence in time, or at most a relatively modest lag between the "consumer revolution" and the beginnings of the

Industrial Revolution in 1760 or 1770, would be necessary for that to be plausible. Yet recent studies suggest that a consumer revolution, if one can be said to have occurred at all, probably peaked between 1680 and 1720 in both England and France. The lag in England seems entirely too long for consumer demand to have been a primary force in technological change. Similarly, the absence of a contemporary Industrial Revolution in France downgrades the role of consumerism as a sufficient cause since it failed there to have this effect.[34]

For a more differentiated, penetrating, and probably more credible account than those of the visitors cited earlier, we are fortunate in being able to turn to Adam Smith in the mid-1770s. Beginning by noting a rise in money wages in which England was substantially in advance of Scotland, he goes on to comment that:

> The real recompence of labour, the real quantity of the necessaries and conveniencies of life . . . , has, during the course of the present century increased perhaps in a still greater proportion than in money price. Not only has grain become somewhat cheaper, but many other things, from which the industrious poor derive an agreeable and wholesome variety of food, have become a great deal cheaper. Potatoes, for example, do not at present, through the greater part of the kingdom, cost half the price which they used to do thirty or forty years ago. The same thing may be said of turnips, carrots, cabbages; things which were formerly never raised but by the spade, but which are now commonly raised by the plough. . . . The great improvements in the coarser manufactures of both linen and woolen cloth furnish the labourers with cheaper and better cloathing; and those in the manufactures of the coarser metals, with cheaper and better instruments of trade, as well as with many agreeable and convenient pieces of household furniture. Soap, salt, candles, leather, and fermented liquors, have, indeed, become a good deal dearer, chiefly from the taxes which have been laid upon them. The quantity of these, however, which the labouring poor are under any necessity of consuming, is so very small that the increase in their price does not compensate the diminution in that of so many other things. The common complaint that luxury extends itself even to the lowest ranks of the people, and that the labouring poor will not now be contented with the same food, cloathing, and lodging which satisfied them in former times, may convince us that it is not the money price of labour only, but its real recompence, which has augmented.[35]

Apparent in Smith's account is a complex interplay of forces. To begin with, technological advances in agricultural practice are said to have substantially lowered the cost to consumers of many agricultural commodities. The shift to the plough is in large part surely a reflection of market-driven changes in land management, not of new technical innovations. Failing to come within Smith's field of view were the impacts both of climate-linked fluctuations in agricultural output and of high, relatively fixed transportation costs. Taking them into account, one may wonder whether the average effects were quite as dramatic as he suggested.

There is no reason to doubt the improvements in the range and quality of consumer goods that were available for purchase as he recorded, along with the qualifying observation that the price of many had become "a good deal dearer." But if soap, salt, and candles are examples of what the "labouring poor" were not "under any necessity of consuming," the "common complaint" that "luxury extends itself even to the lowest ranks of the people" seems, to say the least, overdrawn. Would it not be reasonable to identify in this complaint a fairly familiar example of class-based antagonism that ran far ahead of the real advances in living standards in which most of the population were able to share?

Associated with the idea of a "consumer revolution" appears to be "a studied vagueness in definitional statements and a careful removal of most of the concept from the economic to the cultural sphere: desire, attitude, fashion and emulation furnish the vocabulary of this discourse."[36] Upward trends in demand did occur, but were notably bifurcated in form. Only at the upper social levels, extending down to include a portion of the middle class that remains very hard to define precisely, did rising incomes unquestionably lead to real prosperity and a growing openness to purely discretionary purchases.

In terms of individual attitudes and behavior, Philip Scranton evocatively summarizes the varied meanings rising incomes may have had for a broadened stratum of the population below the aristocracy "from ascetic dispositions toward goods to patterns of expressive possession, from a tendency to save to an appreciation of the duty to consume, and from self-definition in terms of work or occupation to social placement defined largely through reference to commodities held by persons or families."[37]

Along with these essentially self-gratifying tendencies were others that embodied and rationalized a widened gulf between "the wealthier classes—gentry and yeoman alike" and the bulk of the population. Protective walls around estates and parks became common, and a tendency toward social segregation is also reflected in architectural arrangements that stressed domestic privacy.

> In contrast to its predecessors, the rural elite of the eighteenth century began to adopt a studied style of social hegemony based on the majesty of the law and the theatricality of the highly articulated political culture. From the mid-century, in particular, class attitudes toward popular recreations hardened and the game-laws were enforced with increasing ferocity.[38]

But at the other end of the social continuum, the notion of a consumer revolution acquires a much more restricted meaning with the increasing proletarianization of the farm labor force as a result of the enclosure movement. In a lengthy and very critical review of the ideas of the "consumerists," Ben Fine and Ellen Leopold reasonably conclude that "to create the image of a forward-moving society fuelled by demand from above, consumerists have to jettison the interests and contributions of at least three-quarters of its members. The incomes and consumption habits of the labouring and middle classes are left out of the picture altogether, obviating the need to measure the relative impact of luxury spending on the economy as a whole."[39]

Let us turn, then, to a less anecdotal attempt to gauge the condition, and more especially the changing condition, of the greater part of the society that would have contributed little to the swelling demand for luxuries. At around the turn of the eighteenth century, according to Gregory King, some 65 percent of English households were engaged in occupations that were denied the usufruct of the land.[40] Thus they were compelled to purchase goods and services that in many cases their equivalents would have produced for themselves a century or two earlier. For many of them, and especially for those not in a position to benefit from the concentrations of wealth, power, and growth in London, the shift from a household mode of subsistence production to a need to purchase basic goods and services may have carried a meaning that was nearly the opposite of prosperity.

It is difficult to obtain a clear view of the conditions that confronted these social sectors that have left few records. Dietary sufficiency would perhaps be the most directly useful measure. Mortality, which "showed a steady if not pronounced improvement" only in the latter part of the century, may reflect greater concern for public health and improved medical standards as well as higher living standards.[41] Stature, while a seemingly more indirect measure, is probably a more sensitive and accurate indicator of dietary sufficiency. Hence it is noteworthy that at the time of the American Revolution, British soldiers were about three and one-half inches shorter than their colonial counterparts.

Direct dietary information only becomes available for a period in the 1780s and 1790s. To be sure, these "were recognized as especially bad times for workers with population growth, inflation, wars, and urbanization all taking their toll."[42] They are likely to offer only a somewhat distorted picture of what prevailing conditions in the countryside may have been like under conditions three decades or so earlier. But they seem to furnish the only available point of departure that is suitably detailed.

The laboring class household budgets of that later period were collected primarily from agricultural workers and entirely outside the major cities. They provide convincing evidence, at least for this still very large component of the population, that these were anything but good times. Daily caloric intake in the the South had fallen into the 2,100–2,500 range, "which would not seem to provide the energy for hard labor or growth in children and the North's 2,800–3,200 was, at best, barely adequate." Cows, common as peasant laborers' holdings in the late sixteenth century and still held by 20 percent or so in the early eighteenth, were virtually absent.

> By the end of the 18th century, many rural laboring families bought nearly their entire diet, aside from garden stuff, on the market from the baker, the butcher, the grocer, or whoever sold the tea, sugar, butter, milk, flour, bread, and meat in the locality. Apparently, the more prepared the food the better so that fuel would not have to be expended in the cooking. About 10% of the budget went for the new commodities, mainly sugar and tea. . . . There were meat shortages but the steady decline in milk and cheese consumption during the 1700s seems of even greater significance and may help to explain the difference in growth patterns observed between the English and the colonials, who not only consumed more meat but also owned more cows.[43]

The suggestion of considerable privation and hardship that these data provide should not be overgeneralized. It has already been observed that they were collected during lean times, after a rapid population rise was underway in the full thrust of the Industrial Revolution. Carole Shammas has also collected much persuasive evidence of the growing presence of purchased consumer durables—textiles, pewter, brass, glassware, paper, pottery, ready-to-wear apparel, and furniture, in addition to consumables like tobacco, coffee, tea, and sugar—among the possessions of even quite poor individuals when their wills went to probate. To be sure, she has also found that the sharpest rise in the amount of real wealth put into consumer durables was much earlier, in the seventeenth century. But perhaps, since probate inventories only measure stocks at death, the later leveling off of this amount is less a reflection of declining trends in disposable income than of a growing readiness to accept less permanent products that better met the test of stylistic currency.[44]

There seems, in short, no reason to gainsay the general impression that some appreciable rise in *average* living standards did occur as the eighteenth century went on. For perhaps a majority of the population, imported commodities like tea, tobacco, and sugar were moving into a category of fairly regular items of consumption rather than discretionary purchases on special occasions. No less important than the apparently modest upward movement of the average, however, was the intensifying social bifurcation resulting from an ongoing, massive redistribution of income.

What can we conclude from the obviously heterogeneous evidence about whether there was a "pull" from the direction of greater demand on the technology that would presently multiply productivity? Clearly, there were new forms and levels of consumer demand, affecting not only the scale but the structure of production. But we need to disaggregate that demand and in so doing sharply reduce its significance as a force for technological change. Increasing manufacturing productivity would be primarily a response to a widening ability to purchase practical devices and consumer durables on the part of the population at large. The increasing wealth and emphasis on conspicuous consumption at the apex of society, on the other hand, would only have emphasized individually crafted sumptuary, architectural, and other embellishments to express more differentiated styles and tastes.

London's growth as the largest European city of the time precipitated other changes, not only from increases in its own internal division of labor but from the demands it made on a widening zone devoted to supplying its subsistence needs. We have touched on the massive expansion of the coal industry, and on the secondary effects of the growing utilization of coal in industry and the rationalization of transport arrangements it required. Aggregate imports from overseas grew slowly, with the new, exotic consumables from subtropical countries stimulating a widening market and presently finding a permanent place in the national diet. Artisanal products or manufactured goods also found a progressively widening place in ordinary household inventories. And certainly accompanying this intensification of local trade and regional movement was a very substantial growth in the class of merchants, purveyors, wholesalers, drovers, and service personnel and agents of many kinds upon whom it all depended.

However, it is quite another question whether this "pull" extended beyond a perceptible but modest increase in the pace and complexity of economic activities, to serve as a genuine stimulant of the advances in technology that were to come. That position has, indeed, occasionally been affirmed in general but impressionistic terms. Noted by Phyllis Deane particularly for the second half of the eighteenth century, for example, was "a tendency for the demand for British manufactures to exceed their supply. The resultant stimulus to technical change was reflected in the wide interest in innovation."[45] This was left, however, as a somewhat speculative generalization for which no detailed support was offered.

> In theory it is possible to show that technological changes represented a response to rising prices for labour, capital and other inputs contingent upon shifts in consumer demand for particular commodities and services. . . . But when we turn to the case studies of European inventors at work from 1750 to 1850, it is not at all apparent that as a group they can be depicted as a body of men attuned to signals from the market.[46]

Thus at best, if there was a relationship between technological innovation and market forces, it must have been not only diffuse but subject to long lags and other intervening variables. John Kay's flying shuttle (1733), for example, greatly intensified what was already the substantially greater productivity of cotton weavers than of spinners. A reversal of this imbalance followed only after more than four decades, when Samuel Cromp-ton's mule (1779) ingeniously combined principles of the Arkwright waterframe (1769) and the Hargreave spinning jenny (1770). That in turn could be said to have created a need for great improvement in loom perfor-mance. But while Cartwright's first patents on a power loom were taken out in 1786–1788, its industrial adoption was deferred another four decades.

Sequences like this illustrate a frequently poor and delayed fit of innova-tion to industrial requirement and acceptance. More generally the evidence they provide for technical changes having stemmed from changes in de-mand can only be described as

> disquietingly slim, . . . unobjectionable but unsatisfying. Why, after (say) Kay's flying shuttle, could not a producer simply bring more and more spinners into action to supply the quicker weavers? Transport charges would certainly rise . . . but cotton was not a bulky material, and transport costs were only a small portion of its final costs of production. . . . Thus the imbalance resulting from the flying shuttle lay as much in the structure of the family and the community as in the structure of production. In that manner, supply-side considerations emerged in a supposedly demand-side model, as the long-run supply curve turned more steeply upward.[47]

Other examples of delay are even more damaging to the case for de-mand-induced innovation. As we shall shortly see, Abraham Darby's coke smelting process (1709) entered into substantial use only from fifty to seventy years later. Henry Cort's puddling process for converting pig iron to wrought iron (1784), while it eventually revolutionized the iron indus-try, was not substantially taken up for a quarter century. And all these examples embody, in any case, only demands for improvements within manufacturing processes rather than substantially more distant consumer demands for end products that would have had to be communicated through a market.

It seems, in short, that rising consumer demand was likely to provide no more than a modest and diffuse inducement at most to the technological advances that lay at the heart of the Industrial Revolution. Market "pull," in that direct sense, appears to have little explanatory power. But it has another, indirect aspect to which we shall need to return in greater detail. However deeply increases in purchasing power may have penetrated into the populace at large, all the sources are in agreement that considerable increases in disposable wealth came into the hands of a substantial elite during the latter part of the eighteenth century. Perhaps it is not so much their diversified and growing desires as consumers that quickened the pace of technological advance, but rather the increasing supply of potential ven-ture capital for which this elite was beginning more aggressively to seek new avenues for profitable investment.

Patents as an Index of Innovation

There is nothing surprising or counterintuitive about the proposition that,
with regard to technology, market signals can primarily take the form of
queries and initiatives from potential investors—investment "push" rather
than consumption "pull." In fact, it has long been almost axiomatic among
economists "that invention and technical change are the major driving
forces of economic growth" and "that much of technical change is the product of relatively deliberate economic investment activity which has come
to be labeled 'research and development.' "[48] This inclusion of technology
within the framework of deliberate investment activity constitutes a vital,
if in modern times increasingly blurred, contrast with basic research in
science. Journal publication has become the classic form of public dissemination of the latter, exposing new findings to the criticisms of one's peers
and contributing to a collectively pursued advance whose sole purpose is
unspecified new knowledge. Technological contributions, on the other
hand, have always tended to have a proximate economic end in view. They
therefore constitute a more calculably profitable form of new knowledge
and as such are today an important corporate asset.

These considerations of course underlie the existence of a patent system.
The extent of patenting activity and the changing substantive directions it
took would seem to provide an excellent indicator of the pace of inventive
activity as it may have contributed to the onset of the Industrial Revolution. Equally significant would be a quantitative analysis of the changing
orientation of patenting activity, as it presumably reinforced the main directions of technological change that lay at the heart of that great epoch of
change. But while studies of patenting have indeed provided some tantalizing clues along both of these lines, they are beset with a number of serious
ambiguities.

Patents for inventions have been regularly granted in England since the
mid-sixteenth century. But for our purposes, the early English patent system had important limitations. To begin with, it "was one of simple regis-
tration. Extensive scrutiny was not expected of the law officers administer-
ing it. . . . unless a patent was challenged, all the stages except one were
mere formalities. This was . . . to determine whether the patent would be
legal under the Statute of Monopolies. Neither the viability nor the utility
of an invention was their concern, while novelty was generally taken on
trust."[49]

Detailed studies support the characterization of the early patent systems
as a "rag-bag institution" with many anomalous inclusions and such major
exclusions as the Newcomen engine, Huntsman's crucible steel, and the
Crompton mule. But in addition, there are deeper questions to be asked as

to what any measures of the rate of patenting actually represent: "Can technical achievement be put in the same scales as economic potential, for example? Which was more important, the initial breakthrough or the host of later, minor improvements that made it technically and economically feasible? How far was that breakthrough an illusion, the cumulation of many small steps? How does one count failed prototypes or the multiple inventions of the same product or process?"[50]

As this suggests, a consistent relationship cannot be taken for granted between patent series and total inventive activity. Consciousness of the rising pace of economic activity during the early years of the Industrial Revolution, for example, may have intensified competitiveness among inventors or otherwise given greater salience to the patent system and thus increased the propensity to patent.[51] Beyond this limitation, patent series fail to distinguish between fundamental technological breakthroughs, marginal improvements—and, indeed, a significant proportion of patents that, for one reason or another, never entered into the commercial stream at all. With all these necessary caveats, however,

> the graph does exactly what our historical "common sense" tells us it should. It shows a marked upward trend from the third quarter of the eighteenth century. From an average of 60 patents per decade in the century after the Restoration, the decennial total jumped to over 200 in the 1760s, to nearly 500 in the 1780s, and continued doubling every two decades to the mid-nineteenth century. The obvious explanation is that this rise expressed a sudden blossoming of inventive talent; it was the "wave of gadgets" that notoriously swept the country in the late eighteenth century.[52]

A more precise evaluation of the patent statistics has suggested that 1757 can be identified as the critical upward turning point in the curve. From the perspective of the transformative effects of change seeming to occur most rapidly in fields like textile production, earlier it had been generally agreed that "critical innovations were concentrated at first in a small sector of the economy."[53] No such concentration has been detected, however, when scrutiny is directed not at industrial consequences but at the patents themselves. Instead it turns out that the heightened rate of patenting was very diffuse and without apparent focus. Moreover, there is no indication of a change in administrative regulations to which the upward turning can be attributed. Thus the increase in inventive activity seems to have been genuine as well as general. And since it substantially preceded the radical increases in productivity in the textile sector, it cannot be attributed to some sort of demonstration effect from that source.[54]

But while a focus on particular domains of productive activity cannot be discerned in the patent series, there is some evidence for a cross-cutting concern of a different kind. Christine MacLeod, having identified an

interest in patenting new advances in labor-saving technology that remained consistent throughout the eighteenth century, nevertheless suggests that this

> was always subordinate to the desire to save capital. To speed up a mechanical,
> chemical, or metallurgical process promised reductions in the cost of trade credit;
> it might also cut fuel costs, overheads (or even wage payments). These ends were
> also sought in the drive for greater regularity of products and reliability of pro-
> cesses and machinery. In the nineteenth century many "labour-saving" inven-
> tions were prompted not by a desire to economize on the costs of labour, but to
> circumvent the power of organized labour and the fallibility of workers.[55]

The bearing of the evidence of prior patent series on the origins of the
Industrial Revolution is, in short, tantalizingly suggestive but unspecific.
Prior to 1760 there was little to demonstrate that markedly increased returns would accrue to inventive activity—let alone to what we today would
construe as "basic research" from which more practical applications might
presently emanate. Equally significant is the apparent absence of a concentration of patenting in fields that presently would move to the leading edge
of industrial development. Something was in the air, we may conclude, that
carried the suggestion of emergent opportunities for profitable growth and
change. If MacLeod is right, the key may lie in a new outlook on capital
investment not as static and needing little attention but rather as a strategic
resource to be husbanded, managed, and strategically re-deployed whenever favorable combinations of circumstances could be identified.

As has already been discussed in chapter 1, much debate continues on
whether the inherent possibilities of further developments within particular
areas of science and technology furnish a kind of "push" for the technology
enterprise as a whole. The only alternative usually identified, seemingly in
rather sharp contrast (if not opposition) with this, is the "pull" of rising
demand. But discernible in the patent series coinciding with the Industrial
Revolution as the leading example of a great economic transformation is
another, at least partly independent, driving factor. At a time of significantly rising disposable wealth, the appearance of a new and more far-
seeing class of entrepreneurs, bringing with them a new sense of how to
marshal capital resources and put them to more productive use, can be an
important force in its own right. And it is noteworthy that Adam Smith,
with his usual perspicacity, discerned the presence of just such a group at
the appropriate time. Clearly highlighted in his account of them is an impression of their stance and objectives that could be applied with almost
equal relevance to their modern equivalents:

> All the improvements in machinery, however, have by no means been the
> inventions of those who had occasion to use the machines. Many improvements

have been made by the ingenuity of the makers of the machines when to make them became the business of a peculiar trade; and some by that of those who are called philosophers or men of speculation, whose trade it is not to do anything, but to observe everything; and who, upon that account, are often capable of combining together the powers of the most distant and dissimilar objects. In the progress of society, philosophy or speculation becomes, like every other employment, the principal or sole trade and occupation of a particular class of citizens. Like every other employment too, it is subdivided into a great number of different branches, each of which affords occupation to a peculiar tribe or class of philosophers; and this subdivision of employment in philosophy, as well as in every other business, improves dexterity, and saves time. Each individual becomes more expert in his own peculiar branch, more work is done upon the whole, and the quantity of science is considerably increased by it.[56]

Some may regard the presence of such entrepreneurs as no more than a somewhat more visionary or speculative subdivision of the "pull" of market forces. But if theirs was a genuinely creative contribution, as Smith perceived and as seems clear almost by definition, much of its originality lay in the capacity to *anticipate* a new convergence of consumer preferences and technological posssibilities. And where then, other than in the entrepreneurial group itself, can we look for the originating, primary force?

Technological Contributions and Contributors

The issue of "push" versus "pull" is an abstract and probably overgeneralized one. It seems to imply that causal primacy for an epochal turning point in human affairs like the Industrial Revolution lies ultimately in large, impersonal forces beyond the control, and perhaps even largely beyond the conscious understanding, of human protagonists.

The new products and productive processes that began to be spawned at a rapidly increasing rate by a tide of technological inventiveness in the later years of the eighteenth century could be regarded as one of these forces. Was there an inanimate multiplier effect in their potentialities for meeting old wants and creating new ones, such that the flow of new inventions and innovations was itself somehow stimulated by its own success? Were the basic scientific and engineering insights behind the great technological advances simply parts of a broad climate of public knowledge, or did they depend on direct, personal stimuli?

Market demand, obviously overlapping with the former but also deserving some independent status as an alternative, provokes a different set of questions. In the absence of a capital market and institutions assuring the relatively unrestricted flow of accurate economic information, how was

consumer demand apprehended by investors and entrepreneurs, and ultimately communicated to inventors? To what extent did these three categories of individuals overlap? How did they establish contact with one
another, and with what relations of authority and cordiality, if they did not?
What accounts for a striking, and apparently mutual, recognition of opportunities for long-range growth in the development of new producer goods,
rather than in more immediate responses to transient consumer tastes?

These and similar questions are mostly still without adequate answers.
Decision-making processes, seldom well recorded for technology-related
matters even today, are individually creative acts on which older historical
documents are misleading or silent. Those we would like to know more
about in this case balanced ill-defined opportunities against equally ill-
defined risks and uncertainties. Our concern, after all, is for how the atmosphere gradually became receptive to change and growth in the last third
of the eighteenth century. We may lose a sense of how the world of perception and aspiration shifted if we approach it only through impersonal
aggregates.

We can begin to move toward an understanding of the gradually accelerating pace of change, however, by concentrating initially on only two industrial sectors. Textiles and iron-making, by general agreement, lay at the
heart of the Industrial Revolution when it arrived in full flower toward
the end of the century. For textiles, the first of the advances on which the
enormous later growth of the industry was based came only in the 1760s
(and hence will be dealt with only in chapter 4). An analysis of baptismal
records for the Lancashire region shows that a large proportion of fathers
were employed in some branch of the textile industry by mid-century. In
fact, the Manchester environs had seen "the evolution of a capitalist class
and an experienced workforce for nearly two centuries before the first
water-powered cotton mills were built in the area." It was only after that
time, however, that superior technology rather than careful market cultivation became the key to British achievement.[57]

The timing of fundamental inventions and innovations in the iron industry was somewhat different, but with delays in their substantial implementation the overall effect is similar. Ultimately of the greatest importance
was the discovery by Abraham Darby I (1678?–1717) of a practical means
to substitute coke for charcoal in the smelting of iron ore. This was first
successfully carried out in 1709, and the delay in wider adoption of the
practice has long been puzzling. It now appears, however, that neither the
special quality of Darby's coal nor his secretiveness about the process was
responsible for this. Instead it was merely a mundane matter of comparative cost, compensated for in Darby's case by counterbalancing savings on
a by-product about which he was indeed successfully secretive.[58]

It was only in the 1750s that rising and falling trends in the prices of wood and coal, respectively, reversed the cost advantage of charcoal. Probably accompanied by greater efficiencies in furnace management, this initiated a trend that presently led to the virtual elimination of charcoal as a fuel.[59] After considerable controversy, it now appears that an imminent general shortage of charcoal was not a constraint on iron production at its mid-eighteenth century levels.[60] On the other hand, the vast expansion in the production and use of iron that followed could not have been maintained on a charcoal base. Rising demand for iron plainly was an important force in speeding the transition to coke once the process had begun. But there is little to suggest articulations of demand pressure—which might have taken the form, for example, of stimulating supplementary improvements and/or efforts to secure more favorable factor prices. In terms of its social and economic impact, it is thus instructive that comparatively modest cost differentials kept a highly promising technology essentially dormant for more than four decades after it first came into existence.

A second, somewhat later innovation in this industry was of only slightly less importance. Benjamin Huntsman (1704–1776), a clock maker and instrument maker, introduced crucible steel in the 1740s, vastly improving upon the quality and homogeneity of the cementation and blister steels that were then the best available. By 1750 or so, a uniform (if still costly) material was becoming widely available from several competing producers for watch springs, cutlery, high-quality metalworking tools, files, hammers, rollers, and the like. Its superiority was quickly recognized as indispensable by the nascent engineering profession, in sharp contrast to the delay incurred in the case of coke-smelting. But was this prompt exploitation (perhaps even hastened by industrial espionage) a matter of greater demand for this singular product, or rather of timing? With its later invention, after all, crucible steel's exploitation turned out to be roughly contemporary with the delayed spread of coke-smelting. Hence it is equally plausible to suppose that there was a generally rising demand for iron products of all kinds that began only in the 1750s, and that it accounts for the simultaneously positive reception at that time to both crucible steel and the coke-smelting of iron.

So let us turn briefly to iron output as a further clue to the demand for iron products.[61] The British market in the 1720s is estimated to have consumed annually 28,000 tons of domestic pig iron plus around 20,000 tons of imported bar (approximately equivalent to the British tonnage in less refined pig iron). Foreign imports, in their pig iron tonnage equivalence, had climbed to 38,000 tons by the early 1750s, which, together with an almost static level of domestic production as well as 2,000 tons of the new coke pig, brought aggregate consumption up only slightly to 66,000 tons of pig equivalent. By contrast, in 1799 domestic pig production alone had risen to 170,000 tons, at a time when foreign imports were beginning to decline. According to an estimate of 1802, British iron production had multiplied about eleven times since the second decade of the eighteenth century, and about eight times since mid-century.

What the available figures seem to suggest is a level of demand that increased only at a very moderate rate until the 1750s. Then, however, a break becomes rather abruptly discernible, with the trend of growth in annual output turning sharply upward and maintaining its steep slope well into the nineteenth century.[62] The coincidence of this development with the other indications of rising demand that were mentioned earlier is, to say the least, highly suggestive.

Another way to contextualize technological advances begins by dealing individually with a few leading technological protagonists or groups of protagonists, concentrating on those whose achievements, contacts, and influence made them not merely arbiters of opinion but creative forces in their own right. The more fully they are described in individual terms, it is fair to warn in advance, the more quickly they escape categorization. But a minimal characterization that applies to all of them is that their breadth of experience and contact was very wide. On the other hand, it is seldom clear how fully, or even whether, they apprehended either the signals of the market or the state of organized, scientific knowledge of their times. Perhaps the vagueness and ambiguity of the sources that motivated and stimulated them may itself be a clue to the complexity of the changes in which they were primary participants.

The early ironmasters deserve to be the first group to be briefly characterized, since they were the first group of professional industrialists to emerge. They were from very diverse backgrounds, including landed proprietors with an interest in developing their own natural resources, manufacturers of iron products, merchants, and others. It may be significant that those responsible for the major technological breakthroughs tended not to be individuals who previously had been well established in the industry. Instead, for the most part they entered it from diverse other business pursuits. Generally, however, they do appear at least to have had some acquaintance with other branches of metalworking trade and industry. Nonconformists formed the largest contingent of the group as a whole, with Quakers alone said to have owned or operated half of the functioning ironworks. Yet in spite of the general Quaker reputation for prudent moderation, qualities of dynamic aggressiveness came to be associated with the leading ironmasters.[63]

Engineers are another group deserving closer scrutiny, and we can gain some insight into them by considering a man who was probably their outstanding individual representative. John Smeaton (1724–1794) made major improvements in the efficiency of waterwheels, and by all contemporary accounts stood astride both the science and the technology of his day. Having begun as an apprentice to a scientific instrument maker, much of his early experience lay in making models of air pumps and other working models of devices for scientific lecturers. "In connection with his model construction he developed an interest in empirical studies of engine efficiencies, much as Watt did. In fact, during the 1760s and 1770s, just before Watt's separate condenser engine came into use, . . . Smeaton was able to construct the most efficient commercial steam engines then operating, nearly doubling the efficiency of Newcomen's engines."[64]

Pioneering as Smeaton's contributions were, they should be understood not as an unprecedented eighteenth-century development but as falling within an already well-defined intellectual lineage. Newcomen himself had begun to develop rudimentary general rules for calculating the power of his engine, and Henry Beighton (1636–1743) greatly improved the tables for predicting engine performance in pumping water. Beighton's interests also anticipated Smeaton in extending to waterwheel construction, land surveying, and mechanical invention, all displaying a similar concern to unite theory with practical experience.[65] But while he had productive progenitors and contemporaries, Smeaton's own contributions were of outstanding significance for his generation.

> Smeaton sought out the company of scientists, becoming a member of the Royal Society of London and a frequent visitor at the meetings of the Lunar Society of Birmingham, whose members included Joseph Priestley; the zoologist, Erasmus Darwin; and the potter, Josiah Wedgwood, as well as both James Watt and Matthew Boulton. By any reasonable measure, Smeaton was himself a scientist, publishing a number of papers in the *Philosophical Transactions of the Royal Society of London* on mechanics and astronomical instruments in addition to the major paper on water and wind power. . . .
>
> Smeaton's greatest claim to fame as a scientific engineer and as a dominant figure of the early Industrial Revolution came from his innovations in water wheel design, which were among the most important engineering achievements for the conduct of eighteenth-century industry. They allowed for an effective fifty percent increase in power extracted from England's rivers and streams at a time when there was a developing crisis of power needs. The development of steam power was even substantially delayed because of the dramatic success of Smeaton's new water wheel designs, which were widely emulated throughout Britain between 1759 and 1800.[66]

Running through many of Smeaton's experiments on steam and water power was a systematic reliance on what later became a fundamental engineering methodology, parameter variation. Absent a useful, quantitative theory, as is frequently the case with difficult engineering problems of structural design or the complexities of turbulent fluid flow, this is often still the best available approach.[67] Smeaton deservedly holds an important place in the emergence of engineering as a profession.

Josiah Wedgwood (1730–1795), whom we have already encounterd crossing paths with Smeaton, was a no less wide-ranging and influential figure, although his primary talents lay more in the direction of entrepreneurship and there are grounds for considerable skepticism about his qualifications as a scientist. He was, on the one hand, a Fellow of the Royal Society who published several papers in its proceedings, an avid experimenter, correspondent with other researchers, and the possessor of an important chemical library, and the inventor of the pyrometer. On the other hand,

> Doubts about the scientific value of his papers to the Royal Society are as well founded as is scepticism of the scientific significance of being a Fellow in the eighteenth century. Doubts about the importance of proposals which do not develop are equally justified. And such doubts can be multiplied. Wedgwood's pyrometer was certainly industrially useful, but his empirical approach to the problem and his failure to calibrate its scale with that of Fahrenheit raises doubts over the level of scientific achievement involved in its invention. Wedgwood was certainly familiar with the language of chemists but that does not necessarily invest his use of it with any scientific significance. Wedgwood certainly learnt and borrowed experimental techniques from the chemists he corresponded with and read, but that does not necessarily raise his experimental practice above the level of empiricism. . . .
>
> It is this contrast between promise and performance, between suggestion and reality, which has led to such firmly contradictory views about Wedgwood's status as a scientist. It is the same contrast which has fed the general controversy over the role of science in the Industrial Revolution.[68]

Not accounted for in this handful of thumbnail biographies is the indispensable infrastructure of skilled artisans and craft specialists to which there was much reference in the preceding chapter. The capabilities and importance of this large and heterogeneous group, about the details of whose recruitment and individual backgrounds there is little but anecdotal information, should be emphasized. It was, in fact, essential for the successful development and spread of the great inventions that followed later in the eighteenth century. Scaling up from small, hand-powered equipment and softer materials like brass with which they were most at home,

> clockmakers such as Hindley and Wyke had been making wheel- or gear-cutting engines, as well as lathes and other tools, which they had supplied to Smeaton, Watt, Boulton, Wedgwood, and others; such clockmakers and instrument makers played a key role in the development of precision engineering, especially with

what was referred to as "their more accurate and scientific mechanism," which could be used in the "clockwork" of textile and other machinery.[69]

In many largely anonymous hands, principles and practices evolved in this way for cutting, boring, shaping, planing, finishing, and other forms of metalworking, and for coupling these to inanimate sources of power. Ultimately all this would help to pave the way for machine building and the launching of a machine tool industry. The self-reliance and effectiveness of such artisan-practitioners can be recognized if we consider their essential role in installing Newcomen engines as those engines found a widening market all over Europe.

The rapidity with which Newcomen's design was put to use independently by other engineers and craftsmen is impressive. Modern replication of the original machine has shown that many months of skilled effort and patient adjustment must have been necessary to bring each of these new engines into operation, for they lay at the very frontiers of engineering technology of the time.[70] Yet scores were very quickly brought into use in England, and within four years after the first was introduced in 1712, they could be found in eight countries. An "unavoidable conclusion must be that this would have been quite impossible without the prior existence, from the beginning of the century, of a substantial and well-distributed population of skilled, energetic artisans—millwrights, blacksmiths, instrument makers, etc.—who were prepared to erect new and revolutionary machines, and of men who were willing to risk their capital on such ventures."[71]

Still a further personnel requirement is suggested by the fact that most installations of engines on the European continent depended heavily on British engines and engineers. "One might have thought that the New-comen engine, generally regarded in Britain as a much cruder device than the Watt engine, would have spread easily to other countries without the aid of British skills since it worked with much broader tolerances. The facts seem to have been otherwise."[72]

Thus we come to James Watt (1736–1819), whom once again we have encountered as a member of the same elite circle of scientists, engineers, and entrepreneurs. By background a technician and instrument maker, his circle of friends in and around Glasgow and its university included Adam Smith and the chemist Joseph Black. It was once thought that Black had provided a crucial scientific stimulus to Watt's discoveries. Watt himself flatly denied this, however, and in any case Black's views on "latent heat" could only have impeded what Watt undertook. The source of his contribution thus seems to lie within the realm of his own, technologically oriented thought.

As in Smeaton's case, Watt apparently retained something of the outlook of his former background as a craftsman. Testifying to this is his advice in 1781 to a mother on the education of a young would-be civil engineer: "When he is 14 put him to a Cabinet Maker to learn to use his hands and to practise his Geometry—at the same time he should work in a smiths shop occasionally to learn to forge and file. . . . before he attempts to make a theodolite let him be able to make a well joined Chair."[73]

But if an immediate stimulus from a scientific direction was apparently lacking, an entrepreneurial one was not. Watt's close and cordial collaboration with Matthew Boulton played a vital part in his success. Without him, in Watt's own words, "the invention could never have been carried by me to the length it has been." It was Boulton who brought to the partnership not only needed capital but the single-minded vision he once candidly expressed to Watt by letter: "I was excited by two motives to offer you my assistance—which were, love of you, and love of a money-getting, ingenious project. . . . It would not be worth my while to make for three counties only; but I find it very well worth my while to make for all the world."[74]

By all reckonings, Watt's multiple improvements in the Newcomen engine ultimately came to occupy a place at the very core of the Industrial Revolution. His first (1769) patent added a separate condenser and made related changes greatly to increase its pumping efficiency, immediately making possible the deeper and cheaper mining of coal. Later patents in the early 1780s provided vital adaptations of it for driving machinery of all kinds, stimulating more continuous operations and an expansion of scale in virtually every industry. The city of Manchester, widely regarded as the first citadel of the Industrial Revolution, powered its largest spinning and weaving factories with Boulton & Watt engines.

Fundamental as Watt's contributions proved to be, it is important to recognize that, even with Boulton's joint efforts, their commercial success was neither immediate nor assured. The economy and reliability of his steam engine as a power source took time to establish, and waterwheels continued for many years to provide a highly competitive alternative. As has often been the case, cumulative smaller improvements in older and competitive technologies substantially delayed the adoption of the new and seemingly superior one. In particular, advances in waterwheel design for which Smeaton had been responsible greatly extended the aggregate reserves of water power available to accommodate new industrial growth.

The design of the Newcomen engine, too, had not remained static. Its original "duty" (the number of pounds of water raised one foot by a bushel of coal) of about 4.5 had been nearly trebled by 1770, again with a significant contribution from Smeaton. Watt's separate condenser initially raised this only from 12.5 to about 22. It was 1842 before Watt-style engines eventually reached a duty of 100.[75]

With these gradual, rather than abrupt, processes of power succession and improvement, it should not be surprising that Newcomen engines, too,

TABLE 3-1
Numbers of New Eighteenth-Century
Steam Engines, by Decade

	Cumulative Total (Known)	Possible Total
To 1710	4	4
1711–20	30	40
1721–30	83	100
1731–40	136	150
1741–50	232	250
1751–60	319	370
1761–70	510	580
1771–80	711	800
1781–90	1,099	1,300
1791–1800	2,113	2,500

Source: Kanefsky and Robey 1980: 169, table 2.

were slow to disappear. In modern eyes, the patenting of the idea of a separate condenser had rendered them instantly obsolete. But that had little to do with the calculations of contemporaries. As Nathan Rosenberg has observed, Newcomen's

> atmospheric steam engine was not only technically workable, it was commercially feasible as well. To be sure, it experienced great heat loss in its operations and was a voracious consumer of fuel, but it nevertheless survived the market test and was widely used in the eighteenth century. . . . even in 1800, by which time Watt's patents had all expired, Newcomen engines not only continued to be used but, due apparently to their low construction and maintenance costs, still continued to be *built*.[76]

Approaching the effects of successive modification in steam engine design from another direction, the total numbers that were introduced in successive decades of the eighteenth century have been tabulated. Allowing for some residual (and probably irreducible) incompleteness, table 3-1 provides a reasonable approximation.

Plotted on a semilogarithmic scale, the straight line formed by these successive figures testifies to a steady rate of expansion in numbers of engines until almost the end of the century. There was an appreciable acceleration only in the final decade.

Even though we have largely confined ourselves to no more than a representative handful of leading figures, several complementary aspects of early technological creativity seem to emerge. Success was quickly recognized and often well rewarded in material terms. Inventors and entrepreneurs, seeking one another out and working closely together, either closely attuned their directions of effort to market demand or focused on recognizably crucial and potentially profitable areas like the generation of power. The steep rise in patenting suggests that a spirit of innovation was in the air, and that competition over ideas, resources, and market access had become fierce enough to justify increasing recourse to this modest degree of protection. Yet the means of communication, decision making, and amassing and deploying financial, human, and other resources were still so opaque and primitive that—at least by modern standards—the actual rate at which demonstrably superior inventions found their way into use rose only relatively slowly.

Was Natural Philosophy a Technological Stimulus?

Numerous accounts of individuals like the foregoing convey a common, eighteenth-century theme of mutual acquaintance and contact involving gentlemen-devotees of natural philosophy and engineers or nascent entrepreneurs of a more practical bent. Not apparent at any point, however, is the transfer of a directly enabling idea in either direction. All seem to find places within relatively small and loose, self-identified circles. No doubt their informal exchanges of views established a fairly wide and common domain of discourse. But the contribution of what was only beginning to take shape as a set of scientific disciplines to the practical guidance and inspiration of the pioneers of the Industrial Revolution remains elusive.

Is the glass best viewed as half-full or half-empty? Authorities are sharply divided on the matter. To be sure, there is not much apparent difference over the known facts between those who are more and less skeptical of a significant contribution from natural philosophy. Stronger differences emerge in appraisals of the importance of what may have been a shared world view emerging from the ongoing, informal exchanges between the three, intercommunicating groups of participants.

To begin with the skeptical position, crucial technological advances associated with stimulating the Industrial Revolution invaded power-driven textile machinery, improvements in the steam engine and in the conversion of iron to steel, the utilization of coke in smelting ores, and various advances in agriculture. On the other hand, contemporary progress of a more scientific character lay primarily in

> the progress of taxonomy in botany and zoology, the theory of combustion, the foundation of exact crystallography, the discovery of the electric current, the extension of the inverse-square relationship to magnetic and electrostatic forces,

the analytical formulation of mechanics, the resolution of the planetary inequalities. . . . and it is immediately evident that no apparent relationship existed between these two sets of achievements except the vague and uninteresting one that both occurred in a technical nexus. . . .[77]

A different but related skeptical position also has been frequently argued. In modern hindsight, scientific theories of the time may have been manifestly untenable. What credit should be given to science when new technologies supposedly based on scientific discoveries in fact reflect misinformation or false analogies? That applies, for example, to the invention of the seed drill by Jethro Tull, in spite of his belief that air was the best of all manures, and to considerable progress in dye, bleach, and soda fabrication in spite of the unsupportive weight of the whole edifice of contemporary chemical theory.

Turn now to those who are more supportive of the existence of a productive connection with activities of a more scientific character. First to be noted are reservations with which Charles Gillispie, the skeptic quoted above, accompanied his own, otherwise forceful position. "It cannot very well be supposed," he readily conceded, that men of great and deserved scientific eminence "had no more in mind in their frequent references to the utility of science, than to win for it public esteem, and that no real substance lay behind their belief in its practical contribution to the arts" (p. 400).

A more basic statement of the contrary position involves an affirmation that natural philosophy did not so much supply a body of confirmed theory and fact as exemplify a persuasive readiness to pursue new problems with an open, investigative stance. Unaffected by the adequacy of the explanations that were offered, moreover, was the usefulness of new information about the properties of materials and their alteration that had been turned up as a result of investigations of a more disinterested, scientific character.

> We might then readily admit that scientific *theories* were relatively unimportant in connection with technological innovation until well into the nineteenth century at the same time that we insist that *science* was nonetheless a major contributor to innovations that greatly increased agricultural and industrial productivity in eighteenth-century Europe.[78]

As so often occurs with grand questions of historical determination, no simple resolution of, let alone choice between, these two interpretive stances is apparent. Some sense of the power and potential of the new outlook almost certainly was a stimulus to many attempts to discover its practical utilities of which we remain largely ignorant. Even if its quantitative basis may in most cases not have been more than dimly apprehended, for example, a powerful rationale was provided for more systematically introducing simpler modes of calculating into technological and business applications.

In fact, the new philosophical outlook was being enthusiastically propagated by the early eighteenth century to growing, willing audiences that included tradesmen as well as gentlemen and aristocrats. Pulpit lectures quickly diversified into more secular subscription series held in masonic lodges and coffeehouses, emphasizing potential applications to such practical concerns as mining and smelting, water supply and transport, fen drainage, and steam power. As well as providing dramatic replications of classic experiments, these lectures typically used "mechanical devices of increasing complexity, especially air and water pumps, levers, pulleys, and pendulums, to illustrate the Newtonian laws of motion and hence simultaneously their applicability to business, trade, and industry."[79]

It seems clear, in short, that the balance between natural phenomena that could only be passively experienced and those that could be understood and manipulated had shifted. Analytically decomposed aspects of everyday experience turned out, like Newton's beams of light refracted by a prism, to reveal marvelous, wholly unsuspected regularities. Techniques, instruments, and modes of experiment and observation that had become identified with natural philosophy thus were surely seen as offering possible leverage on a widening range of industrial and commercial opportunities. So the utility of holding related, if loosely defined, outlooks must have been engendered in which practical men as well as entrepreneurs and natural philosophers began to perceive that—with technology as their meeting ground—they were advancing along different flanks of a common enterprise.

To the Threshold of a New Epoch

This chapter has sought to describe a number of streams of change that began to converge at an accelerating rate, and with mutually reinforcing effect, in the second half of the eighteenth century. Although at first getting underway only slowly, foremost among them was a significant economic quickening of pace. Its initial focus was concentrated on the continuing growth of London as a great city and port, whose circle of influence spread out over the surrounding countryside that supplied its needs. As time went on, economic interactions deepened and intensified. Far-flung conquests, maritime supremacy, and consequently burgeoning commerce led to rapidly rising, more diversified imports. Coming into use first in London as the setter of styles and fashions, these imports generated not only new tastes and wants but intensified efforts on the part of the middle and upper strata of the population throughout the realm to be able to afford and emulate them. Improvement of transport networks became a target of private investment. Agricultural activity was increasingly oriented toward producing for markets, with increasingly specialized services extending into rural regions to facilitate this end. For rural villagers as well, as enclosures reduced access to their former commons and to subsistence agriculture, there appears to have been a general move toward increasing reliance on markets even for daily consumables.

To speak of this as a "consumer revolution," however, fails to take account of the narrowness of the emergent market for "luxuries" and even "decencies." For the elite, this was indeed a time of rapidly increasing, ostentatiously displayed wealth and luxury. For perhaps one-fifth of the population (this proportion slowly growing), there was a perceptible widening in the practical horizons of choice. This coincided with the appearance of new, more creatively entrepreneurial manufacturers and arbiters of style like Wedgwood, who were ready to stimulate and orchestrate the swelling demand. But especially as the general and painful rural dislocations of the enclosure movement got underway, there is little to suggest that the arrival of the threshold of the Industrial Revolution would have been perceived by the greater part of the population as a time of plenty.

It is very difficult to identify a technological impetus that can be credited with more than a marginal influence on any of these developments. A substantially better case can be made for the reverse—for attention having turned to investing in new industrial processes and products, and to more effectively exploiting existing technologies, as new concentrations of disposable wealth appeared and the general level of economic activity increased. Awaiting a diverse range of new applications were the newly demonstrated efficiencies of concentrating production in larger facilities; the potential for profit and growth inherent in correspondingly more disciplined and repressive forms of industrial organization; and the gradually swelling mass of dislocated urban immigrants seeking some form of employment. By entrepreneurs and inventors alike, in roughly the 1750–1770 period technology began to be seen as a ready means to exploit all these opportunities.

4

England as the Workshop of the World

THE ONSET of the Industrial Revolution, while an epochal transformation in any longer view, involved no sudden or visible overturning of the established order. Many contributory streams of change had converged and unobtrusively gathered force during the first two-thirds or so of the eighteenth century. Leading elements of technological change presently began to stand out—in engines and applications of rotary power, in the production of iron and steel, and most importantly in new textile machinery. But their sources and significance lay primarily in interactions with a wider matrix of other, earlier or contemporaneous changes.

As we have seen, these other changes were largely of a non-technological character. While highly diverse, the most important can be briefly summarized. Among them were the consolidation of large landholdings and the commercialization of agriculture; the widening web of international trade; unprecedented urban growth; rising, increasingly differentiated internal demand and the development of markets to supply it; intensified applications of oversight and discipline in prototypes of a factory system; growing concentrations of wealth and the socially sanctioned readiness to invest it in manufacturing; and, not least, the discovery of common interests and meeting grounds by inventors, engineers, scientists, and entrepreneurs.

Not as anyone's conscious intention, it was out of such diverse elements—and the even broader and more diffuse shifts in cultural predispositions that underlay many of them—that an era of profound and irreversible change was fashioned. This chapter will sketch the main outlines of how it happened and the role that technology occupied along the leading edges of change. But no less important than England's taking of commanding leadership was how it slipped away in the later nineteenth century. We will discover that this was no internally determinate, quasi-biological cycle of youthful vigor, maturity, and decline. It mostly had to do instead with the diffusion and further development of core features of the Industrial Revolution itself. External competitors could quickly assimilate the English example while escaping some of its natural shortcomings as a pioneer.

Uncapitalized, the term industrial revolution denotes a significant rise in manufacturing productivity and an ultimately decisive turn in the direction of industrial growth, anywhere and at any time. So used, it applies to all of the sequential phases of modernization on a technological base that, at a still-accelerating pace, have cumulatively transformed the world since the late eighteenth century. On the other hand, when stated as the Industrial Revolution, it is widely used, and is used here, to refer specifically to the founding epoch of machine-based industrialization in England that began around 1760 and lasted there for something less than a century.

There is no dispute that this was a time of sustained, cumulatively substantial change. Within a human lifetime of seventy years or so England moved forward without serious rival into the status of "workshop of the world." As such, it came not only to dominate world trade but to transport much of that trade in its ships. Industrialization directly and profoundly altered the fabric of life for many, drawing a rural mass not only into urban settings but into factories and other new, urban-centered modes of employment. Population, having risen only slowly until late in the eighteenth century, turned sharply upward by the early nineteenth and rose by 73 percent (from 10.5 to 18.1 million) between 1801 and 1841. Markets proliferated, and popular dependence on them both deepened and widened. Travel beyond one's local community or district came within an ordinary wage earner's reach. National income rose precipitately, although the significance of this for any notion of "average" income or well-being is undermined by growing inequities of distribution.

Important as these largely material changes were in their own right, we must not overlook their intersections—but also their differences—with others in the realm of ideas. Introduced in the wake of the American and French Revolutions were new aspirations for individual rights, equality before the law, and widened political participation. Echoing great themes of popular unrest that had risen to the surface in the seventeenth century and never been completely submerged, they gave ideological form and content to the protests of working-class movements as they emerged to meet the new challenges of the factory system. Similarly energizing were the great nonconformist religious movements that from the mid-eighteenth century onward began to challenge the established Church of England.[1] And differentially directed against against each of these streams was a backlash, chronicled at length by Edward Thompson who saw it as a "political *counter*-revolution," that continued for four decades after 1792.[2]

The very currency of the example and idea of political revolutions in the late eighteenth and early nineteenth centuries tends to project a single model of what a revolution must be. Seemingly implied is an overturning—abrupt, tinctured if not characterized by violence, and only occasionally linear in its development and permanent in its effects. Whether capitalized or not, industrial revolutions are profoundly otherwise. The quickening of pace is relative to the preceding and prevailing pace of economic change, not to the much greater volatility of political processes. Perhaps the defining feature is an upward tilting of the ramp of economic, or more particularly industrial, activity. And the lesson of history seems to be that this change not only is permanent and cumulative but inexorably diffuses outward until it assumes global proportions.

New ideas have always tended to generate their own coherent testimonials, but the effects of industrialization were so many-sided and diffuse that for a long time they failed to do so. Not so clear, therefore, is when and to what extent those living through some part of the Industrial Revolution became fully cognizant of the economic transformation in which they had participated. Adam Smith's penetrating, truly encyclopedic work, appearing in 1776, embodied no apparent sense of anticipation of what was just getting underway. But far more surprising is the fact that something of the permanent significance we attribute to the Industrial Revolution was only weakly recognized in its own country until the 1880s.

Perhaps it was natural that a comprehensive overview emerged only after the return of a slower tempo of change, and also an uneasy awareness of challenges that the United States and Germany had begun to make to England's lead. But making the delay more puzzling, the characterization had meanwhile begun to circulate in francophone Europe in the 1820s and 1830s, thereafter finding its way into Friedrich Engels' *The condition of the working class in England* in 1845. For whatever reason, as an important subject of scholarly concern in England itself, the Industrial Revolution took its place only in 1884, following publication of Arnold Toynbee's *Lectures on the Industrial Revolution of the Eighteenth Century in England.*[3]

Pervasive as the changes accompanying the Industrial Revolution were, it is thus clear that we must not overstate either their novelty or their suddenness. Growth in the scale of manufacturing facilities is a case in point. Large-scale manufacturing establishments in the iron industry, in particular, had their origin much earlier. With them, as Thompson noted, "we are entering . . . , already in 1700, the familiar landscape of disciplined industrial capitalism, with the time-sheet, the time-keeper, the informers and the fines."[4] Although a full mapping of the shifting proto-industrial landscape is probably beyond reach, Maxine Berg has also called attention to an emerging, fluid mix of cottage and workshop industries in the earlier eighteenth century countryside, within which larger factories were beginning to occupy a place.[5] All that can be said with confidence is that increasing scale created a need for new, more centralized forms of control.

The factory-based shift toward tightened time discipline and an accompanying routinization and intensification of labor might precede, accompany, or follow the introduction of new production technologies.

More important is the fact that, as Rosenberg and Birdzell are at pains to point out, "it was not mechanical change alone which constituted the industrial revolution. . . . The spirit of the times was centralizing management *before* any mechanical changes of a revolutionary character had been devised."[6]

At the same time, perhaps contrary to the expectations engendered by what we perceive today as the persuasive logic of the new synergies of organizational and technological advance, this was still a society in which change was very uneven. Through mid-century, the activities of only about 30 percent of the labor force had been radically transformed by new technical innovations, and the factory and large town

> were far from being the dominant spheres of work and life. . . . Half the population still lived in the countryside in 1851, and another 15 per cent lived in towns with less than 20,000 inhabitants. Agriculture still employed over 20 per cent of the occupied population, and domestic or small workshop industries based on handicraft trades like shoemaking or tailoring ran the factory based cotton industry close in terms of numbers employed. Whilst it would be wrong to ignore the fact that agrarian or handicraft work had been substantially affected in many ways by the industrial revolution, it is nevertheless essential to grasp the small-scale, local nature of much social change between 1780 and 1860.[7]

Not entirely clear, in other words, is just how preponderantly "industrial" and disjunctively "revolutionary" the Industrial Revolution really was. In spite of the virtual universality of the use of the term, controversies continue over its distinctiveness and content. Imposing a specific, relatively narrow time frame may overemphasize the abruptness of the discontinuity. It also tends to reify and isolate the technological component of a broader transformation in socio-economic life as a whole.

To overstate matters, speaking of this epoch as the Industrial Revolution may tend to identify technology as the driver, the initiating or primary motive force. Yet in any more nuanced view, the problem is precisely to try to disentangle times, places, or respects in which technological change may indeed have had some independently initiating role from others in which it is better understood as secondary, dependent, or derivative.

The imputed causal role of the technological component may be further exaggerated if the focus is narrowed to those industrial sectors, especially iron and textile manufacture, where process improvements were indeed the most profound. Even within the industrial sector innovation frequently took the forms of improvements in hand technologies and work organization, rather than more dramatic advances in mechanization. And a preoccupation with the industrial sector can lead to the "misleading over-simplification" that significant contemporary changes in agriculture, transport, and commerce were no more than secondary reflections of the dominant, driving forces behind the growth of industrial production.[8]

These are all necessary qualifications. Adding to their weight are important quantitative revisions (although they are still far from precise, and are contingent upon numerous assumptions) in the scale and rate of change of economic activity, and of the role of industry within the larger whole. Not merely a capitalized Industrial Revolution but even an uncapitalized one is under sharp challenge in some quarters. Yet recognizing the pluralism inherent in the concept of revolution itself, I can only concur wholeheartedly with David Landes's judgment that the name Industrial Revolution is "consecrated by clarity and convenience" and "will not go away."[9]

Whether or not the cumulative effects of change were fully sensed by contemporaries, at least in retrospect they are awesome. Not surprisingly, technology-based increases in the productivity as well as scale of the cotton textile industry are the best illustration of this. However unrepresentative that industry may have been of the larger industrial aggregate, the rise in its consumption of raw cotton from 4 million pounds in the the 1760s to 300 million pounds in the 1830s, by which time it had come to dominate world markets, was absolutely without precedent. But more generally also, "in the eighty years or so after 1780 the population of Britain nearly tripled, the towns of Liverpool and Manchester became gigantic cities, the average income of the population more than doubled, the share of farming fell from just under half to just under one-fifth of the nation's output, and the making of textiles and iron moved into the steam-driven factories."[10]

No less important than the technical discoveries that will next be discussed were changes in the locus of production and the thrust of innovation. Centralized sources of power—with steam engines only gradually displacing waterwheels—were essential for machine-dependent production at increasing scale. Together with certain other economies of scale and continuous operations, this led to a concentration of the more easily routinizable aspects of production in larger and larger facilities, and to the repressive supervision of the work force in the factory system.

The factory system also demanded an intensified effort to take command of its own supply of innovations. By fairly early in the nineteenth century mechanical engineering had become recognizable as a profession directly associated with improvements in production processes. Iron and steel replaced wood, especially in the construction of textile machinery. Inventions such as Henry Maudslay's all-metal lathe (1784, 1797), James Nasmyth's steam hammer (1838), and other important advances in cutting, forging, and shaping technologies established the basis for a machine tool industry. Gradually the design and construction of production machinery became a specialization of its own. With growth in the scale of production came not only further specialization but hierarchical organization of the work force on the basis of skills. Mutually reinforcing shop floor modifications and improvements by craftsmen directly engaged in supervising aspects of production significantly supplemented and sometimes even outweighed the contributions of individual inventors working on their own.

Critical Areas of Techno-Economic Advance

Retrospective judgments about the impacts of particular inventions or innovations require caution. It is difficult to avoid a teleological element, seeing them not in the light of their initial capabilities but of what they later became. Particularly within a larger framework focused on the Industrial Revolution, there is a natural tendency to highlight the abrupt disjuncture of their introduction, and to assume that their potentials for inducing fundamental change were immediately recognized.

There is little need for that cautionary note in the case of the steam engine. It is clear that our retrospective sense of the importance of James Watt's introduction of the principle of a separate condenser, as well as his other, subsequent contributions to its improved performance, was very quickly shared by well-informed contemporaries.[11] By astutely having joined forces with a visionary entrepreneur like Matthew Boulton, he not only acquired the capital to go into production but moved at once to establish the credibility of his plans and efforts. Even at the outset, of course, the credibility of steam as a source of motive power was already well established.

But it must be recognized that the technical difficulties facing Boulton and Watt were daunting. Widespread, effective introduction of their engines into use—as distinguished from acceptance in principle—required the slow and painful surmounting of many unforeseen and challenging difficulties:

> Boulton had envisaged making engines for the whole world, probably without fully realising all the problems in design, in control over manufacturing particularly when sub-contracting was involved, in erecting complex machinery far away from the parent company, in training skilled men to both erect and then to run the machines, in the continued maintenance and in an after-sales service. The world of mechanical engineering was then in its infancy and the Boulton & Watt rotative engine demanded higher standards in its manufacture and maintenance than anything else in its scale at that time. It was, after all, the most advanced piece of engineering and technology of its day.[12]

The first and fundamental Watt patent, it will be recalled, was taken out in 1769. Yet even in the textile industry, "the *locus classicus* of the steam-powered factory system,"[13] waterwheels were not fully displaced by steam until after the end of the nineteenth century. It was only in the 1820s that steam finally became the primary power source for the industry. This is partly because its requirements were initially very modest until ballooning textile production began seriously to overreach earlier power capacity. Thus there was at first no immediate shortage of water power to hasten the transition to steam as an alternative.[14]

A shift to steam power could almost immediately bring some important advantages. Removing the constraint of a riverside location with adequate hydraulic "head," it simplified labor recruitment problems that will be discussed later. Moreover, it permitted greater flexibility in planning for future growth.

The case for breakthroughs as a result of newly invented machines in the textile industry is similarly mixed, although dramatic impacts on national economic performance unquestionably were earlier. The innovations in cotton spinning by Hargreaves, Arkwright, and Crompton that have long been identified as central to the Industrial Revolution were quickly recognized as capable of generating major improvements in quality and reductions in labor costs. Each either responded to others made earlier in other parts of an integrated series of production processes or anticipated demands that would be made shortly. The overall result, therefore, was a succession of sharp increases in productivity, leading to steeply rising output and declining costs. By 1799, as Donald McCloskey has observed, "the real price of cloth compared to its level in 1780 had halved; but it was to halve twice more by 1860."[15]

David Landes has accurately captured the striving, pragmatic spirit that was repeatedly manifested in this highly competitive process. He describes it as a "sequence of challenge and response, in which the speed-up of one stage of the manufacturing process placed a heavy strain on the factors of production of one or more other stages and called forth innovations to correct the imbalance. . . . The many small gains were just as important as the spectacular initial advances. None of the inventions came to industry in full-blown perfection."[16]

Nevertheless, there still remains a danger of overemphasizing the suddenness and single-mindedness of the transition. We highlight the cotton-spinning innovations that were ultimately successful, neglecting both the decades of experimentation pointing in different directions that led to them and the long lag before power looms successfully complemented power-driven carding and spinning to complete the mechanization of textile production. The transition from cottage and small workshop to factory was largely consummated for spinning already in the 1770s, but the full relocation of textile production was only partial until hand-weaving also had been superseded. And relocation in factories had, in any case, long been foreshadowed by the advent of water-powered silk throwing mills in the

TABLE 4-1
Total British Exports and Exports of British Cotton (£ million)

	1784–86	1794–96	1804–6	1814–16	1824–26	1834–36	1854–56
Total	12.7	21.8	37.5	44.4	35.3	46.2	102.5
Cotton	0.8	3.4	15.9	18.7	16.9	22.4	34.9
Percent	6.0	15.6	42.3	42.1	47.8	48.5	34.1

Source: Chapman 1987: 43.

1750s. Also to be kept in mind is the uniqueness of cotton and cotton textiles, a result not only of the elasticity of supply of the raw material of which North America proved capable but of the virtual insatiability of worldwide demand.

Nevertheless, as John Walton concedes while cautioning against exaggeration, "there is something cumulatively impressive to explain. Nothing like it had been seen before."[17] Textile production, to be sure, increased only slowly at first. But from the 1790s onward, the pace of growth, and the massive output levels to which it was thus carried, were remarkable by any standards. Those levels are worth noting before we turn in more detail to some of the major technical steps that made this achievement possible. As well as any single set of statistics can, table 4–1 makes tangible the full magnitude of the transformation that the Industrial Revolution represented.

Initially the focus of technical innovation was on the labor-intensive process of hand-spinning, introducing power so that "workers became machine minders rather than machine operatives." First to be introduced was the spinning jenny (1764, pat. 1770) by James Hargreaves (d. 1778). With very modest power, space, and capital requirements, it merely replaced the work of several spinners without forcing a reconfiguration of the putting-out system. Richard Arkwright (1732–1792) followed shortly with his somewhat more demanding water frame. Originally designed at a scale intended for home manufacture, his 1769 patent restricted its use to mills with a thousand-spindle capacity, and thus constituted a significant step along the way to the factory system.[18] The most significant and lasting contribution, however, was developed by Samuel Crompton (1753–1827). His "mule" (1779), especially after steam power was applied to run it, not only added substantially to throughput and further hastened the transfer from rural workshops to urban factories, but also fundamentally improved the fineness and uniformity of the yarn.

This is best measured in hanks of yarn per pound of cotton. Hand-spinning seldom could do better than sixteen to twenty hanks, as did Hargreaves's jenny since it essentially replicated hand motions. With Arkwright's water frame, the level rose into the sixties. Crompton's mule soon was producing eighty hanks per pound, and it attained the three hundreds by the end of the century.[19] This level "allowed Britons to spin the fine yarns for muslins and the like by mechanical means, whereas previously such qualities had had to be imported from the hand-spinners of India."[20]

> Success at the spinning stage of the production process immediately created the need for an increase of output at the earlier stages, and all the important inventors of spinning machines were compelled to divert their minds to preparation machinery. . . . The outcome was the perfection of a system of continuous (or flow) production in which the cotton was mechanically handled from the moment the bales were hoisted from the drays to the top floor of the mill to that at which it was dispatched, in carefully graded yarns, from the ground-floor warehouse.[21]

Concomitant steps of mechanization and productivity enhancement were underway in other phases of textile production. Bleaching, a hand process utilizing sour milk, initially required many months. Commercial production of first sulfuric acid and then bleaching powder reduced this to little more than a day by the end of the century.

Power looms were longer delayed by their technical complexity. Patents were awarded to Edmund Cartwright in 1785–1788,[22] but their first, and still modest, commercial success did not come until early in the nineteenth century. They did not begin to replace significant numbers of handloom weavers until several decades later still, for only then did it become "possible to produce an efficient machine at a cost which a manufacturer would be prepared to incur in order to enjoy the power loom's expected advantages." In fact, it was only the choking off of cotton supplies during the American Civil War that caused cotton handloom weaving to disappear altogether.[23]

There is no question that cotton handloom weavers ultimately were displaced by power looms. They had suffered from periodic crises of unemployment during the earlier course of the Industrial Revolution as well. But although hundreds of thousands of individuals were affected over the course of several decades, this should not be taken as necessarily representative of a generally negative impact of technological advance on the ranks of skilled workers. "Cotton handloom weaving, from its earliest days, was an unskilled, casual occupation which provided a domestic by-trade for thousands of women and children, whose earnings were normally quite low."[24]

Let us turn next to iron-making. Just as in textiles, there was a "see-saw of challenge and response" in which

> small anonymous gains were probably more important in the long run than major inventions that have been remembered in the history books. . . . Iron manufacture was essentially a kind of cookery—requiring a feel for ingredients, an acute sense of proportion, an "instinct" about the time the pot should be left on the stove. The ironmasters had no idea why some things worked and others did not; nor did they

care. It was not until the middle of the nineteenth century that scientists learned enough about the process of converting ore to metal to provide a guide to rational technique and measures for testing performance.[25]

The iron industry played a part in the Industrial Revolution that was no less crucial to it than the contributions of steam and textiles. Iron, not steel, was the primary ferrous material of the Industrial Revolution. Steel, while crucial for specialized uses, was too expensive to be widely employed in spite of its advantages. Cheap steel came to the fore only with the introduction of the Bessemer and Gilchrist-Thomas processes in the 1850s. Until then, there were two major categories of iron as a raw material: pig or cast iron, the immediate product of the smelter and consequently cheap, very high in carbon content and thus hard but brittle; and wrought iron, low in carbon content and consequently tough but more malleable.

As already discussed in chapter 3, Abraham Darby's coke-smelting process was the major advance in iron technology of the earlier part of the eighteenth century. At around the time that coke-fueled furnaces were beginning to be widely adopted in 1760, their annual output was still relatively small, on the order of seven hundred tons. Less than fifty years later this had more than doubled, and some individual furnaces were producing more than three thousand tons. But there had been important other advances as well. Watt-designed steam engines were introduced to supply forced air for the smelting process, freeing the furnaces from the limitations (and seasonal variability of flow and therefore power) of riverside locations. That also removed upper limits on power available at a given plant site, permitting economies of scale as additional blast furnaces were built to take advantage of the same infrastructure.[26]

For wrought or bar iron, England had continued to depend heavily on relatively costly Swedish imports. Typical forges (where cast iron was converted to wrought iron) of the early eighteenth century had also been small, producing on the order of 150 tons of bars annually. Drawing on earlier procedures but integrating them in a new way, this bottleneck was decisively overcome by the *puddling and rolling* process (patented 1783–1784), which remained the internationally recognized way of converting coke pig to wrought iron for as long as any demand for wrought iron continued.

Henry Cort (1740–1800), the originator of this process, was originally a fairly well-to-do supplier of iron products to the navy. Bequeathed a forge, his primary insight was that the carbon content of pig iron could be more economically reduced by a combination of heating and rolling. Stirred with iron rods, carbon monoxide rose from the melted iron mass and burned away. Then the spongy iron mass could be rolled between waterwheel-driven rollers, squeezing out the slag and consolidating the grain of the iron—and eventually enhancing the rate of production twenty-five times.[27]

With the advent of cheap iron, demand multiplied and new uses proliferated. Forges increased even more strikingly in size than blast furnaces, averaging five thousand tons of output annually by 1815 and extending up to thirteen thousand tons.[28] Some use of cast iron for "plateway" rails for intra-industrial transport had begun already in the 1760s. Cast iron pillars were first used in construction in the 1770s, and the first iron-framed building dates from 1796. By the early years of the nineteenth century cast iron gas pipes as well as water pipes were in use. Cast iron was increasingly pressed into durable service for purposes requiring strength and great rigidity, such as steam engine cylinders, gears, machine bases, and moving parts. Wrought iron, meanwhile, played a dominant part in the rapidly growing production of nails, locks, hardware, and agricultural implements.

By the mid-1830s, British pig iron production had passed the million-ton mark. By 1847 it had doubled again. Approaching a hundredfold increase in a century, these figures convey the same sense of prodigious growth as do the earlier figures for textile exports. Britain can appropriately be described as "ironmaster to the world" during the middle decades of the nineteenth century, its production of pig iron as late as 1873 probably being fully equivalent to that of the rest of Europe and the United States combined.[29]

Steam engines, iron, and cotton textiles lay at the core of the Industrial Revolution. They illustrate its rate of change and scope most vividly, but naturally should not be taken as representative of contemporary changes in the industrial economy as a whole. But selectively concentrated as these clusters of advancing technology were, they make it plain why the term Industrial Revolution remains so sturdily in place in spite of continuing criticism. Furthermore, the mechanization of these industries was in no sense a narrowly isolated and divergent development. Instead they became engines of expanding trade and capital growth that enormously invigorated the entire economy.

> The cotton-spinning pioneers, hoping (one assumes) to gain modest wealth by cheapening one process in the manufacture of cloth, could have had no notion that they were going to make cotton the largest single industry in Britain within a generation, that this minor textile would by 1815 account for 40 per cent by value of domestic exports, that it would demonstrate the possibility of creating a world market in the cheapest of cheap consumer goods, that it would provide a dramatic object lesson in the profitability of mechanisation, inspiring hundreds of inventors to attempt the like in other fields, and thus providing one of the main psychological stimuli of technological advance for a couple of generations.[30]

Railroads are the fourth, and in their transformative economic impact the last, of the great technological complexes that distinguish the Industrial Revolution. Their origins, however, lie much earlier. Already by the mid-sixteenth century the great study of mining technology by Georgius Agricola (Georg Bauer, 1494–1555) described and illustrated the use of hand-propelled trucks to move ore, their passage guided by an iron pin passing between wooden rails.[31] By the early seventeenth century flanged wheels had made their appearance on larger, horse-drawn mining railways in Britain, some of considerable length and used to convey coal from pitheads to waterside loading piers. Transverse wooden ties or sleepers came into use early, and cast iron cladding or straps were presently added to reduce wear on the longitudinal wooden rails or "plates." By the end of the eighteenth century all-iron rails had begun to be installed, and there were substantial clusterings of private ventures of this sort in mining areas such as the hinterlands of Newcastle.

Gradually developed and put in place over a considerable period of time, in other words, was an infrastructure of roadbed facilities and operating expertise that merely awaited the addition of steam power. That, too, did not become instantly available even with Watt-designed engines. Needed first were incremental improvements in thermal efficiency, attained in part by high-pressure engines of lighter design. Drawing (although not commercially) "10 tons of iron, 5 waggons and seventy men," Richard Trevithick (1771–1833) first brought the two technologies together in a cogged-wheel, steam-driven locomotive that he demonstrated in Wales in 1804.

Further improvements and experimental variations naturally began to follow in quick succession. Engines relying on their own weight for traction by adhesion were introduced by 1813. George Stephenson (1781–1848), renowned as an early railroad builder, opened a 60-kilometer-long public railroad (although for some years passengers were carried in horse-drawn coaches) in 1825. What can best be regarded as the opening of the railroad era is probably the inauguration of the Liverpool and Manchester Railway in 1830. A public carrier of goods and passengers in the fully modern sense, it was followed by a virtual explosion of effort that saw the completion of over two thousand kilometers of intercity railroad connections within little more than a decade.[32] A passenger trip from London to Exeter in 1841, a distance of 280 kilometers, already could be completed in six and one-half hours, at less cost than stagecoach fare and in less than one-third of the time.

From this point forward, railroads rapidly transcended the modestly supplementary role that had been originally envisioned for them. Improvements in speed and reliability, almost indefinitely expansive carrying capacity, and reductions in cost were of a different order of magnitude than had been imagined at the outset. In addition, and perhaps even more important, "they made competition more effective, squeezing out inefficient units that had hitherto been protected by isolation, raising the general technological and managerial level. They made demands for coal, iron and steel that stimulated the rapid growth of these industries. . . . The huge capital demands of railway construction raised investment rates to quite new levels."[33]

As with the introduction of new spinning technology in the early cotton mills, the railroads can be seen as a natural outgrowth, in many small steps rather than great, revolutionary insights, of converging needs and economic processes of their time. Once set in motion, however, they unleashed a cascade of new energies and economic synergies, radiating in many unforeseen directions. As Samuel Lilley rightly goes on to observe, "The feeling that the inventions were the *cause* of the Industrial Revolution, though not a historical truth, is almost justified by this outcome."

No doubt it is precisely the unforeseen yet transformative character of technological advances like these that principally accounts for the common tendency to treat technology itself as an autonomous, self-generating force.

Last, we need to return in fuller detail to an issue briefly addressed earlier—the progress made in machine-building and machine tools. After all, it was here that the crucial technical means were developed, and eventually mass-produced, without which it would not have been possible for any of the preceding, economically and socially transformative changes to have been introduced on the necessary scale. Among the many contributing innovations, at the hands of inventors and technicians in widely dispersed foundries and workshops, were lathes, drilling machines, planers, and shapers, as well as gauges and templates to standardize production. The cumulative effect was greatly to extend the variety, scale, and speed of metalworking operations.[34]

This differs in some respects from the foregoing accounts of textile machinery, steam engines, and ironworking. Those sequences of development were more clearly punctuated by a small group of major new machines or processes. In the nascent machine tool industry some new machines also played a decisive role. But the emphasis lay more heavily with a diversified, cumulative increase in metalworking capacities, in which many devices and improvements made the decisive contribution through the complementarity of their functions. Short of describing the whole of this complex process, we can perhaps visualize its importance and character by concentrating on two individuals who made central contributions to it.

Henry Maudslay (1771–1831), of a working-class background, was apprenticed as a youth to a leading lock manufacturer before striking out on his own. He is best known for the all-iron, "self-acting" lathe and slide rest in whose introduction he played a major part, and for later improving the slide rest through the addition of change gears and a power-driven lead screw. Lathes themselves had their origins centuries earlier, but his innovations are largely responsible for their applicability to the shaping of metal in an industrial context. Opening a business in the design, manufacture, and sale of machine tools, he is credited with having " . . . developed not only machine tools, but also accurate measuring machines (used as bench micrometers), standard gauges, true planes, and standard screw-threads, thereby laying the foundation for standardized production, which were improved upon both in London and in the provinces by other famous engineers, many of whom had been trained in his workshop."[35]

However, this list does not begin to exhaust his versatility. He was the first to realize the crucial importance of accurate plane surfaces for guiding tools, and the first to make them available as a standard. Beyond this, he was the builder of many stationary and marine engines and was also responsible for introducing methods for printing calico, and for desalinating water for ships' boilers. Finally, he was the builder of a series of 43 special-purpose machines at the Portsmouth naval dockyard that mechanized the production of 130,000 pulley blocks a year, at an estimated 90 percent saving of labor. In operation by 1808, decades before somewhat comparable advances were made in American arsenals for which they became justly famous, "here is the American system in full cry." But in England it remained a largely isolated instance, not an example that soom became widely followed.[36]

James Nasmyth (1808–1890) was one of those engineers, later famous, whose early training came in Maudslay's workshops. Born in Edinburgh and returning there for a time after serving with Maudslay in London, he moved on to Manchester in 1836 and chose a strategic location for his Bridgewater Foundry near the junction of the Bridgewater Canal and the newly opened Liverpool and Manchester Railway. He designed this for "something resembling modern assembly-line production" of a wide array of planing machines, lathes, drills, boring machines, gear- and screw-cutting machines, grooving, shaping, and milling machines, all on a "ready to supply" basis as advertised in printed catalogs. Central to this inventory was his widely celebrated steam hammer (patented in 1842), originally designed to solve the exceptional difficulties of forging a steamship drive shaft. The standardized, if not quite "mass produced," equipment that emerged from this "all in a line" foundry also included steam pumps, cranes, and locomotives for various railway companies.[37]

Nasmyth retired at the early age of forty-eight in order to pursue his hobby, astronomy. Still an extraordinarily versatile and creative figure, he went on to win a prize for lunar cartography in the Crystal Palace Exhibition in 1851. A contributing factor in his abandonment of manufacturing was said to have been frustration over the inhibiting effects of trade-union restrictions and craft conservatism. Real and perceived constraints on entrepreneurship are questions to which we will return later in this chapter.

It should again be acknowledged that there tends to be a persistent bias in such accounts toward the accomplishments of leading technological protagonists. Limitations of space, and of the existing records, make this virtually unavoidable. But the progress of the Industrial Revolution was at least as dependent on the existence of a skilled and creative human infrastructure of whom we generally know little:

> John Wilkinson, it is often remarked, was indispensable for the success of James Watt, because his Bradley works had the skilled workers and equipment to bore the cylinders exactly according to specification. Mechanics and instrument makers such as Jesse Ramsden, Edward Nairn, Joseph Bramah, and Henry Maudslay; clock makers such as Henry Hindley, Benjamin Huntsman, and John Kay of Warrington . . . engineers . . . ironmasters . . . chemists. . . . were as much a part of the story as "superstars" Arkwright, Cort, Crompton, Hargreaves, Cartwright, Trevithick and Watt.[38]

Sectoral and Regional Disparities in Growth

The definition given to the Industrial Revolution at the beginning of this chapter was concerned with manufacturing productivity and industrial growth. This focus on the impacts of mechanization may seem narrow. Concerning itself only secondarily with changes in the texture of daily life experienced by the population at large, it concentrates on only one sector of economic activity. From another perspective, the emphasis now shifts to the long-range transformative potential of a cluster of specific advances for which this epoch was the seedbed.

This is not to imply that the Industrial Revolution was not accompanied by massive changes in ordinary British lives, for indeed there were many. By the middle of the nineteenth century, Britain was well advanced along a path of rapid urbanization and an accompanying decline in agricultural employment. A grid of railroads was rapidly unifying the country. Factories had become commonplace, and were also making voracious demands on increasingly efficient capital markets. For commodities like textiles, they were successfully producing at an enormous scale for world markets.

Differences of interpretation remain over the extent of the increase in national and per capita income. They include disputes over the residual uncertainties in existing statistical aggregates, as well as "uncomfortably wide margins of error" even for the few industries (cotton, woolens, worsted, and iron) for which productivity growth can be estimated.[39] Within the bounds of what aggregated statistics purport to show, the newer, "revisionist" school can claim to have demonstrated that change was considerably more gradual than previously believed.[40]

It should not be a surprise that the national economy grew more slowly than its industrial sector, and more especially so than the production series of its vanguard industries might suggest. The production of pig iron, for example, doubled (from 90,000 to 180,000 tons) in the single decade after 1790, while coal output rose by 47 percent (from 7.5 million to 11 million tons) in the same period. Such extraordinary increases could in no way be representative of the performance of any economy as a whole. But even on the revised, aggregate basis, gross domestic investment more than doubled (from 5.7 to 11.7 percent) in the seven decades after 1760, while the rate of growth of national income for 1780–1801 was almost twice what it had been earlier in the century. The latter then continued to gather speed, so that rates of industrial advance and productivity growth for 1801 to 1861 were, at the revised rates, three to four times what they had been a century earlier.[41] "There was a break, then, and this kink in the curve, however modest to start with, was indeed revolutionary in its import."[42]

But the dispute remains intense over whether aggregate growth accounting can provide an adequate measure of the rate of economic progress associated with the Industrial Revolution. Insofar as it does, the downward revision it is said to document in this rate may almost seem to substitute a mere process of accelerated change of considerable duration for one of truly revolutionary dimensions. But what do the estimates leave out, critics respond, such as the role of small start-up firms that have left few records? What is the effect of changes over time in the proportions and categories of industrial activity that were recorded? Does the revisionists' interpretation of progress having been narrowly confined to a few industries do justice to the multiple ways in which "traditional" and "modern" firms and techniques actually defied these categories and intersected?[43]

Full reconciliation of these differences seems unlikely, for they rest on different encompassing visions of "reality." On the one hand, an aggregative view purports to establish average rates of change in manufacturing scale and productivity with increasing confidence and accuracy. These indices, moving with a significantly less disjunctive change in slope than had been accepted earlier, highlight changes that occurred within the whole of the national economy as the overridingly most significant historical variable. The opposing view focuses in the first instance on the vividness and particularities of testimony that forms the substance of traditional historical scholarship. But ultimately its subject is change itself, demanding that attention be directed primarily toward identifying its most dynamic developments and sources. As Schumpeter once put this approach to the analysis of technological change, "it is disharmonious or one-sided increases and shifts *within* the aggregate which matter. Aggregative analysis . . . not only does not tell the whole tale but necessarily obliterates the main (and only interesting) point of the tale."[44]

But at the same time, and probably more important, it does not appear that any real disagreement exists about either the "kink in the curve" to which Landes referred or that the resultant, cumulative increases represented a significant historical discontinuity. While ostensibly antagonistic, the two approaches are not so much irreconcilable as concerned with two different kinds of illumination—one derived from macro-economic analysis, and the other from micro- and local case studies. Discrepancies between the two testify to the diversity of the British economy and to the apparent lack of anything forcing a closer articulation of developmental trends within it.

The selectively focused effects of external trade are a case in point. Varying in the range of 10–18 percent of GNP between the mid-eighteenth and mid-nineteenth centuries, it would seem to have made only a modest contribution to the growth of the national economy. But the stimulative effect on the most dynamic sectors of industrial growth was in fact very large since "exports consisted overwhelmingly of manufactures, many of them with the potential for being mass-produced. In no other European economy were manufactures such an important proportion of exports so early and at such a low export level. At times during the late eighteenth and early nineteenth centuries, as much as 35% of industrial output was exported."[45]

Any appreciation of the relative status of leading and laggard sectors, let alone of their respective importance within the larger, national economy, is perforce somewhat impressionistic. But table 4–2 provides an appraisal of how profoundly their approximate standings changed over time, relative to one another. It appropriately emphasizes the leading role of cotton and metals.

Just as there were substantial differences between sectors of Britain's economy in their rates of industrialization, there were comparably wide regional disparities. The Industrial Revolution, like all economic growth, was an inherently uneven process. Resource constraints—coalfields, seaports, access to waterpower—played an obvious part in this. But societal considerations like access to markets were no less important. Urbanization, in particular, was a primary factor, both attracting and reflecting the ongoing concentration of industrial production in new factories.

London was already the first city of Europe in 1750, a ranking it retained throughout the Industrial Revolution and still retains today. But of the forty largest European cities in 1750, only Edinburgh, far down in thirty-fifth place, joined London in that listing. With fifty-five thousand inhabitants at the time, Edinburgh had less than 9 percent of London's population. In 1850, by contrast, Liverpool, Manchester, Glasgow, Birmingham, Edinburgh, Leeds, Sheffield, Wolverhampton, and Newcastle all had joined London (in that order) among the leading forty, with the first three also among the largest ten. Manchester and Liverpool had quadrupled in size in a half-century, and none of the ten largest cities in England had less than

TABLE 4-2
Indices of Output, Various Industries (1841 = 100)

Industry	1770	1815	Industry	1770	1815
Cotton	0.8	19	Metal	7	29
Wool	46	65	Food and Drink	47	69
Linen	47	75	Paper/Printing	17	47
Silk	28	40	Mining	15	46
Clothing	20	43	Building	26	50
Leather	41	61	Other	15–50	40–60

Source: Harley 1993: 181, table 3.1.

doubled its population during that interval. "Nowhere was this new order of society more clearly visible than in the rapidly growing industrial towns of northern and midland England."[46]

Many examples of the disproportionately heavy, regional impact of industrial growth could be given, but three illustrations may suffice. One, already underway before the Industrial Revolution, involved the production of wool textiles. For the country at large, this grew by a fairly modest 150 percent during the entire course of the eighteenth century. Were that growth uniform in its incidence, Pat Hudson has pointed out that it probably "could have been achieved simply by gradual extension of traditional commercial methods and production functions." But undermining that smoothly evolutionary outcome was the fact that Yorkshire's share in the national aggregate in the meantime rose from 20 to 60 percent. This "necessarily embodies a veritable revolution in organisational patterns, commercial links, credit relationships, the sorts of cloths produced, and (selectively) in production techniques."[47]

Similarly transformative were the impacts of the introduction of iron production into the West Midlands after 1766:

> Within a single generation, the coal-fired blast furnaces with their accompanying puddling furnaces and rolling mills revolutionised not only the south Staffordshire economy but also its settlement pattern and its landscape. The output of pig iron in Staffordshire rose from 6,900 tons in 1788 to 125,000 tons in 1815 and the Black Country's share of national output rose from 9.8 per cent to 31.6 per cent. The population of the south Staffordshire parishes increased three and fourfold, and new settlements sprang up near the furnaces. Agriculture became progressively more difficult, the night sky was illumined with flames and the day darkened with smoke, and the district began to be called the Black Country.[48]

The third example, industrial Manchester, is of a different kind. It was famous as the center of the cotton textile industry, and this must be largely credited with its extraordinarily rapid growth. From a population of only around fifteen thousand in 1760, a commercial center of merely local importance, it rose to a quarter of a million seventy years later, growing at more than twice the rate of London itself. In the mere thirty-year span between 1790 and 1820 the number of its textile mills had grown from just two, employing 1,240 workers, to sixty-six with 51,800 workers.

A further recitation of statistics on its growth would merely recapitulate this picture in more detail. Of greater interest is the composition and orientation of Manchester's elite. Large, independently minded, and newly prosperous, it tended to be "in conflict with existing social structures. . . . Remnants of eighteenth-century elitism existed alongside a more thrustful provincialism."[49] In a penetrating study, Arnold Thackray has provided details of how science somewhat unexpectedly became the beneficiary of their constraining circumstances: "Given their social isolation, political emasculation, and tumultuous surroundings, Manchester's new and increasingly wealthy elite understandably sought cultural means through which to define and express themselves."[50]

The Manchester Literary and Philosophical Society became their organizational as well as cultural focus. It was representative of a wide array of provincial and metropolitan disciplinary bodies that had sprung up to supplement the position that "until 1781 the Royal Society of London had enjoyed [in] lonely splendor as the sole institutionalized, enduring English organization devoted to the pursuit of natural knowledge." As to why science became the focus, possible explanations include "its possibilities as polite knowledge, as rational entertainment, as theological instruction, as value-transcendent pursuit, and as intellectual ratifier of a new world order."[51]

More positively, Thackray recognizes that new organizations like the Manchester "Lit. and Phil." articulated "fundamental qualitative shifts in the meaning of science as a cultural activity," as well as the rise of the associated neologism "scientist." It is surely not coincidental that this occurred in a rising provincial city that was a preeminent center of industrialization. But specifically with regard to the role of science in the process of growth, his conclusion is more cautionary:

> [T]he legitimation, the institutionalization, and the growth of science itself was more nearly a by-product of the society rather than the reason for it. . . . it becomes evident that the interaction between science and technology within the society's walls has assumed for historian commentators a degree and kind of importance it never possessed for contemporaries, whether manufacturers or men of science.[52]

At least insofar as Manchester can be taken as representative, there is little to add to the account given in chapter 3 of the contributions of science and scientists to the onset of the Industrial Revolution. Recognition of the potential contributions of science, and more especially of the scientific attitude, played an essential part in an emergent entrepreneurial world view. Few recognizable and specific contributions found their way into new industrial products or processes out of an elite circle of friends and acquaintances that included scientists. But more was expected, and perhaps eagerly awaited. And the warm acceptance of what was perceived as the scientific attitude was of no small importance in itself.

Industrialization and the Working Class

Dominating popular understanding of the Industrial Revolution is an impression of sustained, transformative growth. In the preceding account also, attention has been focused on technological and economic advances. But what has been said may leave the sense of a rising tide that, if at somewhat different rates, surely was managing to raise all boats. Is this impression accurate?

The gross disproportion in the levying of the tax burden might suggest otherwise. It had almost doubled by 1815 (to more than 18 percent) as a share of national income, as a result of war expenditures that had spiraled upward from the time of the American Revolution. Excise duties on domestically produced goods and services were principally called upon to sustain this increase. Although they also impacted on the cost of decencies and luxuries for those with discretionary incomes, these duties fell more heavily on price-inelastic necessities for factory workers. By contrast, with relatively minor and temporary exceptions, "There was no effective tax on wealth holders throughout the period."[53]

How disproportionate, in fact, were the respective allocations of the benefits of growth to different groups within British society? What are we to make of the great waves of popular protest, including especially the Luddite movement that aimed at the destruction of textile machinery in the second decade of the nineteenth century, and the working-class movement for parliamentary reform in the late 1830s and 1840s that was known as chartism? Were they only transitory episodes of conspiratorially incited unrest, as was repeatedly proclaimed by the courts and aristocracy in suppressing them with considerable severity? Or were they instead manifestations of deep-seated, broadly felt grievances over declining living standards and a loss of security that was in increasingly sharp contrast with growing national wealth?

In impressive detail, Edward P. Thompson's massive work on *The Making of the English Working Class* (1963) argues the "pessimistic" case for what he regards as the preponderantly negative impact of the Industrial Revolution on England's working people. Beyond this, Thompson also records the repressive legal climate with which virtually any form of individual or organized speech or action to obtain redress of political or economic grievances was received.[54] Drawing upon "political and cultural, as much as economic, history" across the half-century or so after 1790, he argues that this forged a working class able self-consciously to identify and struggle for its own strategies and interests. That outgrowth of resistance and self-discovery was, for him, "the outstanding fact of the period."[55] But Thompson's emphasis on the centrality of class formation is sharply disputed by some other authorities and is, in any case, not one on which it can be said that any consensus has emerged.[56]

More immediately relevant to our concerns is a further judgment reached by Thompson. The working class did not come into existence, he insists, as a spontaneously generated response to an external force—the factory system, or the technological advances of which factories were an outcome. That would imply that the class itself was composed of "some nondescript undifferentiated raw material of humanity," lacking common, historically derived aspirations and identifying symbols of its own solidarity. "The working class made itself as much as it was made."[57]

Legitimate differences on the cumulative impacts of the Industrial Revolution on working-class well-being remain, and they are wider than Thompson's righteous anger allowed him to concede. To some degree, they hinge on unresolved uncertainties over issues like the inroads of unemployment on the average income of the industrial work force. But they also involve the credence placed in indirect indicators of distress and in eye-witness accounts of idleness, as distinguished from more systematic efforts to compute average pay rates and periods of employment. Reinforcing Thompson's "pessimistic" appraisal, to begin with, is the account of Sidney Pollard:

> It is generally agreed that total national income per head increased substantially in the period 1750–1850; it is also beyond dispute that the relative share of labour fell. . . . Few indices are quite as striking in this period as the stagnation in *per capita* food consumption and the increase in the numbers of domestic servants. . . . Throughout a period of nearly a century, wages remained somewhere near a level which had come to be accepted as subsistence. This betokens an economy operating essentially in conditions of abundant labour, and it is clear that an elastic labour supply at low cost and a transfer of income from labour to capital were two basic features of the British industrial revolution.[58]

It is important to note Pollard's reservations about the limitations of the available statistics. Secular trends in working-class earnings, he maintains, in aggregate may have been either positive or negative in sign. All he can conclude is that apparently the differences were comparatively slight.

There is also a considerable variety of eye-witness testimony as to trends in working-class well-being. Thompson highlights that of Henry Mayhew, "incomparably the greatest social investigator in the mid-century." Mayhew's vivid attestation that in the 1840s and 1850s the proportions of the working population that were un- and underemployed each were comparable to the fully employed proportion does indeed lend support to the idea that the first two categories must have had a large depressive effect on the last. Pollard similarly estimates that "in the trough of the depression, employment in the industrial centres ran at about one-half of labour capacity only, and in the worst cycles at one-third." "An estimate of a loss of employment of 15–20 per cent of capacity averaged over good and bad years together," he concludes, "does not seem too pessimistic."[59]

Analyses like these emphasizing the severe depressive effects of un- and underemployment as late as the 1840s contradict earlier views like those of T. S. Ashton, who, however, is less selective and more balanced in the choice of eye witnesses he cites. While having less to say about proportions of un- and underemployment, he recognized that the real income of most factory workers had declined during the Napoleonic Wars. By that criterion, however, he maintained that conditions for the majority began to improve steadily after 1820 or so, excepting only a narrowing circle of those whose skills became obsolescent with factory modernization.[60]

Some detailed recent studies broadly support the Ashton position. Peter Lindert and Jeffrey Williamson, in particular, have attempted comprehensively to review pay rates, days per week worked, annual earnings, and costs of living for all wage-earning sectors of society for the century beginning in 1750. Some of their findings are not inconsistent with those of Pollard, such as that "from 1815 to the middle of the nineteenth century . . . the gap between higher- and lower-paid workers widened dramatically." But their broader finding, supported by a number of remarkably parallel data series, is not: "There was general real wage improvement between 1810 and 1815, and a decline between 1815 and 1819, after which there was continuous growth. After prolonged wage stagnation, real wages . . . nearly doubled between 1820 and 1850. This is a far larger increase than even past "optimists" had announced." Conceding that they "cannot answer questions about living standards, but rather only questions about real earnings," their "best-guess" estimates call for "impressive net gains in the standard of life: over 60 per cent for farm labourers, over 86 per cent for blue-collar workers, and over 140 per cent for all workers. The hardships faced by workers at the end of the Industrial Revolution cannot have been nearly as great as those of their grandparents."[61] "The great majority of these human gains," they are careful to add, "came after 1820." The earlier picture is one of "stagnation," leaving room for much human suffering across many decades.

For all periods, the picture certainly becomes more mixed if we consider individual fields of employment. Loss of employment security and downward mobility or "de-skilling" may have been the dominant trend. On the other hand, some groups like engineers were able to introduce high entrance standards based on skill and training, limiting their numbers, and raising their professional level of acceptance—and presumably incomes. Mechanics, too, were noteworthy for seeking to elevate their position through advanced training. The notion of an aristocracy of labor, widely referred to at the time, perhaps had a special association with artisans engaged in hand-producing costly luxuries for the elite, and in this niche, too, they held their own without difficulty. But in general, "it is not even possible to say with any certainty whether the proportion of skilled and semiskilled workers as a whole rose or fell in this period; all one can say is that the nature of skill and the sources of privelege were different in the new conditions, both within the factory and without."[62]

Even Thompson does not dismiss the possibility that there may have been a "slight improvement in average material standards" over the 1790– 1840 period. But more importantly, over the same period, he believes there was "intensified exploitation, greater insecurity, and increasing human misery. By 1840 most people were "better off" than their forerunners had been fifty years before, but they had suffered and continued to suffer this slight improvement as a catastrophic experience."[63]

It appears that, with the growing array of statistical findings, a consensus is beginning to take shape at least with regard to the later, markedly positive course followed by factory incomes. However, there are continuing differences over whether the evidentiary base genuinely supports the weight of interpretation that the optimists have placed on it:

> On closer inspection, the real wage data is found to suffer from a number of rather serious defects. One is that they cover only limited data points and that the choice of the end year (1851) by Lindert and Williamson is unfortunate, because that happened to be a year of unusually low prices. The nominal wages changed very little in this period, so that the rise in real wages came almost exclusively from falling prices. Hence, the optimist conclusion is highly sensitive to the correct specification of the price deflator, and its deficiencies weaken the optimist findings even further.[64]

Other differences between the optimists and pessimists also persist. The latter, for example, place greater stress on the markedly increasing intensity of labor over the course of the period, and on the social costs of the shift of womens' and childrens' labor from domestic or rural settings into factories.[65] And problems in assessing the significance of cyclical unemployment remain substantially unresolved. Womens' incomes, to begin with, are not taken into account in most compilations of data like that by Lindert and Williamson. If there were increasing interruptions in their continuity of employment, on a family income basis this could offset many male gains. Also left open is the likelihood that the incomes of domestic workers and independent artisans continued to fall. Agricultural income estimates, too, are felt to be particularly fragile because of seasonality factors, growing redundancies, and the erosion of noncash subsidies such as access to commons.

It is strongly implied in most contemporary accounts that there was widespread aversion to factory employment. Those whose skills made them members of the "aristocracy of labor" no doubt constituted an at least partial exception. For the mass of less skilled and unskilled workers, however, factory jobs may have been seen as the only alternative following loss of rural livelihoods as a result of enclosures and the consolidation of smaller workshops.

No doubt people often entered the mills in the (usually illusionary) conviction that such work was only a temporary expedient and hence could be briefly tolerated. Symptomatic of this is a textile factory owner's complaint in the 1830s about the "restless and migratory spirit" of his mill workers. Long, closely supervised shifts were, after all, a starkly unpleasant departure from periodically demanding but on the whole far more intermittent agricultural labor schedules. Moreover, the factories themselves often bore a disturbing resemblance to parish workhouses for pauper women and children—from which, in fact, not a little of the early factory labor force had been involuntarily recruited.

The pauper apprenticeship system, barbarous as it was in terms of "children working long hours for abysmal wages," was in any case short-lived. The need for it was greatest prior to the general shift of spinning factories to the more advanced Crompton's mules driven by steam power. For as long as the availability of waterpower was a requirement, the necessary location of some mills in remote settings had isolated them from the rapidly growing potential work force that was congregating in the new industrial cities.[66]

Some, but not all, of the deterioration that had occurred in the material conditions of life can reasonably be laid at the door of the conscious discretion of industrial owners and managers. That applies, for example, to the sometimes almost unbelievably harsh conditions of exploitation of child labor that Thompson cites, and perhaps to some deliberate manipulation of skilled and unskilled, male and female groups of factory operatives in order to depress wages by maintaining an unemployed but dependent reserve.[67]

Almost entirely outside the control of employers, on the other hand, were boom-and-bust aspects of the business cycle that had been greatly intensified by an era of market-oriented expansion of production to meet a breathtakingly growing, world demand:

> There had been years of unemployment and distress before, resulting from wars, bad harvests, interruptions of overseas trade, or the secular decline of individual industries. But it may be doubted if these were as regular and persistent as the cyclical unemployment now superimposed on the evils of casual work and structural unemployment; and above all it is most unlikely that in the past there had ever been such a large proportion of the population exclusively dependent on income from market-oriented industry.[68]

Still other aspects of general deterioration in living conditions also must have been essentialy outside anyone's control—or consciousness, for that matter. Surely one was the rapid growth of population. Peaking in the 1811–1821 decade, we have already seen that the cumulative increase between 1801 and 1841 for Britain as a whole reached 73 percent. For the great industrial cities growing up in the Midlands, I have noted earlier that their rate of growth in the decade of the 1820s alone exceeded of 40 percent. Clearly, as Thompson acknowledges, this was an enormous influence on economic conditions, forcefully if unevenly at work throughout the Industrial Revolution.[69]

We must not forget that the science of public health was in its infancy. The effects of urban congestion on endemic and epidemic diseases were only beginning to be understood. No body of experience made it possible to gauge the combined effect of malnutrition, pollution, and extended hours of daily employment on expectant mothers and especially on the health and growth of children. It has remained for late twentieth-century analysts to discern such indirect consequences as declining human stature, and persistent high rates of infant mortality, rising illiteracy, precisely in the latter part of the Industrial Revolution when the incomes of male factory operatives are supposed to have risen. The general impression thus is left that there was little or no improvement in living standards before mid-century, "despite the optimists' evidence of rising real wages."[70] To whatever degree a product of innocent or willful ignorance, the cumulative effect was that "wherever comparison can be made . . . staggering differences in life expectancy appear, amounting in the worst decades to an average of twenty years of life expectancy lost by the average male urban wage-earner; and whatever horrors the English statistics showed, the Scottish were invariably even worse."[71]

More needs to be said about the employment of women and children, especially but not exclusively in the textile mills. Such employment was not an unprecedented development, to be sure; most members of farm families had typically participated in various ways in the seasonal cycle of agricultural work. But mill employment was unremitting and of much longer daily duration. Hence womens' participation in the factory labor force constituted a new and major disruption of family life. And the fact that women were regarded as especially suitable for some factory tasks also had the effect of almost doubling the size of the population seeking work.

First to be affected in this way was spinning. But after 1820, as power looms began to be introduced in significant numbers, skilled men were replaced in weaving also by less skilled women and children. Within two decades, some 46 percent of the work force of 420,000 in cotton factories were less than eighteen years of age, and only 23 percent were male adults. The proportion of women and children was even higher in some other branches of the textile industry.[72] The number of skilled handloom weavers, having peaked at 240,000 in the decade of the 1820s, had declined to 110,000 by 1841 and to 40,000 a decade later. Three times as productive as handlooms in a quantitative sense, power looms had also been improved to match them in the quality of the textiles produced.[73]

There were, however, important restraints on the sweeping replacement of more skilled workers. The prevailing ideology was that factory employment for women was no more than a temporary expedient. For young, single women, marriage was expected to follow in a few years at most, while also for married women domestic responsibilities were regarded as permanent and primary. Elaborately rationalized by beliefs that women were not fit for many, more skilled types of labor, this became the justification for strict, gender-based job classifications as well as differences in remuneration. "As a result of the gradual acceptance of these interlocking values and beliefs, many historians now argue that the changing nature of the work . . . served only to redefine and revitalize *patriarchy*."[74]

The depressive effects on living standards of the enormous net growth of the urban population seeking employment, for all the reasons given, surely were supplemented and intensified to some extent by the technological displacement of semi-skilled and unskilled labor from the factories. But how much did each factor contribute to the deterioration of working-class income and well-being? Only in an exceptional case like the explosively growing cotton textile industry does it seem reasonably certain that technological advance was the dominant factor: "within two or three decades, the formerly respected and privileged occupations of weaving, framework knitting, or calico-printing had been reduced to virtual unskilled status, to be entered by any untrained outsider."[75]

More generally, however, I do not believe a rigorous or even plausible way has yet been found to assign accurate proportions of overall responsibility for labor redundancy as a whole. In other words, there is no compelling evidence that would gainsay Thompson's impression that the formation of the British working class, while proceeding hand in hand with a host of new innovations in leading industrial sectors, was a process with immensely deeper and wider roots than technological progress alone.

In just the same way, labor came to terms with the results of the Industrial Revolution—once it had begun to stabilize—by a process proceeding on many levels:

> It took a generation—which was a lifetime in industrial Britain—to learn how to deal with industrialism, but in due course it was done. Workers learnt by bitter experience, and after experimenting with all kinds of organisation they ultimately evolved the most viable types of trade union. New forms of mass agitation achieved some protective legislation. Hours began to be reduced, by Act of Parliament and by union power, so that the twelve-hour day became common in the 1820s, the eleven-hour day in the 1840s, and the ten-hour day thereafter. Men came to accept the factory discipline; children were taught new skills; housewives learnt to make the best of urban shopping and cooking facilities. Education, introduced in order to improve obedience, also promoted independent thought. Labor not only learnt the "rules of the game" of capitalist society: it also helped to make its own rules.[76]

Ownership and Management

Leading sectors and regions of techno-economic advance, as well as the concurrent development of the working class, have now been identified. But little reference has been made heretofore to the enterprises in which both took shape. Some account is needed of the financial and managerial arrangements sustaining the new factories in which the Industrial Revolution was principally embodied.

Little can yet be said about the origins of entrepreneurs as a group, save that they came from very diverse occupational and status backgrounds and tended to be of Quaker and Nonconformist religious persuasion.[77] Perhaps this diversity reflects what apparently were prevailingly low threshhold requirements for initial entry. This was especially so in the textile industry, which may well have offered the most attractive opportunities for rapid capital accumulation. Fixed capital requirements there are described as having been comparatively modest, and were still further eased by the availability of credits for raw materials and machinery, and of rental space in existing plants. The real challenges lay instead in the later financing of expansion, in surviving financial crises or sudden contractions of the market, and in avoiding a widespread tendency toward "feverish overproduction."[78]

As the development of the factory system went forward, the enlarged industrial settings opened the way to centralized supervision and new standards of labor discipline—more exacting in the eyes of supervisors but merely harsher in the eyes of employees new to any form of sustained oversight. No less unprecedented than the capability to monitor individual workers' performance was the new capability to safeguard and schedule stocks of material inputs and to increase the rate of production itself.[79]

There were complex problems to be solved if factory management was to profit fully from these new capabilities. More effective tools of planning and accounting were obviously necessary, as were more effective means of delegating responsibility for ongoing improvements in manufacturing methods. The putting-out system of cottage and workshop production had been able to function with little more than periodic visitations of management and infusions of working capital. With the factory system came a greater requirement for management-in-place, as sharply distinguished from formerly absentee ownership.

> In a dynamic sense, changes in control helped new technologies to achieve higher output speeds, and in turn higher speeds, new materials and quality shifts eventually demanded changes in control. Tensions between technological change and organization, between organizational change and finance, and so on, had arisen on many occasions before the Industrial Revolution; but now all were changing in response to the 'exogenous' changes in one another. . . .[80]

Few if any of the salient technologies of the Industrial Revolution can be thought of as science based in any direct sense. They can better be described as craft based in important ways. Reference has already been made to the quasi-managerial role of highly specialized craftsmen, "the aristocracy of labor," and this group was indeed an essential contributor to the success of the factory system. In the latter part of the nineteenth century, the need for many master craftsmen to occupy these wider roles receded with further advances in mechanization. But during the years of the Industrial Revolution they frequently made an indispensable contribution:

> Masters of their techniques, able to maintain their tools as well as use them, they looked upon their equipment as their own even when it belonged to the firm. On the job they were effectively autonomous. Most of them paid their own assistants, and many played the role of subcontractors within the plant, negotiating the price of each job with management, engaging the men required, and organizing their work to their own taste and convenience. The best of them "made" the firms they worked for.[81]

Thus a characteristic of the first great wave of British factories was, as William Lazonick points out, a reliance on the externality of "an abundant supply of skilled operatives who performed many of the day-to-day organizational functions that in later times and other places were taken to be the prerogatives of management." As Peter Payne, too, has observed, it was possible for owners virtually to ignore many of the problems of labor management "as long as managerial responsibility as well as risks of managerial failure fell to the lot of subcontractors."[82]

Coupled with other external economies—not least the market penetration and stabilizing influence of the emergent British Empire itself—this "made it unnecessary to develop managerial structures that could generate economies internal to the enterprise." Only under another set of conditions would this organizational underdevelopment, as well as the closely related horizontal and vertical fragmentation of ownership, cease to be a source of adaptability and strength. In the latter part of the nineteenth century, as "technological requirements made for an increase in size beyond that manageable by the partners, and capital requirements went beyond the resources of small, often related, groups of men," it became a critical weakness calling for "a new structure for the firm."[83]

The secular trend that might have been expected to lead away from personal proprietorships and partnerships and toward managerially more dense forms of capitalism was the advance of the Industrial Revolution itself. Capital requirements for plant expansion and the introduction of new technologies became progressively greater, straining the capacity of fragmented, familistic forms of ownership to provide the necessary resources. This effect was exacerbated by legal constraints on the prevailing forms of proprietorship, including an absence of limitations on personal liability and difficulties in obtaining parliamentary approval in order to form joint-stock companies not subject to this risk. Yet the basic pattern of fragmentation remained essentially unaltered until long after these legal logjams had apparently been broken. As late as 1885, limited companies still accounted for at most between 5 and 10 percent of the total number of important business organizations.[84] By that time fragmentation of ownership and thinness of managerial resources were becoming serious impediments in Britain's efforts to sustain its industrial leadership against new and larger competitors in Germany and the United States.

Taking and Losing the Lead

This book is primarily concerned with an accelerating succession of waves of innovation. Of less relevance, therefore, is the slowing pace of technological advance in England from the later nineteenth century onward, as a maturing industrial economy became less path-breaking and competitive. Still, the processes of decline are of great topical concern in their own right, as the United States struggles to maintain its own position of leadership a century later. Is the organismic parallel—a natural, internally dictated cycle of birth, growth, maturity, and decline—a valid and inescapable one?

Or alternatively, as I would suggest, is intensifying competition the decisive force, with later centers of creativity and growth presently devising improved paths of their own for superseding the former leader?

I cannot adequately deal with these questions in the present framework, but also cannot entirely avoid them. At the very least, a further, derivative question arises. To what extent must subsequent American success be attributed to some internal dynamic in the new setting, as opposed to an exploitation of the special opportunities that Britain's own, faltering leadership opened to younger competitors? Herein lies the issue provocatively introduced long ago by Thorstein Veblen under the heading of the "penalty of taking the lead."

Inveighing against pervasive features of "conspicuous waste" associated with British capitalism, he condemned with a broad brush investors' "inhibitory surveillance over technological efficiency," unemployment-inducing business cycles, provisions for future revenue in the form of deferred spending and insurance, the "nearly net waste" of expenditures on marketing, and "wasteful consumption" that included not only every accretion of style and custom but virtually all sports. Entirely absent from his understanding is Schumpeter's later demonstration that capitalism generates new wants, new technologies, and new markets as well as new business opportunities even as it ruthlessly sweeps away old ones. Yet a powerful kernel somehow remains as to the "efficacy of borrowing." Technologies that are quickly taken over ready made in a new national setting need be accompanied by few or none of the obsolescent features that attended their original growth and have become lodged in law or tradition. This kernel, at least, remains viable eighty years after Veblen first propounded it.[85]

In its acknowledged position of leadership among industrializing nations, England had no standards of measurement by which to evaluate her performance along the path she had taken. Nor was there an energizing vision of a yet more perfect model of progress to strive for. It is understandable that few specific technological innovations are inherently predictable in detail. But had England's relationship to the Industrial Revolution as a whole been "emulative" rather than "originative," it is likely that supportive accommodations in the institutional setting would have occurred more quickly and comprehensively. And not a few of those accommodations— in corporate structure and management, in capital investment, in the stimulation of research, and in marketing—might well have left British industrial leadership unchallenged for a much longer period.

As it was, having to pioneer in the devising of new institutional contexts as well as in how to implement novel technologies, the pace of England's effort was modest. Assuredly it was rapid in comparison with other countries that had not yet begun industrial revolutions of their own. But by hindsight, according to the standards of later industrializers, Britain's rate of growth instead appears to have been relatively restrained, if hardly as slow as the "snail's pace" on which one commentator insists.[86]

The initial phase of the Industrial Revolution was an age of textiles, preeminently cotton. The mechanization of weaving continued longer, but the direction was well established by the early 1840s. Already getting underway by that time was a subsequent phase, deserving to be identified as an age of railroads and steel. England still led the way at first in both of these respects, the high-water mark of its supremacy appropriately celebrated in the great Crystal Palace exhibition of 1851.

As we shall see, however, there were already hints of what was to come that ruffled the triumphant complacency of that occasion. American success in beginning to master interchangeable-part manufacturing technology, bursting into public consciousness at that exhibition, was immediately troubling to only a few. But by the 1870s, the threatening significance of new initiatives elsewhere could no longer be denied. The increasing bargaining power of skilled workers, as for example in holding back the acquisition of new, ring-frame spinning technology to replace the now-obsolescent mule, kept labor costs uncompetitively high.[87] England found its industrial leadership increasingly subject to cost-competitive challenges emanating from continental Europe as well as North America.

Although it was not apparent to anyone at the time, Britain's trend of growth in industrial output had begun to slow already by mid-century. More significant, and probably also unnoticed, were declines in the rate of growth of total factor productivity that began somewhat later. According to the estimates of C. H. Feinstein,

> while in 1870 GDP/hour worked in the United States was only 90 per cent of the British level, by 1890 it was 5 per cent and by 1913, 25 per cent, higher. Over the period 1870–1913, American total factor productivity growth appears to have been about three times the British rate. . . . J. G. Williamson, noting growth in skills per worker over 1871–1911 at only 70 per cent of the American rate, concluded that "it may well be said that the 'failure' of British industry in the late nineteenth century can be laid to the doorstep of inadequate investment in human capital . . . compared to her main competitors in world markets."[88]

Of course aggregates like these, as we noted earlier, tell only part of the story. Accordingly we will touch briefly on a few specific features of the epoch of gradual decline that began in England during the last third of the nineteenth century. Its most general, seemingly almost inevitable, characteristics have been well summarized as follows:

> Unlike many of their international competitors, who had access only to much more confining markets, Britain's international marketing structure meant that British firms could get enough orders of similar specifications to reap economies

of long production runs, and had a large enough share in expanding markets to justify investment in (what were then) up-to-date and increasingly capital-intensive plant and equipment. But the tables were turned by the spread of tariff barriers and indigenous industrialization.[89]

A second aspect of decline has been perhaps too sweepingly characterized as "entrepreneurial failure." It involved an apparent unreadiness "to confront institutional constraints innovatively." If the objectives of this study were extended to include all aspects of international competition for industrial leadership, detailed attention would need to be given to the steps by which ascendancy in the field of organic chemicals and dyestuffs passed to Germany in the later nineteenth century. But briefly to summarize, the heavy chemical industry was initially one in which Britain had been dominant. Its lead was based in part, however, on enormous caustic soda manufacturing capacity using the LeBlanc process. This may have been unwisely retained in Britain after having been superseded in terms of cost and efficiency by the introduction of the Solvay process on the European continent.

Careful retrospective analysis has indeed tended to confirm that this was "costly conservatism . . . profit-losing attachment to continuity and . . . reluctance to admit a major mistake."[90] The argument on this point is complex, however, and the rationality of historical hindsight may not accurately reflect the balance of factors involved in actions and decisions at the time they were made. Initially introduced from France in 1818, there were terrible deficiencies in the alkali industry based on the LeBlanc process that became "*the* chemical industry" in Britain during the first half of the nineteenth century. Extremely wasteful of materials and labor, it was also a source of serious atmospheric and soil pollution and a direct threat to human health. By contrast, the Solvay ammonia-soda process, while more capital-intensive, was at least 40 percent less costly, required only more readily available raw materials, and was environmentally much more benign as well. That much is clear. Yet it is also true that British manufacturers made some progress with process improvements and with shifting their emphasis to the conversion of former waste materials into useful by-products. They may have felt a heavy responsibility, too, not only for the enormous capital tied up in existing plant but for the livelihood of forty thousand employees.[91]

Perhaps even more complex, but also no less suggestive of management weaknesses, was Britain's failure to retain the advantage of its initial priority in synthesizing an aniline dye from coal tar residues in 1856. Henry Perkin, the brilliant young discoverer, had even followed up this achievement within little more than a year by making arrangements to produce the dye on a commercial basis and almost instantly finding a ready market for it. Perkin's was a largely serendipitous discovery, and his greater achievement lay in overcoming costly, technically deficient engineering obstacles to the production of a mauve dye.

The synthesis of alizarin, in which Germany won the race by hours in 1869, led more directly to a wide family of further discoveries and had a much greater economic impact in its own right. More important still, it demonstrated that the newly emergent field of structural organic chemistry could be deployed in the pursuit of commercial success. Only by German industry was the lesson quickly grasped that "the secret of commercial success lay in continuous chemical research."[92]

Already by 1873 Britain's decline in the field was evident, and by the late 1880s Germany had completely seized the lead in organic chemicals. Recruiting few salaried managers and relying on outsiders for marketing, British firms were equipped to exploit potentialities of neither scale nor scope. Vigorously investing in both of these directions, a handful of German firms reduced the price of a kilo of dye from well over 100 marks in the early 1870s to 23 marks in 1878 and then to 9 marks by 1886. By 1913, in consequence, out of a world total of 160,000 tons of dye, Germany produced 140,000 and Britain only 4,400. Nor was comparative performance much different in pharmaceuticals, films, agricultural chemicals, and electrochemicals. And by 1913 also, two-thirds of the electrical equipment machinery made in British factories had been developed not in its own research laboratories but in those of Siemens in Germany and General Electric and Westinghouse in the United States.[93]

As Anthony Travis has shown, this was more nearly a cause than a consequence of the subsequent passing of scientific leadership. Germany's emergence as the dominant power in the dye industry had its roots in business principles that are hauntingly familiar to students of Japanese economic performance today. They included careful study and detailed improvement of the English technology that was initially copied; a greater readiness to assume the risks consequent upon long-term investment; greater effectiveness in meeting the needs and tastes of diverse foreign markets; and painstaking insistence on maintaining product quality.[94]

The British empire, as well as its areas of traditional industrial strength, seemed to promise higher returns with fewer risks. On the other hand, "the German investors, as latecomers on the industrial scene, had to look elsewhere for new opportunities and found them in domestic industries based more heavily on scientific research and newer technologies." In a word, Perkin and his British contemporaries simply "did not build the organizations that could manage the resources needed to do this kind of work, so the world's leading textile-producing country lost its dye industry."[95]

Even greater challenges to Britain's industrial leadership emerged soon afterward in the United States. First to appear was the commercial generation of electrical power, with large-scale grids making it economically available for domestic as well as industrial use. Following almost immediately, and indicative of an accelerating pace of change, was the introduction of the internal combustion engine. The revolution in transportation that it promised was quickly visualized, although it was not fully consummated for several decades. Later still were vast new industries devoted to electronic consumer products and, most recently, microelectronics and the computer. The United States played a preeminent role in all of these, so that a fuller account will be found in chapters 6 and 7.

As illustrated by both the United States and Germany, a growing dependence on sustained research became a characteristic of the industrial age that began in the 1870s. Research in universities that was devoted not to applications but to strengthening the underlying sciences was recognized to be a part of this. But in spite of the reputation and strength of its great universities, the support of both basic and appied research, the according of commensurate status to those engaged in it, and indeed the education of a widening proportion of the population at large to professional scientific or engineering standards, found comparatively much less support in Britain.[96]

A fundamental difference was that, unlike the mechanical innovations of the textile age, chemicals, internal combustion engines, and electrical and electronic machinery and devices all could be introduced only as elements embedded in larger systems. All were relatively complex, open-ended systems themselves, requiring sustained testing and improvement—and frequently innovative marketing arrangements as well. While in some abstract sense small organizations may seem better able to respond quickly and flexibly than large, integrated ones, these new, longer time horizons and heavier capitalization requirements changed the rules of the game: "Here the less competitive German oligopolies, led by a small number of powerful investment banks, and the large firms that emerged in the United States by the end of the nineteenth century may have provided a better environment for technological change."[97]

So it appears that in England there was a progressive loss of vision and nerve. Traditional firms were less ready to transpose themselves into new organizational forms that might have secured new resources and adopted more competitive strategies. A balance was maintained that favored retention of existing consumption patterns rather than greater investment. Veblen's analysis retains its kernel of truth. It is not to the macroeconomics of international competition, however, but to the more particularized contexts of technological change itself that we must next turn—on a new continent.

5

Atlantic Crossing:
The American System Emerges

A SHARED POOL of available technology, its basic features all held in common, was among the many cultural continuities extending from Britain to its American colonies through much of the eighteenth century. Entirely absent at the birth of the new Republic, however, were most of the rapidly crystallizing preconditions for the mother country's impending Industrial Revolution. Slowly at first but with gathering speed, they were selectively stolen, borrowed, or imitated, perforce incorporated into new institutional settings, and then increasingly modified, supplemented, or wholly replaced to meet a different set of challenges. Within seventy years or so, borrowing had become reciprocal. A formidable new competitor could be dimly discerned on the English horizon, its American system of production representing a new and largely original manufacturing synthesis.

Both the rapidity and the originality of this transition are as striking as they were unexpected. They obviously cannot be accounted for solely in economic, let alone technological, terms. Although they fall largely outside our purview here, some of the major, contributing features of the larger cultural and historic setting should at least be mentioned at the outset.

One, surely, was the role of the frontier. It was a pole of attraction for not only immigrants but old settlers seeking a new start in life. The access to land and unstructured opportunity it afforded were powerful stimulants to social and occupational as well as geographic mobility, inhibiting the development of the deep class antagonisms that were simultaneously accompanying the Industrial Revolution in England.

No less ubiquitous as a contextual influence was the successful American Revolution itself, and the spirit of personal independence it enshrined. Of special importance were the constitutional guarantees of individual liberties, equality before the law, and due process that resulted from it. The acceptance of slavery was, of course, a glaring omission from the field of application of these guarantees. No doubt it helps to explain why the American South long remained marginal to the main thrust of industrial and technological progress. But elsewhere these founding documents heavily reinforced a general readiness to innovate and experiment, and an accompanying breadth of economic opportunity, that had no English equivalents.

Transatlantic Transfer

To most contemporary observers, eighteenth-century Britain's transatlantic colonies must have seemed to offer significantly more limited cultural amenities and economic opportunities. Inland from a narrow coastal strip lay a rugged, heavily forested, progressively more sparsely settled landscape, terminating in a slowly advancing but always dangerous perimeter of normal access. In sharp contrast with England, land in America was available for little more than the arduous effort of clearance and readiness to accept the risks and crude necessities of frontier living. But transport costs, kept high by widely dispersed markets, prevented the economies of scale that in the mother country were beginning to encourage the consolidation of workshops into more productive factories. And the constraints of Britain's mercantilist policies were in any case intended to encourage colonial agriculture and natural resource extraction, not manufacturing.

Faced with these conditions, colonial attitudes toward technological change were naturally conservative and "the rate of inventive activity was very low."[1] Taken largely for granted was a continuing dependency on the importation of manufactured goods. The onset of the American Revolution brought a protracted disruption of existing supply lines, affecting civilian necessities as well as vital military hardware, and led to an urgent, frequently innovative development of local substitutes. But while some enhanced industrial capabilities remained, few of these substitutes survived the renewed flow of imports following the end of hostilities.

At least as a matter of national policy, the earlier unreadiness to see industrial growth as a desirable course of action was not immediately displaced. While Thomas Jefferson, for example, was certainly familiar with and receptive to mechanical contrivances up to and including steam engines, there is nothing to suggest that he ever viewed the latter as an agency of economic and social change. Nothing could have been further from his mind than to transform the primarily rural landscapes he knew and loved into the sprawling proliferation of ironworks and textile factories that had begun to darken the landscape of the English Midlands. With an "immensity of land courting the industry of the husbandman," workshops were to him one thing and machines clearly another. John Adams, in the same spirit, even opined that "America will not make manufactures enough for her own consumption these thousand years."[2]

In part, they may have been merely echoing the skepticism about the potential for industrial growth in what were then still British colonies that Adam Smith had, with immense authority, expressed on the eve of the American Revolution. However, Smith's views, consigning an agricultural and natural resource role to the colonies and manufacturing to the mother country, were in any case almost immediately undermined by the simple fact of political independence. Smith also noted that positive avenues of comparatively rapid growth in the North American colonies were already being exploited.

The gross product of the American colonies in the year that *The Wealth of Nations* was published was still only about 40 percent of that of Britain. But as late as the beginning of the eighteenth century it had been a mere 4 percent.[3] Undoubtedly conscious that the gap was rapidly closing, Smith had gone on to observe that

> it is not . . . in the richest countries, but in the most thriving, or in those which are growing rich the fastest, that the wages of labour are highest. England is certainly, in the present times, a much richer country than any part of North America. The wages of labour, however, are much higher in North America than in any part of England. . . . The price of provisions is every where in North America much lower than in England. A dearth has never been known there. . . .
>
> But though North America is not yet so rich as England, it is much more thriving, and advancing with much greater rapidity to the further acquisition of riches.[4]

Even before Smith administered the coup de grace, British mercantilist policies had been subjected to considerable erosion in their application to America. Shipbuilding provides a good illustration of why this was so. Blocked by Dutch control of the Baltic, England had to obtain much of its shipbuilding timber from the colonies. But the American product, having typically been hauled to tidewater over winter ice and then shipped the following summer before it was seasoned, suffered heavily from dry rot. Predictably, all contrary considerations of general mercantilist policy notwithstanding, British merchants and shipowners found ways to build ships in America and then to charter or acquire them.[5]

Timber is only the beginning of the story. To take another example, shipbuilding in England, constrained by fear of corrosion and guild regulations, traditionally employed wooden dowels. In the colonies, more cost-conscious, lacking guild restraints, and prepared to believe that few ships lasted long enough in rough North Atlantic waters for corrosion to be a problem, iron nails were good enough. Locally manufactured iron nails became a growing export product on their own—one of many that help to account for American production of iron having climbed to about 15 percent (30,000 tons) of total world output by the eve of the Revolution.[6]

In other words, even before the Revolution many small and large ways were being found to exploit some of the distinctive advantages the Atlantic littoral offered. The unparalleled abundance of wood was one. Another was the existence of long, navigable rivers penetrating deep into the interior of the country and offering a profusion of opportunities for water power.[7] Cheap land, a magnet for immigrants, was of course a third.

The absence of bituminous coal east of the Appalachians might have seemed, at first glance, a substantial disadvantage. But different considerations were at the fore in the early Republic. Wood was so cheap and almost universally available that it long provided the major domestic fuel. With its much more extensive use on a per capita basis than in Britain, scrap lumber from sawmills, construction sites, and other industrial facilities further reduced cutting and transport charges to charcoal ovens. "Charcoal-smelted and hammered iron was little more expensive at New York or Philadelphia than British imports and was judged by American users such as Mathias Baldwin of locomotive fame to be better for machinery than foreign coke smelted and rolled iron. It was certainly preferred for farm and other metal implements by local blacksmiths."[8]

Meanwhile, anthracite coal from eastern Pennsylvania proved to require only some easily mastered technological changes in grates before being regarded as an acceptable low-sulphur substitute for many purposes. Ways to make use of it for smelting directly, without recourse to the coking process, were worked out in the 1840s, and the resulting pig iron is said to have been preferred for heavy construction.[9] Hence the dominant American trend in this period was from charcoal to anthracite, by-passing coke-based pig iron until the major centers of iron production moved west of the Appalachians. As late as the eve of the American Civil War, coke still was employed by only 10 percent of American furnaces.

In other words, successes came from being ready to take different paths than the highly successful ones pioneered in the country from which the colonies had now detached themselves. The challenge was to adapt, flexibly and without preconceived limits or directions, to what the new setting had to offer.

Equally important were human resources whose further development the new environment encouraged. From the outset of the colonial era, the self-reliance necessary for frontier living led to a familiarity not only with the use of tools but with principles of elementary machinery. Enthusiasm for mechanization accordingly was widespread. Clearly self-selective processes were at work, matching the continuing inflow of vigorous, potentially productive immigrants with these demanding requirements.

A three percent a year increase in population in the early nineteenth century was in itself spectacular, but combined with the fact that immigrants were chiefly adults anxious to buy or build houses and equip farms added to a market demand for construction and crude durable goods probably unknown in world history. The estimate by present-day statistical economists . . . that United States per capita gross national product grew only about half of one percent a year during the early nineteenth century can be misleading compared either to other nations or to modern times. That it grew at all after absorbing a three percent annual increase in people needing transportation and basic equipment for home and farm-making is a mark of high achievement in meeting this growing demand, and also a tremendous stimulant to the improved production of inexpensive durable goods.[10]

Thomas Cochran argues that some historians have a skewed view of American dependence on technology transfers from Britain. American practices began to diverge very early, away from the British pattern of an increasingly specialized division of labor and toward greater reliance on labor-saving machinery. Innovations like factory-made nails amd woodworking machinery accordingly met strong resistance in Britain, he notes. In any case, there was little or no eastward migration to introduce them there. "The net result has been that many American technologists admitted a debt to Great Britain but only a few Englishmen before 1850 recognized any inspirations from America."[11]

Construction, Textile Manufacture, Steam Power

With a million houses built between 1800 and 1820, the construction industry was, along with agricultural machinery, a primary center of innovation. Because of the cheapness and ready availability of wood, its readiness to accept a flood of innovations in woodworking machinery was only natural. Sawmills appeared by the thousands,[12] supporting a per capita consumption many times higher than in England. Again in Cochran's words, it is an example of "gradual industrialization without a factory system," leading to the machine manufacture of

> nails, bricks, and shingles after 1800, the circular saw after 1815, and the balloon-frame house from 1833 on. The latter did away with corner posts and heavy framing by using a cage-like construction of two-by-fours nailed to each other. It also greatly decreased the cost of two-story buildings. Thus this major improvement, from the standpoint of costs, depended on American factory-made nails in place of Old World morticing and tenoning of the members of the frame.[13]

In its innovative reliance on an exceptionally cheap and plentiful raw material, the construction industry illustrates one important pathway to the industrialization of America. The ideas and innovations, coming out of innumerable small carpentry and machine shops, stayed home for the most part, and left comparatively few records. Taken as a model for what followed, it stresses the local adaptability and independence of the course that America took.

There was, of course, another important pathway, perhaps best represented by the New England textile industry. After small-scale beginnings and a lengthy period of rather precarious existence in the shadow of its mighty British counterpart, it eventually prospered and became concentrated in great mills that depended overwhelmingly on British technology, and that of course shared in the use of cotton as a raw material. Taken as a model, this pathway instead emphasizes aspects of the American experience that followed directly from the Industrial Revolution in Britain as a kind of classic prototype.

Borrowing of the new advances in cotton spinning technology began with alacrity. Multiple-spindle jennies were being made in America within five years of Hargreave's 1770 patent. With high transport costs and a limited market, however, as late as 1790 there were still fewer than two thousand in use in the United States, at a time when there were more than a thousand times as many in Britain.

Inability to circumvent British export embargoes on skilled workmen and textile machinery was at best a secondary problem. Machinery, while sometimes successfully interdicted by customs officers, frequently found its way across the Atlantic by well-known smuggling routes. Skilled workmen even more regularly and easily evaded the restrictions, either on their own initiative or under the blandishments of American recruiting agents. As the French minister Calonne acidly explained in 1788: "You can keep a secret in a German factory, because the workers are obedient, prudent, and satisfied with a little; but the English workers are insolent, quarrelsome, adventurous and greedy for money. It is never difficult to seduce them from their employment."[14]

Furthermore, there were fundamental inconsistencies inherent in British policy. While the prohibition against the emigration of artisans was technically universal, there were entrenched interests supporting the profitable export of machinery. Yet as was even more clearly the case with steam engines that the textile mills employed, and the engine-erectors needed to assemble them on-site, how could the one be encouraged with occasionally allowing the other? In the event, the issue was simply evaded and "neither Savery's, Newcomen's nor Watt's steam engine ever came into the machinery clause of the prohibitory statutes."[15]

In any case, as David Jeremy comments, "the new technologies were embodied in the artisan."[16] Most of the immigrant textile workers were only handloom weavers, with relatively obsolescent skills. He estimates that no more than a few dozen, broadly experienced and managerially inclined British operatives played a decisive part in the early establishment of the New England industry. And from that point forward American innovations in the organization and design of cotton manufacturing equipment "fundamentally reshaped the imported technology" during the two decades after the War of 1812.

Expansion of the industry continued to be slow during the long period leading up to the imposition of an embargo in 1807. Illustrating the weight of competition, five-sixths of all U.S. imports from Britain in the early years of the nineteenth century consisted of cotton and woolen textiles.[17] As a result of the embargo, however, there was almost a sevenfold increase in the number of spindles involved in cotton textile production, from 20,000 in 1809 to 135,000 in 1815, and a jump in domestic consumption of cotton over the same period from 500 to 90,000 bales. That response turned out once again to be short-lived, with many American manufacturers having to close their doors upon the return of relatively free trade in 1816.[18] But as the internal market continued to grow because of natural population increase and immigration, there was now a more solidly based recovery and a resumption of growth.

A plausible case can be made that American protective tariffs were a decisive stimulant to the growth of a native textile industry during this critical early period. Certainly what followed was a spectacular epoch of growth: "an average annual rate of 15.4 percent from 1815 to 1833 and a still respectable rate of 5.1 percent from 1834 to 1860."[19] The number of emigrating British textile workers, which had been averaging less than two hundred a year, more than trebled in the late 1820s.[20] Even this heightened level remained modest, however, when weighed against the total of 115,000 employed in the industry on the eve of the Civil War. By 1831 American investments in this area were over one-third the size of those in England, and British manufacturers were beginning to be disquieted by some of the more efficient New England mills.

It is important to remember that aspects of both of these pathways—the local innovativeness of woodworking and the borrowed implantation of textile technology—coexisted and intermingled with one another in virtually every area of economic activity. Much technology was borrowed, much was independently evolved or adapted. What eventually took root on the new continent was neither an act of origination nor a slavish imitation but a fluidly changing mix of both.

Common to both the American and British processes of industrialization was a tendency for transformative developments to cluster in certain strategic sectors of manufacturing. The construction industry was an example not shared by the two countries, while the cotton textile industry conspicuously was. Steam engines provide an example of still a third kind, clearly borrowing basic technology from England but then responding to different conditions by developing the technology in an independent direction.

Knowledge of the existence of devices employing steam to transform heat into mechanical energy had diffused in educated circles well before the American Revolution. Such knowledge, however, was almost entirely academic rather than practical.[21] Remembering that this was prior to the radical improvement in steam power represented by James Watt's condenser, virtually the sole economic use that had been found for Newcomen engines in England was for pumping water, primarily from coal mines. In America, with coal as yet virtually unexploited, steam power stayed in the libraries and drawing rooms.

After the Revolution, both official and private interest in developing uses for steam power slowly intensified. Riverboat transportation had a special appeal in a rugged, sparsely settled country, and by 1807 Watt-improved engines had been successfully adapted to propel them. Both New York and Philadelphia had steam-driven waterworks by 1812. Responsible for the latter was Benjamin Henry Latrobe (1764–1820), America's first great engineer and architect of international stature, who deserves a brief digression as a leading example of "technology transfer" on a different level.

Latrobe, British by birth, had received his professional engineering training under John Smeaton, whom we have already encountered in chapter 3, and then had trained further under the neoclassical architect Samuel Pepys Cockerell. Emigrating first to Virginia and then to Philadelphia, he secured a number of important planning and architectural commissions including, in 1799, the municipal acceptance of his proposal for the Philadelphia waterworks. His greatest engineering achievement, this can be characterized as "the first urban utility in the United States (in the modern sense), and . . . the first convincing display of steam power in the Western hemisphere."[22] Latrobe went on to serve at President Jefferson's behest as Surveyor of the Public Buildings of the United States and later as Architect of the Capitol, and to take an active part in canal-planning and -building operations that would ultimately bring food supplies into Philadelphia from its hinterlands.

Consciously modeling himself after Smeaton as a "learned engineer," Latrobe "became one of the leading steam engineers of his time, using engines in a variety of commissions, and helped make Philadelphia the center of steam engine building and use in the United States."[23] He also brought new stature to this still infant professional discipline through his active contributions to the affairs of the American Philosophical Society. In an inchoate, formative era, his achievements, both in substance and in standard setting, exemplify the small number of individual contributions that belong in the same rank as many shiploads of imported textile machinery and of the immigrant spinners and weavers to assemble and work them.

Steam power rapidly began to fill a growing American niche. By the late 1830s well-established industrial users had taken over the best eastern waterpower locations (and would continue to rely on them through the 1860s for more than 70 percent of their power requirements). Newer, more diversified enterprises, "from printing presses to gunpowder mills," often had only steam to fall back upon. As manufacturing began to move west of the Alleghenies, good waterpower sites were in any case less plentiful.

"The effect on American manufactures was profound. Once scattered by the geographical imperatives of waterpower, industry, now relatively free, was inclined to concentrate in towns. Mills, once limited in size by the waterpower of their streams, were now as large as the engines that ran them."[24]

However, stationary steam engines remained a relatively high-cost source of power. For this reason their use tended to be concentrated in only a few high value-added industries, and where possible their role was kept supplementary rather than primary. In effect, as Peter Temin has observed, steam engines were used "only when the freedom of location gained by using steam was large." In such settings it substituted for cheap transportation, a function that would presently be superseded as steam also revolutionized railroad transportation.[25]

There was a striking departure in the prevailing direction taken by further steam engine development in America as compared with England. Under the cautious advocacy of Watt himself, England continued to rely mainly on low-pressure engines, in which the pressure of the atmosphere alone acted against the vacuum produced by the condensation of steam in Watt's condenser. High-pressure engines, which instead (with Latrobe as an exception) found favor in America, "used steam at higher than atmospheric pressure to push against the atmosphere in the same way that the low-pressure engine used the atmosphere to push against a vacuum." Since the higher pressure was technically more demanding, requiring strengthened boilers and new and higher levels of accuracy in the machining of cylinders and pistons, it is a significant indicator of growing American metalworking proficiency that they were introduced almost simultaneously in the two countries (by Richard Trevithick in England and Oliver Evans in America, in 1803–1804). But by the late 1830s, even though no valid basis for different preferences can be traced that is related either to safety or economy, "almost all stationary steam engines in Britain were low-pressure engines and almost all those in America used high pressure."[26]

In America, in spite of the greater stresses that high-pressure engines had to handle, engine-building quickly became a widely understood skill. American-built steam engines were successfully entering European markets by the later 1820s.[27] An accounting of 1838 provides the names of hundreds of known—mostly very small-scale—builders, responsible for an average of a mere five engines apiece.[28] Only six had produced more than thirty. With long distances between potential purchasers and relatively poor transport facilities, it is likely that most builders were limited to purely local markets.[29]

As both steam power and textile manufacture suggest, the American entry into the industrial age may initially have been slow and uncertain but then underwent a marked acceleration. That American processes and products were beginning to find a place in international competition so soon after the birth of the new nation seems fairly remarkable. It compares well with a similarly impressive acceleration both of innovation and output that had taken place earlier in England. Both involved a conspicuous clustering of transformative events in certain strategic sectors of manufacturing, attracting entrepreneurial capital to them and unleashing new energies as they succeeded. Yet the larger social, economic, and resource environments were very different, and the American course increasingly incorporated independent objectives of its own.

There is considerable controversy over other factors distinguishing early American industrialization. Inelasticities in the supply of labor and higher interest rates on U.S. capital investments both have their proponents.[30] The "confident expectation of rapidly growing markets" has also been put forward as an "extremely powerful inducement":

> [T]he use of highly specialized machinery, as opposed to machinery which has a greater general-purpose capability, is contingent upon expectations concerning the *composition* of demand—specifically, that there will be strict and well-defined limits to permissible variations. Both of these conditions . . . were amply fulfilled in early-nineteenth century America.[31]

Underlying this atmosphere of optimism as an influence on investment strategies was, as already mentioned, an unparalleled rate of population growth that (with immigration included) was more than twice as high as that achieved anywhere in Europe. With an abundance of good land additionally meaning that food prices were prevailingly low, a relatively large margin of average incomes was left for nonfood products. The climate for capital investment in industrial growth was excellent, in other words, and the market for plain but serviceable consumer products must have appeared nearly insatiable.

Parallel trends toward industrialization, differing in the specific mix of measures relied upon, began to appear at irregular intervals in some areas of manufacturing. Already at the outset of the nineteenth century, Eli Terry was shifting craft-based production of wooden clocks in a more industrialized direction, introducing a more highly specialized division of labor to supply a dependable, standardized, substantially cheaper product. As a contrastive example, Elisha Root's efforts culminated some decades later in the establishment of a production line for axe manufacturing.[32] Well exemplified by the textile industry, growing demand frequently led to increasing operating speeds, and that in turn to greater mechanization in order to reduce dependence on a relatively unskilled work force. Other predictable outgrowths were innovations in support of standardized, minimally complex operations such as automatic fault detection, sequential integration of machines, and controls over the flow of materials. All these features presently became characteristically American contributions to industrial operations.[33] New, automatic milling devices, anticipating the subsequent emphasis on continuous flow technology, had already been introduced by the time of the Revolution. But perhaps the most persistent and widely celebrated theme was work directed toward the development of machines and implements employing interchangeable parts.

The "American System" of Manufacture

The success of this "American System," as it came to be called, was more than a mere exemplification of the march of industrial progress. Needing to be recognized, to begin with, are the multiple elements of which it was composed—only a few of them truly determinative of the distinctive course that American experience took. Trends troward standardization and mechanization, for example, characterized the more complex British metalworking industries just as much as those of the United States. The real difference lay in

> the use of special-purpose machinery, very often exhibiting the use of self-acting mechanisms in the workpiece feed and disengage modes of operation. Moreover, this special-purpose machinery was sufficiently accurate so that common piece parts produced by machining would be so close in dimensional tolerance as to permit assembly with a minimum of time for fitting, or mating the connected parts of an assembly by the use of hand labor.[34]

There is no widely agreed-upon, fixed set of elements constituting the American system. At a symposium grappling with the question some years ago, the most that the editors could discern was an agreement that "an accretion of elements" was involved, and that "the system was something transcending mere technology, something suggestive of an underlying managerial philosophy."[35] Clearly, we are dealing here with an evolutionary process, meaning that it changed in content and form over time. Also, what ultimately emerged was a basically new and different configuration than the one characterizing the Industrial Revolution in Britain from which many of the originally constitutive elements had been selectively borrowed. And further still, it is clear that several convergent streams, reflecting opportunities and constraints that differed sharply in different industrial sectors, all came together to shape the final outcome.

Under these complex circumstances, interpretive differences are less disturbing than they might be otherwise. Those primarily concerned with the origin and derivation of the American system naturally stress its dependence on British antecedents. They also tend to rely on British observers' criteria for evaluating how long American products and processes "lagged behind" their nearest British equivalents. By contrast, those seeking to understand the long-term dynamics of American industrial growth look for the earliest seeds of change they can identify that began to distinguish American from British practice. The fundamental point they set out to explain is that "American technology *became* something that British technology was not *becoming*."[36] Their concern, in short, is with the emerging *Gestalt* of a "managerial philosophy," admittedly less tangible than the specific features of products and processes that are regarded as salient by observers who are themselves embedded in a different system.

Paul Uselding has nicely captured the elusive and yet fundamental character of the emergent American-British differences by recalling a distinction that Charles Babbage had introduced in 1832 between *making* and *manufacturing*: "In the *making* system, machines or machine-fabricated components are made one at a time, or in limited quantities. Under the *making* system, the methods of production are not greatly affected by the number of units produced. In the *manufacturing* system, which embraces large-scale production and interchangeable manufacture, the production methods employed are a direct function of the number of units produced."[37]

Having acknowledged that there are significant differences in perspective on this question, it must also be said that there is no disagreement on the basic, initial dependence of American industrialization efforts on British know-how and machines. A distinctive American path is generally seen as first becoming apparent in the early 1840s, confined at that time to certain light industrial areas and especially the American national armories. This position was consolidated early in the third quarter of the nineteenth century, with the entrance into general use of milling machines and turret lathes—the most important new machine tools of American origin.[38] Already by the 1850s "it was widely recognized that the cherished innovations in any particular industrial field were as likely to occur on one side of the Atlantic as the other."[39]

One of the convergent streams referred to earlier lay in those American armories, and it deserves a fuller account. The victorious end of war with Britain in 1815 had brought with it a quiet realization of significant American military failings, not least in the production and procurement of small arms and munitions. Authority over the national armories in Springfield, Massachusetts and Harpers Ferry, Virginia was accordingly centralized, with a legislative mandate stressing "uniformity of manufactures" as a guiding principle. The policies actually followed, however, were rooted in a continuing tradition of reliance on French artillerists and engineers.[40]

Responsive to this advice, and in the face of at times entrenched resistance from the independent, traditionally oriented craftsmen employed in the arsenals, consistent efforts were put in place to standardize the manufacture of American military firearms. Finally, by the mid-1840s, production of fully interchangeable rifles and muskets was successfully launched. This was truly "one of the great technological achievements of the modern era."[41]

There were two, essentially technical keys to success. One was provided by an unprecedented reliance on multiple inspections and an elaborate series of gauges; the other by new levels of precision in working drawings that could assure precise replication.[42] More important than both, however, was sustained emphasis on rigid work rules and "regularized procedures in stabilizing the complex human and physical variables present in the workplace." Only gradually, over a period of decades under close military superintendence and at an aggregate cost of over two million dollars, were workers persuaded and/or compelled to abandon "the task-oriented world of the craft ethos and reluctantly [enter] the time-oriented world of industrial capitalism."[43]

Fully interchangeable parts are a sine qua non of modern mass production, so that there is a tendency to see them as a defining characteristic of the American system. But things are not so simple. As late as 1840, the census of manufacturing shows that "the techniques noted by British observers in the 1850s had been adopted only in government armories and within a small circle of manufacturers producing weapons on contract for the United States government."[44] The overwhelming competitive necessity of interchangeability that we would associate with every branch of industry today was quickly recognized as a technological achievement—but not one that was immediately practical or useful to emulate.[45]

Military considerations assume considerable importance, in other words, in explaining how the American system of manufacturing originated. Somewhat comparable trends toward standardizing products and processes were undoubtedly at work in the private sector as well.[46] But without the "externality" of military policies that could remain in place for an extended period without cost becoming the determinant factor (and indeed, possibly without *French* military authority to sanction the course of action), it is not clear that the United States could possibly have achieved primacy in this emphasis on manufacturing uniformity. As Joshua Rosenbloom observes, this strongly "reinforces the view that the development of new technologies is a path-dependent process in which historical accident can significantly alter the course of economic development."[47]

In a wider sense, then, the American system needs to be identified with increasing refinements of standardization, with mechanization, and with, in the words of a British observer, "the adaptation of special tools to minute purposes."[48] Once its practicability had been demonstrated, it is clear that a different, essentially economically oriented, set of mechanisms gradually began to come into play. American consumers proved to be markedly more ready than their English counterparts to value function over finish and to accept a restricted choice of final products. This made it possible for producers of capital goods to seize the initiative in setting the objective of standardization and suppressing variations in product design.[49] Of course, this is not to deny that it also contributed to American *symbolic* leadership in pursuing the goal of standardized, interchangeable parts in a number of different industrial sectors (however imperfectly it was met at the time), which was a special subject of admiration in the Crystal Palace.[50]

> European comment often gave much attention to the crudity, the lack of finish, the use of wood in place of metal, or the light construction of various articles; but the admiration for their simplicity, originality, effectiveness, and above all their economy and volume of production, led not only to recognition at the world fairs but to increasing entrance of American products and methods into European markets as imports, or through licensing or notably less formal practices.[51]

Considering the originality and distinctiveness of the totality of this pattern, John Sawyer's dissatisfaction with purely environmental or utility-maximizing explanations, although first expressed nearly two generations ago, remains valid:

> Large numbers of observers freely expressed their displeasure or even disgust with American institutions and American ways. But whether they talked of "a noble desire to elevate one's station" or "vulgar dollar chasing," and whether they liked or denounced a society in which business rode high and a wide open social structure fostered mobility, rootlessness, restlessness, and the like, and gave enhanced significance to the visible results of economic success, they were pointing to social values and a social order uniquely favorable to . . . particular patterns of manufacturing. . . .[52]

Sawyer's dissatisfaction was directed toward the apparent inadequacy of explanations of the American system that focused narrowly on consumer preference for products that were rugged and economical, even if with a limited range of choice and lesser quality of finish. As the interest and controversy evoked by the subsequent work of H. J. Habakkuk attests,[53] that inadequacy is now generally recognized. Production processes, management-labor relations, and entrepreneurial and investment strategies are clearly of no less importance in understanding the distinctive directions taken as the American republic began to industrialize.

Looking at the earlier part of the nineteenth century, Habakkuk's starting point was an observation entirely consistent with Adam Smith's for the late prerevolutionary period in America (see the section on "Transatlantic Transfer" earlier in this chapter). Both noted that American labor costs were appreciably higher than their British equivalents. This overall comparison seems to find general agreement even in a highly disputatious field, although with significant exceptions for certain sectors of the labor market.[54]

Habakkuk's estimate was that, as industrialization was getting underway around 1830, industrial money wages in the United States were "perhaps a third to a half higher" than in Britain. He attributed this to the fact that "land was abundant and . . . it was accessible to purchase by men of small means. In the 1820s it could be purchased from the Federal Government at $1.25 per acre, which was well within the reach of a skilled labourer, who might at this time earn between $1.25 and $2.0 per day."[55]

Visualizing an open and attractive avenue of alternative employment in clearing and owning western agricultural land, Habakkuk had no difficulty in explaining this cost differential. Other, time-dependent factors also enter in. The beginnings of massive Irish immigration along the northeastern seaboard in the mid-1840s swelled the surplus of unskilled labor available for factory employment, leaving skilled labor in shortest supply and hence with greatest bargaining power. This and other perturbations of what was never an equilibrium pattern to begin with might be expected to shift the precise direction, although not the general trend, of labor-saving mechanization.

More debatable are at least the details of Habakkuk's underlying assumption, that it served as a spur to labor-saving technological innovation. On the one hand, it is generally in accord with the suggestion of an informed British observer that he found American workers "much more willing than British workers to cooperate in technological innovation. Because they were better educated and because labor was less abundant, there was less fear of labor redundancy, more appreciation of the benefits of mechanization, an absence of apprenticeship restrictions, more social mobility, and less class hostility."[56] But on the other hand, others have drawn attention to the complexity of motivations that have historically accompanied alterations in the capital-labor ratio.[57] In sum, the so-called "Habakkuk thesis" seems to contain an important kernel of truth. But at the same time, it probably attaches too much weight to gross differences in factor prices between the two countries. Hence the extent of its applicability, and the contributive role of other factors, remains to be fully worked out.[58]

Emphasizing that the Habakkuk thesis remains subject to doubt are Alexander Field's macro-economic comparisons of American and British capital stocks and capital-labor ratios in the mid-nineteenth century. On an aggregate basis, he has found that the American economy as a whole, including its manufacturing component, was substantially less capital-intensive than the British economy in 1860.[59] The difficulty with aggregated data, of course, is that it can subsume gross differences between sectors. As with Japanese high-tech industries in relation to that country's whole industrial production today, the possibility remains that the American system is a characterization applying only to a limited part of the whole.

What must be stressed in any account of the American system is the fact that it emerged in an evolutionary process and did not constitute an invariant cluster of attributes. The era of mass production more properly falls in the following chapter, but mass production is quintessentially a part of the American system. Appropriate, therefore, are Alfred Chandler's two perceptive "preconditions" for mass production:

> First, the railroad and telegraphic network had to be built in order to permit large amounts of raw materials to move at a rapid but steady pace into the factories, and the finished goods to be shipped out as regularly and as speedily. Second, there had to be available large quantities of fossil fuel (in this case coal) to provide the essential heat and power. The new production processes, therefore, had their beginnings in the 1850s as the railroad and telegraph networks expanded and as coal supplies became abundant. They reached fruition in the 1880s after the railroads had perfected cooperative arrangements to haul freight cars to any part of the country without transshipment and after the completion of the telegraph and cable network enabled factories to have almost instantaneous communication with any town in the country and indeed in many other parts of the world.[60]

Machine Tools, Railroads

A crucial domain of American industrial advance, providing the basis for many others, was the machinery-producing sector. Beginning somewhat earlier, it had also played a significant part in the Industrial Revolution in England. Similarly, the American system of manufacture as a whole, as well as major, metal-fabricating industries that would develop later in the century, depended on products of the machine tool industry.

Nathan Rosenberg has convincingly outlined the main stages in the evolution of the machine tool industry. As late as 1820 or so, this does not appear to have been a separately identifiable sector of manufacturing activity. Machines needed for production processes instead tended to be constructed in-house, on a largely ad hoc basis. By mid-century, however, a number of adjunct facilities of textile factories had been created that (apart from filling their primary missions in support of their parent plants) were beginning to branch out into machine construction more widely. Among the areas in which they became engaged were textile machinery, steam engines (the Baldwin Locomotive Works is a striking example), turbines, milling machines, and machine tools for other uses. Still other specialized machine producers grew out of the requirements of arms makers, or later appeared as spinoffs of manufacturers of consumer durables.

Evident throughout this development is a reciprocal interaction between the introduction and refinement of new consumer products and further advances in the industrial machinery used in their manufacture. In its rapidly escalating demands for rolling stock, locomotives, and ancillary equipment, the railroad industry re-directed the machine tool industry's emphasis toward large-scale machines capable of high throughput. Bicycle production, beginning on a large scale in the 1880s, falls in the temporal frame of the following chapter rather than this one. But as an illustration of the theme of reciprocal interaction, it played a crucial part in the development of hardened, precision-ground ball bearing components that immediately enhanced the performance of the machine tool industry. Later still, of course, ball bearing assemblies made a vital contribution to the founding of the automobile industry and to virtually every other system for harnessing rotary power. As the recursive cycle continued, the latter further advanced the design of machine tools through the demand for rapid improvements in transmission assemblies and drive shafts.[61]

Rosenberg discerns two, initially distinct sets of machine requirements:

> Whereas the production of heavier, general-purpose machine tools—lathes, planers, boring machines—was initially undertaken by the early textile machine shops in response to the internal requirements of their own industry and of the railroad industry, the lighter, more specialized high-speed machine tools—turret lathes, milling machines, precision grinders—grew initially out of the production requirements of the arms makers. (p. 419)

But while originally responsive to different sets of needs or requirements, as they developed these machines came together to form complementary clusters of growing versatility. Characterizing most of the originally quite distinct areas of manufacturing were a common need for closer tolerances, exacting performance requirements, and higher-strength, longer-wearing materials.

There was an emergent commonality of a relatively small number of processing operations, solutions to technical problems of power transmission, control devices, friction reduction, and the like, all playing essential parts in shaping and finishing metals in increasingly complex ways.[62] This secular process of convergence encouraged the formation of firms that would specialize in producing highly versatile milling machines, turret lathes, and other machine tools to very exacting standards. Finding their places in very diverse industrial settings, these machines immensely broadened future manufacturing capabilities and fully and permanently superseded much more expensive and imprecise hand operations.

Machine tools are capital goods, of great importance but subsumed in productive processes. Railroads played—and still play—a far more direct part in national economic and social life. In the United States during the nineteenth century, it is not too much to say that they knit together the new nation. As expansion of the United States continued west of the Allegheny mountains during the early years of the republic, regional divisions emerged that strengthened the requirement for closer linkages. Production of consumer goods was concentrated along the Northeastern and Middle Atlantic seaboard, while the region of most rapidly growing markets lay in increasingly distant hinterlands. Those hinterlands were producers of agricultural surpluses, sharply deflated in price, or even marketability on any terms, unless they could be economically brought to the swelling populations of eastern cities. Cotton produced in the South, accounting for two-thirds of all U.S. exports, could strengthen the national economy as a whole if carried to New England's textile mills at rates undercutting transatlantic transport.

As noted earlier, the Atlantic seaboard was fairly well endowed with navigable waterways. Further inland, sparse population and rugged landscapes were formidable barries to economical transport. Canal- and railroad-building were two almost simultaneous responses to this stimulus.

Presently railroads carried the day, partly as a result of their greater speed and locational versatility but also facilitated by government-assisted financing arrangements. New York State's construction of the Erie Canal system, linking the Great Lakes to the Hudson River, was merely the emblematic centerpiece of a canal-building effort of imposing scale. Begun in 1817, it was fully opened in 1825 (with later improvements and extensions). By 1840 some 5,300 kilometers of canals were in use, primarily in New York, New Jersey, Pennsylvania, and Ohio.

Compared with wagon freight, canals were an immediate economic success. Even in a westward direction, rates per ton-mile for canal shipping were a small fraction of overland rates. The movement of western grain, lumber, and other high-weight, low-value, bulky goods to eastern markets was even more substantially facilitated, as were the beginnings of western urban and manufacturing centers. But railroads proved to be an altogether more dynamic force. Although three or four times as expensive as freighting by river or canal, they were much less expensive, as well as far more rapid, than wagon freighting or passenger coaches. Directly responsive to passenger demands for greater speed and flexibility of operations, they quickly realized an unanticipated explosion of passenger service— and of community demand for inclusion in service networks.[63] Perhaps the best indicator of the combined economic effect of railroads and canals was a steady, cumulative improvement in terms of trade that is primarily a reflection of declining transport costs: "A composite unit of western output (at a western market) would fetch twice as many eastern goods in the 1840's as it did in the 1820's, three times as many on the eve of the civil war."[64]

Railroad-building's commercial appeal was instantly evident; the technology was almost as readily available for purchase in the United States as by British entrepreneurs; and local capabilities were not lacking to develop that technology in new directions that were responsive to different conditions.

On the other hand, there were greater obstacles to technology transfer in the case of railroads than of textile machinery:

> With the textile industry it was possible to mechanize one step of the process while retaining a handiwork technology in another step. Modern railroads, however, came as a package: they required trained engineers, graded ways, rails and ties, locomotives and cars, inclined planes (in the early years) and stations and other freight-handling facilities. While some experimentation was necessary and desirable, potential transferors generally had to promote the adoption of an entire complex of new technologies.[65]

Horse-drawn freight tramways of local construction, on wooden (or iron-faced wooden) rails, were introduced in the United States not long after the beginning of the nineteenth century. But this larger complex of needs was another matter. Initially locomotives, rolling stock, and rails all were imported from Britain and it was not until 1841 that British-made locomotives disappeared from the American market.

By that time it had also been recognized that every aspect of American railroading faced a different set of challenges. With limited investment capital available and large, low-density zones of service, capital costs had to be kept low on a per-mile basis. That dictated a departure from prevailing British reliance on extensive cuts and fills to provide nearly-level beds, and the construction instead of routes with steeper grades and many curves to follow local topography. And that in turn stimulated a series of innovations by which American-produced locomotives came into favor that more adequately met American conditions.[66] But while eliminating its dependence on British locomotives in this way, U.S. dependence on Britain for iron rails lasted considerably longer. "Only after mid-century did the American rail industry become the major supplier to American railroads."[67]

John Stevens (1749–1838), pioneer in American steamboating and also sometimes called the "father of American railroads," obtained the first state charter from New Jersey for railroad construction in 1815. That project failed for lack of financing, although he is credited with having built the first American steam locomotive ten years later. The Baltimore and Ohio Railroad Company, chartered in 1827, opened its first 21 kilometers to revenue traffic in 1830—the same year the opening of the Liverpool-Manchester line inaugurated the railroad era in England. Scheduled, steam-drawn passenger transportation opened in South Carolina only a few months later, and other projects also quickly came to fruition. Intercity traffic was opened between Albany and Schenectady in 1831, and between Boston and Lowell, Wooster, and Providence in 1835. Reflecting a gradually accelerating coalescence of local initiatives into larger networks, 4,500 kilometers of railroad routes were in service by 1840 and 50,000 by the eve of the Civil War twenty years later.[68]

Agricultural Expansion and Commercialization

As had also been true in England, there was a close association on the western side of the Atlantic Ocean between massive changes in industry and agriculture. The pace of technological change in U.S. agriculture underwent a marked acceleration around the middle of the nineteenth century, and successive transformations continued for at least some decades into the twentieth. Those dates suggest at least a partial reversal of the arrow of causality in the American case, with the production of large agricultural surpluses less a precondition and more a consequence of industrial growth.

Playing a much more salient part in the United States than in Britain were newly founded industrial organizations that aggresssively promoted technological enhancements of American farming practices. By the 1850s they were achieving striking successes in the production and sale of newly invented agricultural machinery. Encountering rapidly rising demand, they were no less successful in extending and improving their product lines to meet newly articulated needs. Capital investment in such machinery on the part of individual farmers unquestionably was responsible for a major share of the ensuing advances in agricultural productivity.

However, much more was involved in this development than a tour de force in marketing and corporate management. The turning from subsistence to commercial agriculture rested more fundamentally on twin foundations. One was the growing ability of railroad networks economically to interconnect lightly settled western regions of concentrated agricultural production with eastern urban centers. The other was the demonstrated readiness of a large part of the farming population to abandon the autarchic, subsistence-oriented conditions of life along the earlier frontier as soon as this became feasible. Far from being hapless victims of urban or industrial initiatives, American farmers actively took part in becoming specialized producers tied to the interdependencies of mass markets.

There are other, no less substantial differences between the British and American instances. In particular, land was limited and labor relatively more abundant in England, while approximately the opposite relationship later obtained in the United States (at least on newly settled lands inland from the former colonies). Associated with this contrast was another, between English farmers who were primarily tenants and Americans who were predominantly freeholders. Tenantry had an apparent braking effect on adoption rates for new technology, but at the same time it helped to generate the great landed incomes that went into English industrialization.

Given the vastly greater geographic sweep and degree of climatic zonation in the United States, the emergence of regional variation is obviously a theme of fundamental importance in any comprehensive account. However, for our purposes it is of secondary interest. The major themes of distinctively American development are better portrayed by focusing on the northern midwestern heartland, extending from central Nebraska eastward through Ohio and from the Dakotas southward through Kansas. Well into the twentieth century, it was here that the cutting edge of agricultural advance was consistently to be found especially with regard to the cultivation of the two principal midwestern crops, wheat and maize (corn).

Steeply rising population and economic activity in early America naturally led to a disproportionate rate of growth of towns and cities. The 1820 census, for example, found a population of 5.7 million people, 90 percent of them rural, living in the northern states. By 1870 the rural population had increased 3.4 times—but the urban population 14.5 times. Where there had been only 880,000 people west of the Alleghenies in 1820 (two-thirds of them in Ohio), there were 13 million fifty years later. By that time some 35 percent of rural agricultural production was finding its way to markets in northern cities alone.[69] Less easy to document, but nevertheless beyond question, is a corresponding improvement in the quality and density of the transport grid (roads, railroads, rivers, and canals) by which the subsistence needs of this population could be met.

As farmers' access to markets improved, they had the increasingly attractive option of directing their production primarily toward meeting market demands rather than supplying their own subsistence needs. In so doing, their own consumption preferences were inevitably altered. Acccess to markets as sellers meant that they increasingly encountered superior, cheaply manufactured consumables that—with growing cash incomes—they could substitute for goods which, earlier and more laboriously, they had produced at home.

Rising expectations of this kind must have contributed to the broad, secular, irreversible shift on American farms toward consistently market-oriented behavior. As this shift went on in much of rural America throughout the nineteenth century,

> the concept of surplus slowly lost its significance, replaced by considerations of marketability, cost, and profit. Family need as a determinant of the application of productive effort became subject to a cost-price calculus. The homestead gradually surrendered many of the functions that had given it a subsistence character

and became increasingly specialized in its activities. The farm became an integrated and dependent unit in the market economy.[70]

Much evidence of these rising levels of consumption can be found in contemporary accounts: the shift from rye and cornmeal to wheat bread, and from lard to butter; the rising consumption of meat, particularly beef, and of purchased rather than locally grown vegetables and fruits; the increasing use of salt, condiments, molasses, and sugar; the displacement of homespun linens and woolens by factory-woven cottons and woolens; and the new preference for factory-made shoes and hats.

Among more permanent investments reflecting the influence of the market may be cited the widely noted transition from predominantly log and rough plank homes to the use of brick, stone, and mill-sawn lumber; and the increasing presence of cabinet-made furniture, factory-made candles, oil lamps, carriages, harnesses, churns, and corn shellers.[71]

Still, it is important not to overemphasize either the extent or the abruptness of the shift toward the market as the source of supply for home consumption requirements: "Farmers cashed in on commercial production when times were good, but relied on their farm's ability to feed the family when the markets turned sour, ... successfully straddling the fence between agriculture as a way of life and as a business enterprise like all others."[72]

Briefly touched on here is the special importance for early American agriculture of the omnipresent issues of risk and uncertainty. No public "safety net" of any kind was in place in the western territories, and the prevailing pattern was for early settlers to operate at or beyond the outermost margin of their own capital and human resources. Inherent in the seasonal and annual variability of farm life was the likelihood of having to *alternate*, with very substantial but irregular and unpredictable swings, between years in which farmers were able to satisfy at least some of their rising expectations as consumers and other years of serious, sometimes overwhelming deprivation. Subsistence autarky and market participation, as this suggests, were not (or not simply) alternate life-styles or value orientations that may confuse the analyst by occasionally coexisting. Indeterminate periods of deprivation were periodically imposed by the larger environment. Investment in technological or other means to reduce their impact became a high priority whenever a sequence of good years made this possible.

It is a widely held stereotype that those engaged in subsistence agriculture tend to be possessed of a set of goals, values, and incentives fundamentally different from the "economic" motives that dominate production decisions in industrialized societies. The nineteenth-century American record strongly belies this popular impression. A first phase of geographic expansion was indeed subsistence-oriented—but constrained to be so by the limited capital and equipment of the pioneers and the virtually total absence of transport linkages to distant markets. The characteristic absence of concern over exploitative standards of land management, and the short time horizons before settlers moved on to fresh lands further west, are hardly what one would expect of value systems centered on self-sufficient, inward-looking family enterprises. And this phase was, in any case, followed within a very few decades by a widespread shift toward technologically based enhancements in productivity. In short, the onset of intelligibly "rational" investment strategies in agriculture paralleled similar patterns in corporate organization and industrial growth with very little lag.

A comparison of the trajectory of technological change in American agriculture with that occurring more or less simultaneously in industry thus is an important one. Family farms may remain a cultural ideal even today, but it has long been almost universally recognized that this is an ideal largely detached from reality. Corporate forms of ownership and organization have unquestionably become the dominant ones, just as farming has become fully dependent on heavily capitalized, mechanized equipment and chemical, biotechnical, and other inputs of industrial origin. For at least a century, agricultural decision making has tended to be linked at all times to news of market fluctuations, not merely locally or even nationally but in other world regions.

The remarkable extent of the apparent convergence between industrial and farming strategies invites further consideration of the processes of change by which, in both sectors, broadly consistent patterns succeeded one another. Consistent with the popular stereotype would be the supposition that the growth of industry was the dominant and dynamic force, and that agriculture was only tardily and involuntarily responsive. This view would have technological and organizational innovations clustered tightly in urban, industrial settings, and diffusing outward from them into the countryside as one system of values simply overwhelmed another. But on closer inspection, it appears that American values were more homogeneous across the entire urban-rural continuum. Transformative technological impulses thus could be, and in fact were, no less characteristic of agriculture than of industry.

It would be conceptually convenient, but inaccurate, to suggest that we can trace the evolution of American agriculture from a more or less stable, geographically undifferentiated baseline in colonial times. Quite to the contrary, substantial regional differentiation began fairly early. New England, with thin soils, soon turned toward livestock-raising on small, fractionated holdings, and perforce toward a movement away from the land into nonfarm occupations. Meanwhile the middle Atlantic states could specialize in producing cereals for export. Prototypes of plantation agriculture relying on slave labor were well established in the South before the American Revolution.[73] Westward movement beyond the Alleghenies, into what later became the midwestern heartland of American agriculture, was at that time only beginning.

An irregular east-to-west gradient gradually developed as the tide of new settlements moved westward, extending from the Atlantic coast into and through the heartland of the Midwest. Westward with it, through the late eighteenth and much of the nineteenth centuries, also moved the focal areas of change and intensification. The pattern finally stabilized only upon encountering a climatic barrier (really, a broad transitional zone) on the western plains beyond which, again, a differentiated regional variant of that pattern was called for, one that emphasized stockbreeding over agriculture. As this subpattern gradually stabilized, the rawness and impermanence associated with the frontier finally began to recede into myth and memory.

As the frontier advanced across the Midwest, a pervasive feature of individual farmers' strategies was to maximize the size of their holdings. While uncleared land was abundant and therefore relatively cheap, the effort of clearance remained a heavy burden for a lengthy period. It has been estimated that, on average, a month's labor with a team of oxen was required to clear an acre: "The usual course was for the family to clear five or ten acres each year, planting crops on land as it became available. This meant that it took five or ten years to build a complete farm. . . . [I]n the pre-Civil War era, it was clearly a serious drain on productive labor resources."[74]

Faced with these heavy demands on available time and energy, and with typically very limited capital reserves on the part of individual farmers, it is understandable that the approach to agriculture was initially a short-term, highly exploitative one. The ideal of staying in place long enough to build a complete farm was seldom put into effect. As an observer noted in 1819, the prevailing intention was "to compel the earth to produce the largest return with the least possible outlay of time and labor. . . . And when their land is exhausted they abandon it for virgin fields, apparently exulting in the thought that there is a best area yet intact in the far West." Similar attitudes prevailed toward livestock, with hardy, slow-moving oxen retained as the principal beasts of nonhuman traction since speed of movement counted for less than utility in heavy land-clearing and timber-hauling. Investments in farm buildings, too, were only those minimally necessary while any discretionary resources were devoted to the acquisition of more land.[75]

Exemplifying the most widely prevailing criticism of farming during this period is an 1833 reference to "an almost universal disposition to cultivate more than could be advantageously handled." As late as the 1850s, the average life of a recorded mortgage even in Ohio was only two years, and about 10 percent of the acreage in the state still was being sold each year.

Customarily, wheat was the first crop introduced. Aimed at maximizing immediate returns, this was frequently grown for a number of years without any rotation. Rapidly declining yields were an inevitable consequence. Prone to crop failure, it was also subject to particularly wide oscillations in market price and its highest labor requirements coincided with the hottest part of the agricultural season. Therefore, it was gradually abandoned in favor of corn across much of the central Midwest, while continuing to find favor in the northern part of the region. Yet it remained, in general, "the preeminent crop for exploiting the fertility of new lands with a minimum of capital, achieving a maximum return in the shortest possible time."

The alternative, corn, was most profitable as feed for hogs and cattle, but in addition, as was noted in 1850,

> it is the main article of food for man and stock, and can be cooked in a great variety of ways, so as to be equally acceptable at the tables of the poor and rich. The cultivation of corn is admirably adapted to the climate and soil of the State [of Indiana], and to the habits of the farmers . . . with proper cultivation the corn does not often suffer fronm cold, deluge or drought, and our laborers prefer to work hard in spring and early summer when the corn most needs it, and then relax exertions in the latter part of the season, when they are not required, and the heat is more oppressive.[76]

A further shift over time, intensifying after 1830, was toward the substitution of horses for oxen. This entailed considerable additional cost for feed and maintenance. As market dependence grew, however, this was more than offset by the increased speed, and hence area, of operations like plowing that were the critical determinant of market income, and by the increasing importance of transport to market.[77]

Small-scale inventive activity relating to new agricultural implements was well underway in the 1820s and 1830s if not earlier. However, long sequences of innovative refinements were needed before substantial industrial production—hence widespread utilization—of any of the potentially important, even revolutionary, new devices could begin. Given heavy use and consequent frequent breakage, standardization of materials (particularly iron and steel) and the availability of replaceable parts were virtually as important as superior design. And in any case, designs rapidly became sufficiently complex so that very few of them could be readily reproduced by local blacksmiths.

Among the inventions of particular importance in growing grain were several independent elaborations of wrought iron and steel plowshares and moldboards in the 1830s, much superior in their capacity to scour cleanly and without sticking. John Deere was turning out ten thousand of these annually by the late 1850s; during the Civil War they were further improved by the introduction of a sulky plow that could be ridden by the operator.[78] The new cast iron and steel plows permitted deeper plowing on deeper, western soils. By 1860 or so they had become almost competitive in price with wooden plows, as well as being much more dependable and predictable in operation.[79]

Largely anticipating a later development, it may be mentioned that disk plows began to be experimented with in the same period. While scouring well and working effectively, especially in light soils, they required a greater pull on the drawbar than moldboard plows and so could not be used economically enough with horses or oxen. Not initially favored for this reason, their ascendancy, especially in multiples, awaited the introduction of the tractor.[80]

Already by 1860, essentially every major operation in the growing of grain had been adapted to the use of newly designed, horse-drawn equipment.[81] Threshing machines were available by the late 1840s, and corn planters came rapidly into use during the 1850s. Horse-powered rakes and mowers were other important developments of the second half of the nineteenth century, with hay-loading equipment, balers, potato diggers, corn shellers, straw and root cutters, hay tedders, and fanning mills of lesser significance.

Most significant of all was the invention of the reaper, for it "replaced much human power at the crucial point in grain production when the work must be completed quickly to save a crop from ruin."[82] A workable machine was patented by Obed Hussey in 1833 and another by Cyrus H. McCormick in 1834. By 1852, McCormick, who had become the dominant figure in the industry, was making one thousand machines a year.

A long succession of further improvements followed. Some involved combines that did both reaping and threshing in the field simultaneously, while others reduced hand labor and speeded the pace of work by mechanical bundling and by tying the sheaves, first with a wire binder and then— an important and difficult innovation—with twine. Jumping ahead merely to indicate the striking advances in output that were in prospect, 250,000 combines were in production by 1885.

One difficulty besetting the farm equipment industry as a whole was a propensity to make annual model changes. Defeating the prospect of great economies of scale that might have followed from mass production, this also led to a continuing succession of production problems. McCormick's success (prior to the comprehensive reorganization of production in 1880) thus is to be attributed not at all to advanced production technology but to "a superior marketing strategy, including advertising and sales techniques and policies."[83] Newly broadening definitions of the American system that had prevailed earlier, marketing considerations make their appearance as a core component!

Reapers and combines were most quickly and widely adopted in midwestern wheat areas. There, finally, the seemingly "uneconomic" strategy of giving highest priority to maximizing land holdings came into its own. Twenty acres of tilled crops in the East and thirty or forty in the West formerly had been about the maximum that a single man could cultivate. Almost from the first appearance of the reaper, these limits were superseded. In fact, admonitions were appearing in Western agricultural journals already by 1846 and 1847 against its purchase by a farmer "who has not at least fifty acres of grain."[84]

It is worth noting, however, that this individualistic, "farm-specific model" of resource allocation, while finding favor among neoclassical economists, provides a poor fit with actual behavior. Matching contemporary Censuses of Agriculture and Population with records of early McCormick reaper purchasers, there is little to suggest a threshold in decision making that follows this advice. Joint purchases, as well as much testimony in farm diaries on the prevalence of teaming of machines, horses, and labor among neighbors, point to other, no less "rational" options.[85]

Coincident with the process of mechanization and industrialization was a marked trend toward fewer but larger manufacturing establishments. Growing rapidly in size and diversification of products, these companies soon found that they required more effective systems of distribution both for their primary products and for the succeeding flow of replacement parts. Independent agricultural retail and wholesale warehouses came into existence to meet this need, while corporate initiatives also led to the appointment of resident local agents and traveling salesmen.[86]

As might be expected, such processes of vertical rationalization took other directions as well. There was a marked acceleration, for example, in the granting of agricultural patents. From an average of around thirty annually in the late 1830s, the number climbed by the late 1860s to an average approaching 1,200—a fortyfold increase.[87] This dramatic change reflects not only the commercial success that producers of agricultural machinery had achieved but the extension of a basically commercial outlook to the agricultural sector as a whole.

As the rapid and substantial demographic changes that were mentioned earlier imply, the scale of agricultural output made immense gains during the corresponding period. To use a different (but suggestive and overlapping) temporal span than that shown in table 5–1, for which calculations are conveniently available, gross output of wheat and corn increased approximately sevenfold between 1840 and 1910, of which the contribution of increased labor productivity was approximately fourfold.[88]

With underutilized land widely available, there was for a long time little incentive to enhance crop yields per acre. In fact, wheat and corn yields remained roughly constant throughout the nineteenth century. Gains in

TABLE 5-1
Nineteenth-Century Gains in Agricultural
Productivity

	1800	1840	1880	1900
Man-Hours/Acre				
Wheat	56	35	20	15
Corn	86	69	46	38
Man-Hours/100 Bushels				
Wheat	373	233	152	108
Corn	344	276	180	147

Source: Rasmussen 1962: table 1.

labor productivity, on the other hand, were in no small part a product of the availability of new agricultural technologies and at the same time a requirement if those new technologies were to remain within the range of farmers' purchasing power. As table 5-1 attests, the cumulative record across the century was highly impressive.

By the eve of the Civil War, a new era in agriculture was clearly at hand. Industrially produced farm equipment was beginning to make major differences in productivity, and an interconnected railroad grid had begun to provide service all the way to the Mississippi River. Midwestern agricultural products were being traded to advantage in the larger urban markets of the Northeast. Predictably, this led to complex new patterns of regionalization. Even if at times badly overexploited, the deep, relatively level soils of the Midwest were exceptionally productive for the cultivation of grain. Fortunately for eastern farmers, the continuing growth and prosperity of cities along the Atlantic seaboard offered diversifying markets for specialized products—dairy farming in the rocky New England uplands and specialized truck-gardening to provide fresh fruits and vegetables in the case of the Middle Atlantic states.[89] These patterns of regionalization proved to be highly complementary, mutually advantageous, and therefore durable.

Scientists, Engineers, and the Shaping of Early Federal Policies

A central issue in the early years of the Republic was the balance to be struck between the powers of the federal government and the powers reserved for the states. That was an overridingly political issue, whatever the economic motives imputed to the contending parties. The young government naturally began with few resources with which to institute even a narrowly defined range of essential functions. As long as ideals of agrarian self-sufficiency remained uppermost, even slight tendencies toward the assumption of new national roles (and the taxing powers they would have required) provoked immediate and widespread opposition. Thus, to speak of federal "policy" as significantly influencing either technology or science would continue for several decades to be something of a misnomer.

Admittedly, Thomas Jefferson's part in this introduces an element of ambiguity. Taking all knowledge as his province, he epitomized for his—and practically any other—time the ideal of the scholar-statesman. Presiding over the American Philosophical Society, finding a place for natural history exhibits in the White House, communicating endlessly with European colleagues across an extraordinarily diverse array of interests and disciplines, he set a personal example of identification with science and humanism. But whether that allows us to see him in the reverse image, shaping the policies he advocated (let alone those he succeeded in implementing) *as a scientist* in some sense that he as well as we would recognize, is very doubtful. And it also must be said that, in the rough give and take of politics, his interests and accomplishments were not entirely an asset.[90]

Even Jefferson himself, dedicated to the twin principles of economy in government and strict construction of the Constitution, opposed moves for the establishment of a national university and national museum. Having taken a leading part in the compromise that established the national capital in the small town of Washington rather than in one of the culturally better-endowed cities of the North, he helped to assure that American science and culture could only develop for many years along regional rather than national lines.

To be sure, the picture is more subtle or complex than these positions of his imply—not so much reflecting politically rationalized passivity as, perhaps, a preference for Baconian empiricism. He played a central part in finding ingenious ways, after all, to secure the acquiescence of Congress for Ferdinand Hassler's Coast Survey and for the Lewis and Clark expedition. Under the latter, for an artfully modest $2,500 appropriation,

> three important precedents at once arose. The Congress, usually so reluctant to grant a penny for science, opened the purse on the authority of the commerce clause. The members must have realized that by using army funds they were also blessing scientific exploration under the military powers. In addition, they authorized the expedition to leave the boundaries of the United States to explore foreign soil. . . .[91]

Having resisted George Washington's proposal to establish a service academy as unconstitutional in 1793, Jefferson's support of the founding of the Military Academy at West Point in 1802 hints at another complex blending of motives. The Academy's initial stress was heavily on engineering, preponderantly influenced by French military traditions and in fact dependent in the early years on French textbooks. Engineering officers long continued to have charge of it, and its graduates, encouraged to return to civilian practice as civil engineers, became the first such cadres to have been trained in their own country.[92] Civil engineers were even more essential than military officers for the growing demands for infrastructure and for the westward expansion of the new country.

In 1816, even after the graduation of a number of early West Point classes, the engineering profession in the United States was still in its infancy. An estimated thirty "engineers or quasi-engineers" were all that was then available. Demand spiraled as canal and railway construction got underway on an increasing scale, but fortunately these projects could largely generate the needed supply through their own recruitment and training activities. Two thousand civil engineers were recorded in the census of 1850, the first to enumerate the new profession.[93]

Similarly casting a long shadow into the new century was patent policy. The Constitution had provided for the issuance of patents "to promote the Progress of Science and the useful Arts," an ambitious phrase that for a time seemed to lend encouragement to subsidizing research under this sanction. Unsurprisingly, funds were not found to take advantage of that possible opening. The first patent law followed in 1790, introducing ill-thought-out administrative arrangements that presently devolved into almost automatic approval, ill-kept records, and consequently endless litigation. Jefferson's attitude had been once again complex and equivocal. While opposed to all monopolies, he in the end became reconciled to encouraging the inventors of real novelties—so long as those novelties had commercial possibilities and were in no sense contributions to science. Meanwhile the number of patent applications began to rise, exacerbating the problem.

Patenting in the United States was at first elastic, declining with heightened foreign competition and rising in periods of domestic business expansion. Numbers of patents awarded accordingly rose during the 1790s, and the rate of rise accelerated during the trade interruptions of the Napoleonic Wars and the embargo that was imposed in 1807. At the end of the War of 1812 there was a corresponding, slow contraction as international trade re-opened. Then there was a steep increase once again during the return of prosperity in the 1820s. More so than in Britain, inventive activity in the early years of the new republic appears to have been volatile and demand-induced. Seeking to explain this contrast, the investigators reporting the difference reasonably suggest that both the comparatively backward technological starting point in the United States and the correspondingly greater fluidity of specialization may have been contributing factors.[94]

It took a good many decades to overcome resistance to centralization in any form in order to straighten matters out. As with the dismal fate of so many other measures on which he sought to effectuate and extend Jefferson's views, John Quincy Adams was fiercely blocked in his attempt to reform the administration of patents in 1826. With the problem continuing to deepen, however, ten years later that reform "appeared a simple and obvious necessity." Fundamental changes were introduced, including a systematic review of applications for novelty and usefulness by a staff of technical experts. While this led to the denial of a significant proportion of them and to a corresponding, temporary drop in the number of applications, it also prepared the way for patent policies that would provide a genuine encouragement to inventors and a stimulant to the new industries that before long would come to depend on them. From 437 patents granted in the year following reform, the total had risen to 993 by 1850 and 4,778 by 1860.[95]

There were a number of other indications of a slow and grudging but fundamental shift of attitudes in the 1830s. All were introduced as practical, ad hoc measures without reference to the new departures in policy they also embodied. Congressional concern over boiler explosions, for example, led the Secretary of the Treasury to contract with the Franklin Institute in Philadelphia to collect information and conduct experiments on the subject. The United States Exploring Expedition was sent out under Lieutenant Charles Wilkes to conduct extensive mapping and other studies in the Pacific under a purely scientific rationale, "to extend the bounds of human knowledge." The Coast Survey was revived by Congress under civilian control in 1832 after a lengthy period of desuetude under Navy auspices. Ferdinand Hassler, its competent but irascible and politically troublesome head, finally had to be replaced in 1843. But fortunately he was succeeded by an outstanding scientific and engineering administrator, Alexander Dallas Bache (1806–1867). Bache, who had also chaired the Franklin Institute during the boiler study and founded the first journal of civil engineering there, was able to develop it from that time forward into the dominant institutional influence in government-supported science.[96]

> Chronometers and mapping led to celestial observations, which led to astronomy. Compasses led to terrestrial magnetism. Natural resources led to geology and natural history. . . . relatedly, *ad hoc* organizations tended to become permanent or require permanent services. The Coast Survey and the Wilkes expedition had set off chains of scientific demands the end of which, much to the consternation of Congress, was not even remotely in sight in the early 1840's. . . . although technological change was beginning to create a sporadic need for government sciences, it was surveying and exploring that set the tone of activity in a period of rapid geographical and commercial expansion.[97]

Closely attuned with these developments at the federal government level were trends in higher education. "By the mid-1840's the stress on practical-

ity produced schools of science at both Harvard and Yale in which a dominant theme was the utility of the sciences."[98]

What would become the Smithsonian Institution was beginning to receive prolonged consideration by the Congress already in the late 1830s.
Funds for this unprecedented—and, in fact, still unparalleled—venture
came to the United States under the will of James Smithson (1765–1829),
an Englishman who had never visited this country. Little was known of his
intentions for the bequest of approximately a half-million dollars (but of
course with a current value many times higher), other than that it was to
support the work of an institution dedicated to "the increase and diffusion
of knowledge among men." Debate in the Congress touched on a wide
array of possibilities, including an observatory, a national university, and
a library, before settling on a fairly open-ended cataloging of possibilities
that included a library, natural history exhibits, a chemical laboratory, and
a gallery of art. Also worked out was a unique mode of governance for it,
with its ruling Board of Regents to be composed jointly of representatives
of all three branches of the federal government as well as distinguished
private citizens.

Joseph Henry (1797–1878), a professor of physics at Princeton and perhaps the leading physical scientist in the United States in his time, was
selected to serve as the Institution's first secretary. Especially noted for
discoveries in the field of electromagnetism, his name is appropriately preserved as the international standard unit of electrical induction. He quite
explicitly and self-consciously identified himself as a discoverer of scientific truths, rather than as an individual concerned with applying that
knowledge for useful ends. Inevitably, therefore, the course he initially
charted for the new Institution was primarily concerned with the pursuit of
basic research and the international dissemination of its findings among
fellow professionals.

Given that emphasis, as well as the modest scale of its activities permitted by the income from its endowment, the Smithsonian of Henry's time
plays at most a marginal part in an account dealing with the role of technology. Indirectly, to be sure, there is no denying that it played a significant,
if largely symbolic, role. One of its functions involved the additional support Henry was able to bring to the promulgation of exacting scientific
standards. Another was the public affirmation of a commitment to basic
research, in the face of considerable congressional and public skepticism.

But Henry's eminence as a scientist, and the importance of the position
he later came to assume at the Smithsonian, make necessary some reference to the strength of his views on the science–applied science distinction.
Henry claimed credit for having discovered the basic scientific principles
embodied in both the telegraph and electric motors, moralistically and
sometimes rather patronizingly deprecating the contributions of those who

ultimately received credit as their inventors by bridging the very large gap between his "philosophical" prototypes and practical applications. And while enthusiastic from the outset over the potential utility of the telegraph, Henry failed to foresee any other possibility than a galvanic cell for generating electrical current and hence dismissed the electric motor as a useless novelty that never could compete with steam for the generation of motive power.[99] Fundamental to his views was an insistence on the priority of the guiding hand of the theoretical scientist, on "a one-way relationship between science and technology: theory gradually worked its way down to application."[100]

We need to take note of the fact that this position was powerfully, if not widely, articulated already at a time when the attainments of American science were still comparatively modest. It is a position that has been repeatedly re-asserted in the history of American science and technology, most prominently in Vannevar Bush's influential 1945 report to the president, *Science, The endless frontier*. Were the Bush report (and of course also Henry's views) a balanced and accurate representation of the nature of the relationship, there might be no place for a book like this. Changes in technology could then be dismissed as essentially derivative and without interesting characteristics of their own.

The telegraph, in any case, became a further indicator of the softening, although by no means the disappearance, of congressional opposition to federal support of at least some ventures in science and technology. Samuel F. B. Morse was able to demonstrate the efficacy of his invention to the House Committee on Commerce in 1838, and finally, four years later, secured an appropriation allowing for a Baltimore-Washington telegraph line to be installed and tested. This further successful demonstration notwithstanding, funds were not forthcoming—at least from federal sources—for the line to be extended to New York. As Dupree concludes, "the uncertain constitutional position of internal improvements generally was responsible for the feeble start and the abandonment of this policy."[101]

We must return to Alexander Dallas Bache as a leading and, in many respects, representative figure in the slowly gaining rapprochement of science and government. Admitted at the age of fifteen to West Point, he subsequently remained in the army for a three-year period of service, first as an assistant professor there and then as a lieutenant of engineers, resigning afterward to become a professor of natural philosophy at the University of Pennsylvania. Outstandingly well connected politically, he was a distant descendant of Benjamin Franklin.

Bache's scientific interests were especially directed toward terrestrial magnetism and meteorology, but his accomplishments in these directions were relatively modest. It was as a creative and broadly effective scientific organizer and administrator that he made his primary contributions. He also carried on educational research in Europe during the 1830s and organized a model system of public education for Philadelphia upon his return. As noted earlier, the Coast Survey both grew and prospered under his exemplary direction, receiving the largest budgetary allocation yet devoted to a scientific bureau. Not least, as a regent of the newly organized Smithsonian Institution, he contributed to maintaining its standards and shaping its policies. His was probably the guiding hand in the selection of Joseph Henry as its first secretary.[102]

As this account suggests, Bache was indeed a pivotal figure in a time of change. Bridging the roles of scientist and engineer, he moved smoothly back and forth throughout his career between the conduct and advocacy of basic research and the taking of a high level of responsibility for its practical applications. His Coast Survey budgets during the peak years in the 1850s dwarfed those of the Smithsonian by a factor of more than a dozen times, nearly trebled those of the railroad surveys, and greatly exceeded even those of the Wilkes expedition. Before the Civil War it was, as Dupree rightly judges, "the best example of the government in science."[103] But to that it is not unreasonable to add engineering as well.

Finally, it should be noted that Bache was the leading figure in an informal but highly influential group of (mostly) physical scientists known as the Lazzaroni. Devoted in their own minds to raising the standards of scientific professionalism in the United States, they were widely perceived by others as a factional, self-aggrandizing, and prematurely centralizing influence. Bache himself quite naturally became the key organizing figure and first president of the National Academy of Sciences upon its chartering in 1863, but it would be some time before that title was invested with the unchallenged content it claimed.

The election of President Lincoln in 1860 and the oncoming Civil War that it signaled provided an historical conjuncture in scientific and technological terms as in many others. As Hunter Dupree observes, in spite of new departures like the Coast Survey, the exploring expeditions, the Naval Observatory and the Smithsonian Institution, "the United States still hesitated to embrace the theory that the government should have a permanent scientific establishment. The concept of the Union as a federation of states was still a powerful argument against a forthright commitment to the support of science. . . . The attempts of the government to use science in regulation and aid of technology had been timid and intermittent."

Yet the new departures, "theoretically temporary and disguised institutions" though some of them had to be, would prove to be forerunners of a slow but inexorable pattern of federal patronage.[104]

More especially with reference to technology, the continuing—although soon to be compromised—preeminence of sectional interests and rivalries left initiatives in most areas almost entirely within the hands of private entrepreneurs. Only three salutary instances of federal encouragement of industrial development stand out sharply against this prevailing, only slowly abating passivity. Originating earliest was the slow refinement of what later became enshrined as the American system of production in military armories. Both a practical set of near-term process innovations and a distant vision of industrial mass production based on the rigorous attainment of the standard of complete interchangeability of parts, it would ultimately constitute the fundamental contribution to America's technological coming-of-age. The second was the reform of the administration of patents, its importance well attested by the more-than-tenfold rise in the number of patents annually issued over the two decades prior to the Civil War. And finally there were the many forms of direct and indirect encouragement given to the headlong growth of American railroads. Although full achievement of the objective was once again distant, only in this way would the potentialities of a vast internal market for American manufacturers later be realized. Isolated as they were, these were no small achievements.

6

The United States Succeeds to Industrial Leadership

Impacts of the American Civil War

The Civil War defines the beginning of a new watershed in American economic and technological history, a thunderous entry upon the uncharted landscape of modernity. In an unexpectedly sanguinary, all-out struggle, its logic and momentum forcefully submerged, without entirely displacing, earlier, regionally concentrated opposition to growing national integration. Similarly anticipated by few at the outset was the abolition of slavery. Meanwhile, having made a major contribution to victory, the North's superior industrial base and logistical mastery helped to precipitate a confident new, national pursuit of international industrial and technological ascendancy.

So much is common knowledge. Most historical accounts of this great national turning point concentrate on the gathering political tensions that led to it, the conduct of military operations, and the impacts on the South of the era of Reconstruction that followed. However, more relevant for our purposes were the ensuing extensions of national authority by statutory enactment and constitutional amendment. With the Confederacy's defeat, strict constructionists and advocates of the primary rights and powers of the states had permanently lost key citadels of their strength.

So of primary concern here are a different set of outcomes, focused primarily instead on the victorious North. The contrast is overdrawn, but is also not without some underlying accuracy, between the plantation-based leadership of the South that retained an important national role only until the war and the more complex and dynamic amalgam of business leaders, a growing industrial work force, and the yeoman farmers of the North and West. In particular, the general-welfare clause of the Constitution, its favorable implications for federal government initiatives previously encountering entrenched opposition and hence largely unutilized, now could take on a more active meaning within the coalition that formed the Union and retained its primacy after the Union victory.

Commencing over a sectional clash that had dominated and constrained political life almost since the founding of the Republic, the Civil War thus had the ironic effect of helping to clear the way for the requirements of a modern industrial state. The reduced importance of traditional sectional balances meant that the federal government's intervention and assistance could be more openly sought as an investment in national unity and economic growth that would benefit all groups in society. With population rapidly flowing into national territories that had bridged the continent, increasingly urgent needs could be federally addressed for a more effective fabric of administration and communications. As an enormous railroad transportation grid rapidly materialized, the prospect of an integrated national market of unprecedented size was sure to stimulate correspondingly ambitious plans for industrial and commercial growth. Although Alexander Hamilton's *Report on Manufactures* had been simply ignored by the Congress when it was submitted in 1791, his far-sighted advocacy of affirmative support of industry by a strong national government thus began to receive a belated measure of vindication.

Along with the historic legislation associated with the emancipation of slaves, the 37th Congress passed a series of measures embodying this outlook. Considered together, they may well deserve the label "revolutionary."[1] Common to these measures was a readiness to treat portions of the vast public domain in the still largely unoccupied West not as a permanent asset but as a strategic resource—using land grants as an incentive to attract new settlers, new schools, and the railroad-building activity that would ultimately tie the nation together.[2]

The Morrill Act that led to the establishment of land-grant colleges carried with it the seed of federal support for research as well as education in the field of agricuture. Also established in 1862, with some of the same objectives, was the Department of Agriculture. Its duties included acquiring and diffusing "information on subjects connected with agriculture in the most general and comprehensive sense of that word, and to procure, propagate, and distribute among the people new and valuable seeds and plants."[3] Before long those responsibilities would require the assistance of a staff of scientists to assist in carrying them out.[4] As Hunter Dupree observes, "Although opponents could and did invoke states' rights against federal scientific activity, the outbreak of the Civil War had ruled in favor of Alexander Hamilton's interpretation of the general-welfare clause as clearly as it presaged the triumph of Hamilton's vision of an industrial nation."[5]

No doubt the exigencies of the war effort contributed at least a greater sense of urgency to both of these undertakings. With military recruitment, able-bodied farmers declined substantially in number. Nevertheless, "northern farm production continued its consistent increase even as the war progressed." Hog production is perhaps the most notable example, with the number of hogs slaughtered in Chicago alone increasing from 270,000 per year at the beginning of the war to a level of 900,000 at the end.[6] Federal civil spending in the decade after the war (now including veterans benefits and interest payments on the debt) was approximately twice as high in real terms as it had been in the 1851–1860 decade.

Increases of this magnitude carry a significance of their own. We should not lose sight of the transformative effects of the sheer scale of the Union war effort. During the period of the war itself, federal expenditures rose ten- to fifteenfold in real terms.[7] The army's Quartermaster Department, growing from a skeleton force at the outset and independent of expenditures by other departments on ordnance and troop rations, was by the end of the war "consuming at least one-half the output of all northern industry" and spending at an annual level of nearly a half-billion dollars.[8] While the initial weakening in the ranks of states' rights advocates must be associated with the secession and the congressional acts and constitutional amendments that followed, the irreversibility of the trend away from those views must also owe something to increasing familiarity with, and consequently acceptance of, this previously undreamt-of scale of federal expenditures.

Achieving such increases was more than a matter of simply ratcheting up the scale of existing practices. Intimately associated with them were technological advances associated with railroads and the telegraph, for example. These were absolutely indispensable in facilitating military planning, maintaining command-and-control, and executing logistical movements, throughout an enormous theater of war that was being conducted at new levels of complexity. The experience thus gained would be readily translatable into the increasingly national scale of corporate activities at war's end.

The supplying of army ordnance was technologically conservative but in other respects well managed. Difficulties with fusing impeded what might otherwise have been significant changes in the design and effectiveness of artillery, and repeating rifles, already known earlier and also of great potential importance, were introduced into service only belatedly and on a very limited scale. Relatively new technologies—steam power, screw propellors, armor plating and iron ships—impacted much more heavily on naval warfare. In Dupree's judgment, the Navy's record in supporting these new technologies with research and qualified management was an excellent one: "in no important way did they further the naval revolution, but to keep pace with it was a major accomplishment."[9] Again, it may well be that the power of a successful example of quick and resourceful innovation at a hitherto unanticipated scale was the key lesson that was later passed on to peacetime industrial enterprises.

A qualified appreciation of these scientific and technological aspects of war was no longer to be found either in the White House or at cabinet level. However, three senior scientific administrators who were deeply familiar with the ways of government were fortunately already on hand. Along with Joseph Henry of the Smithsonian Institution and Alexander Dallas Bache of the Coast Survey was Charles Henry Davis, who became head of the Navy's new Bureau of Navigation in the reorganization of 1862. With them science and technology were for the first time represented by "professionals who had emerged at the level of bureau heads. From this unusual historic position they were able to undertake what had always failed or missed the mark before—the coordination of the government's scientific policy."[10]

In short, for all of its long-term consequences, the course of the Civil War was itself unaffected by any major scientific or technological breakthroughs. Many new ideas were advertised for and briefly, rather haphazardly considered. Advice was no doubt often sought from qualified individuals like Bache and Henry on an ad hoc basis. But in spite of the war's considerable duration and the massive mobilization of effort it led to, sustained research efforts of the kind that have become so familiar in our own time did not play a part in it. On the other hand, the chartering of the National Academy of Sciences may reflect some anticipation of that possibility. The Congress took this action early in 1863, while the war's outcome still seemed to hang in the balance. Although not specifying military demands, the charter stipulates that the Academy's unremunerated advisory services could be "called upon by any department of the Government to investigate, examine, experiment, and report upon any subject of science or art." However, the creation of such an advisory body had been an old dream of Bache's, and little was said of national need in the maneuvering that led to its realization. Moreover, the government inquiries directed to the Academy during the remaining years of war proved to be as unrelated to military exigencies as they were scientifically undemanding.[11]

But we must not lose sight of the forest for the trees. Fundamentally incorporated in the conduct of the Civil War—one might perhaps even say in its outcome—were recent advances in telegraphic communication, railroads, and shipbuilding. Placing enormous demands on the existing technological base in agriculture as well as industry, it called forth prodigies of large-scale planning and management. There is little doubt that national policies, interests, and networks of distribution would have presently displaced sectional ones even without the years of protracted conflict and destruction. But it seems equally certain that the war not only accelerated the process but placed a decisive stamp on the directions it subsequently took.

Trends in Industrial and Corporate Growth

Growth is best understood as a long-term process, even if the technological stimuli that spur and help to influence its course tend to occur in punctuated bursts. But does sustained growth act to stimulate those irregular technological impulses, or does the causal arrow fly in the other direction? The relations between the two are, in either case, of fundamental importance. Hence we turn first to a brief overview of the cumulative results of industrial growth across the long period from about the time of the Civil War until World War II.

Real national product increased nearly seventyfold in the United States between 1840 and 1960. Population growth alone made a substantial contribution to this, however, the aggregate having increased by about 10.5 times while the size of the labor force increased nearly thirteen times. Product per worker made a relatively more modest contribution, having increased over five times while per capita real income is likely to have risen more than six times.

Simon Kuznets, who is responsible for these calculations, found it "puzzling" and "somewhat of a surprise" that this rate of growth of productivity was "no higher than in the large European countries (except, moderately, compared with England) and in Japan, despite freedom from destructive impacts of the major wars." Calling upon some distinctive features of the American experience, he was able to offer only some suggestive and plausible, but as yet untested, hypotheses by way of explanation.[12] What this clearly highlights is a comparatively modest but impressively long-sustained performance, receiving great reinforcement from immigration and native population growth but relatively little advantage from some group of special technological stimuli.

The abundance and diversity of natural resources within the continental United States adds to the puzzle of a relatively modest rate of growth of productivity. There is much to suggest that rich resource endowments should have provided an additional, powerful incentive to approaching industrial ascendancy in the late nineteenth century, although the reported rate of growth hardly corroborates their overall impact. It is clear, in any case, that by early in this century the United States had become by a wide margin the world's leading producer of copper, coal, zinc, iron ore, lead, and other minerals at the core of industrial technology for that era, in addition to oil and natural gas.

> In an era of high transport costs, the country was *uniquely* situated with respect to almost every one of these minerals. Even this understates the matter. Being the number one producer in one or another mineral category is less important than the fact that the *range* of mineral resources abundantly available in the United States was far wider than in any other country. Surely the link between this geographical status and the world success of American industry is more than incidental. . . .
>
> Resource abundance was a background ingredient in many other distinctively American industrial developments.[13]

Gavin Wright goes on to show that the preeminence the United States attained in this respect was more a matter of its great geographical extent, and of its early development of these natural resources and the efficient

TABLE 6-1

Percentage Shares of Agriculture and Industry, Value Added in
Constant Prices

	1839	1859	1889	1919	1949
Agriculture	78	61	41	22	16
Industry	22	39	59	78	84

Source: Gallman and Howle 1971: 25–26.

transportation that made them accessible, than of an especially rich resource endowment on a unit area basis. It also should be noted that a nation's possession of natural resources has tended to decline steadily in importance over the last century or more. Today, with the high and growing
substitutability of synthetics and with supposedly scarce natural resources
tending to become commodities that move freely and more cheaply in a
unified world economy, their domestic availability confers a less striking
advantage on our economy. Energy imports, to be sure, offer a significant
exception to this trend.[14]

No less important than the long-term, aggregate gains referred to above
were progressive changes in the basic character of the producing economy.
These are given in table 6–1.

As table 6–1 illustrates, it was during the early 1880s that the United
States became preponderantly an industrial rather than an agricultural
country. Also by 1880 or so, it surpassed England as the world's leading
industrial power. But more important than this conjunction of the mid-
points of trends more than a century ago is the fact that both trends have
continued on the same course ever since.

Other indices corroborate the unleashing of a wave of rapid industrial
growth during the decades following the Civil War. The number of steam
engines in industrial use doubled between 1860 and 1880, for example, and
doubled again in the following twenty years before reaching a maximum in
1910. Since water did not begin to be substantially displaced as a source of
industrial power until toward the end of the nineteenth century, these numbers are a useful surrogate for the early rate of growth of investment in
industrial plant.[15]

A further index reinforcing the case for industrial growth measures
changes in the proportion of workers employed in various sectors of manufacturing. From 1870 to 1910 there were broad declines in the percentages
devoted to various categories of consumer goods, and corresponding, even
more substantial increases in the percentages for durable consumer goods
and producer goods. Especially significant was the aggregate decline for
cotton and woolen textiles from 16.85 to 10.88 percent, and the corresponding doubling for iron and steel products from 7.58 to 15.19 percent.[16]

TABLE 6-2
Percentage of Steel Produced by Various Methods,
1870–1940

	Bessemer	Open Hearth	Crucible	Other
1870	54.5	1.9	43.6	—
1880	86.1	8.1	5.2	0.6
1890	86.2	12.0	1.7	0.1
1900	65.6	33.4	1.0	—
1910	36.1	63.2	0.5	0.2
1920	21.1	77.5	0.2	1.2
1930	12.4	86.1	—	1.5
1940	5.5	91.0	—	2.6

Source: Niemi 1980: 156, table 10-2.

The same pair of opposing trends continued, and even accelerated, until the onset of the Great Depression in 1929.

Given the closeness of the association between iron and steel products and both the production of producer goods in particular and industrial plant expansion in general, we need to look further into the increasing efficiency and—at least for a time—ruthless driving-down of costs in the iron-and-steel sector. Making this especially noteworthy is the fact that increasing labor productivity was nevertheless accompanied by a marked increase in employment. Explaining this in part was the progressive replacement in steel production of Bessemer converters, which represented an enormous advance at the time of their introduction in the 1850s, by open hearth furnaces, the next major stage of technological advance (see table 6–2). Not only more efficient, open hearths offered the even greater advantage of controllability of the product's uniformity and performance characteristics as demand began to rise for these new and increasingly valuable requirements in steel products.

There were also economies of scale as corporate mergers led to integrated operations of blast furnaces, rolling mills and finishing mills, in giant works capable of producing a full array of specialized steel products.[17] Most of these new plants were located west of the Alleghenies and in the Midwest, accessible to economical Great Lakes shipping of Mesabi iron ore from Minnesota and closer to great bituminous coal reserves than to the anthracite of eastern Pennsylvania. Hence the shift to bituminous coke as the primary source of heat and energy for steel production, largely complete by early in this century, was another significant contribution to greater efficiency.[18]

Still other advances in productivity and efficiency took place simultaneously, based on improvements in the design of plant layouts, greater attention to maintaining throughput at capacity levels, and intensified use of energy. In this manner, Andrew Carnegie reduced his costs for producing steel rails from nearly $100 a ton in the early 1870s to $12 in the late 1890s.[19] As Peter Temin has vividly recorded:

> The speed at which steel was made was continually rising, and new innovations were constantly being introduced to speed it further. Steam and later electric power replaced the lifting and carrying action of human muscle, mills were modified to handle steel quickly and with a minimum of strain, and people disappeared from the mills. By the turn of the century, there were not a dozen men on the floor of a mill rolling 3,000 tons a day, or as much as a Pittsburgh rolling mill of 1850 rolled in a year.[20]

The essence of Carnegie's strategy, as Chandler records in detail, was aggressively competitive and investment-oriented. His policies were to invest heavily in growth and the substitution of energy and machinery for human labor, to integrate backward into ore extraction and forward into steel fabrication in order to capture economies of management and transportation, and further to drive down his costs (and hence steel prices) by operating his mills "steady and full."

As a strategy dominating the industry, this was replaced by an oligopolistic trend after Carnegie was bought out by J. Pierpont Morgan in 1901. An ensuing series of mergers and acquisitions led to the formation of the United States Steel Corporation as essentially a holding company of hitherto unprecedented size. Its policies, no doubt heavily influenced by the possibility that federal antitrust action might be directed toward the breakup of the firm, turned substantially away from the pursuit of market dominance through growth and what might be characterized as predatory pricing. Quite to the contrary, higher, "artificial" prices for steel products were instituted and maintained even at the expense of sharp reductions in output and temporary shutdowns. Predictably, this led to lowered productivity, higher unit costs, and higher prices for steel products—but also increased competition within the industry as significant parts of U.S. Steel's market share were lost to newer, smaller producers.[21]

Similar advances in efficiency, leading to corresponding cost reductions, occurred in the production of energy fuels. After consolidating its production of kerosene in three new and efficient refineries, the newly formed Standard Oil Trust was able to reduce the price of a gallon of kerosene from 2.5 cents a gallon before 1882 to 1.5 cents in 1885, while increasing its profit even at this reduced price level to slightly more than one cent a gallon. Apart from improved manufacturing facilities, a major part of these economies once again depended on "steady and full" operation. The large new refineries had to maintain a daily throughput three- to fourfold higher than before, with two-thirds being directed to overseas markets.[22]

TABLE 6-3

Energy Sources as Percentage of Aggregate Consumption, 1850–1950

	Bituminous Coal	Anthracite	Oil	Wood	Gas	Hydro
1850	4.7	4.6	—	90.7	—	—
1860	7.7	8.7	0.1	83.5	—	—
1870	13.8	12.7	0.3	81.2	—	—
1880	26.7	14.3	1.9	57.0	—	—
1890	41.4	16.5	2.2	35.9	3.7	0.3
1900	56.6	14.7	2.4	21.0	2.6	2.6
1910	64.3	12.4	6.1	10.7	3.3	3.3
1920	62.3	10.2	12.3	7.5	4.0	3.6
1930	50.3	7.3	23.8	6.1	9.1	3.3
1940	44.7	4.9	29.6	5.4	11.7	3.6
1950	33.9	2.9	36.2	3.3	19.2	4.6

Source: Niemi 1980: 159, table 10-4.

More generally, as economies of scale were gradually added to those of processing, petroleum products began a long-term trend toward market dominance in replacement of coal (table 6–3).

Across a wide range of industries, developments comparable to those in steel and fossil fuels were occurring in approximately the same period of time. Growing, mass markets, partly as a result of steeply reduced transportation costs, were for the first time making substantial economies of scale an attainable goal. They required in particular, however, massive investments and readiness to accept long time horizons. In addition, they called for entrepreneurial vision to be extended in a number of new directions: toward backward and forward integration, as we have noted; toward improved selection and training of managerial personnel; toward stability of full-production levels as a key cost-saving measure; toward greater and more continuous emphasis on product improvement; toward improved labor relations; and, not least, toward unprecedented attention to marketing considerations.[23]

Price reduction, on the other hand, for the most part was not a dominant goal. Instead, the industrial giants "competed for market share and profits through functional and strategic effectiveness. They did so *functionally* by improving their product, their processes of production, their marketing, their purchasing, and their labor relations, and *strategically* by moving into growing markets and out of declining ones more quickly and effectively, than did their competitors."[24]

The enormous size of the internal American market clearly presented special challenges and opportunities. This is nowhere more evident than in the chemical industry. Still of very minor proportions at the end of the Civil

War, it was largely limited to explosives and fertilizer manufacture. Growing rapidly to meet the demand, it was comparable in size to German industry, the dominant force in international trade, already by the time of World War I. In recognition of the size and rapid growth of the U.S. market, especially in petroleum products, U.S. companies naturally focused on the special problems of introducing large-volume, continuous production and finding ways to take advantage of the potential economies of scale they offered. As L. H. Baekeland, the discoverer of bakelite, said of the transition he had to make from laboratory flasks to industrial-scale production of this synthetic resin not long after the turn of the century, "an entirely new industry had to be created for this purpose—the industry of chemical machinery and chemical equipment." With heavy capital investments for new plant in prospect, fairly elementary laboratory problems of mixing, heating, and contaminant control became much more difficult to handle with precision and assurance of quality. This largely explains the distinctive existence in the United States of chemical engineering as a university-based discipline.[25]

Still, it needs to be recognized that profitability, not absolute market dominance, remained the major consideration. Chandler quotes a revealing letter advocating the policies that the E. I. du Pont de Nemours Powder Company would presently follow. It urged that the company *not* buy out all competition but only 60 percent or so, since by making unit costs for 60 percent cheaper they would be assured of stable sales in times of economic downturn at the expense of others: "In other words, you could count upon always running full if you make cheaply and control only 60%, whereas, if you own it all, when slack times came you could only run a curtailed product."

Thus "in the United States the structure of the new industries had become, with rare exceptions, oligopolistic, not monopolistic. This was partly because of the size of the marketplace and partly because of the antitrust legislation that reflected the commitment of Americans to competition as well as their suspicion of concentrated power."[26]

American railroads deserve to be considered as a final, and in many ways most significant, example of industrial and corporate growth. Alfred Chandler has demonstrated that "the great railway systems were by the 1890s the largest business enterprises not only in the United States but also in the world." Thus they were unique in the size of their capital requirements, and hence in the role that financiers played in their management. Ultimately even more important, because they were widely emulated, were organizational advances in which the railroads "were the first because they had to be," developing "a large internal organizational structure with carefully defined lines of responsibility, authority, and communication between the central office, departmental headquarters, and field units." Railroad man-

TABLE 6-4
Aggregate Miles of Railroad Trackage

1840	2,818	1900	167,191
1860	30,625	1910	249,992
1880	93,261		

Sources: Winter 1960: 298; Schlebecker 1975: 166.

agers became the first such grouping to define themselves, and to be generally recognized, as career members of a new profession.[27]

Railroads had quickly become the central, all-weather component of a national transportation and communication infrastructure. Around them, depending on their right-of-way and facilities, grew up further, closely allied elements of that infrastructure: telegraph and telephone lines and the corporate giants controlling them; a national postal system; and, with frequently monumental urban railroad stations as key transfer points, the new and rapidly growing urban traction systems. Added to this mix by the end of the century was railroad ownership of the major domestic steamship lines.[28]

The total length of the networks of track that were placed in railroad service over successive periods of time (table 6–4) can serve as a convenient index of aggregate railroad growth. Extensions of track naturally accompanied growth in areas calling for railroad services as incoming population claimed new lands for settlement. But no less importantly, the increasing density of the networks reflected the essential role that railroads had begun to fill in the movement and marketing of growing agricultural surpluses, internationally as well as in the cities of the eastern seaboard. The major epoch of railroad expansion came to an end at about the time of World War I.

Albert Fishlow has shown that productivity improved on American railroads at a faster rate during the latter half of the nineteenth century than in any other industrial sector. He adduces many factors that were involved: economies of scale and specialization as trackage lengthened and firms consolidated; industrywide standardization; technological improvements (heavier rails and locomotives and larger rolling stock); and the growing experience of the railroad work force and management. By 1910 real freight rates had fallen more than 80 percent and passenger charges 50 percent from their 1849 levels.[29]

The multiplicity of factors accounting for the marked growth in railroad productivity parallels that in the steel and petroleum industries and is, in fact, a general characteristic of the period. As Robert Fogel has succeeded in demonstrating for American railroads (vis à vis canals in particular), "no single innovation was vital for economic growth during the nineteenth century." For the American economy more generally, he has reached the conclusion that "there was a multiplicity of solutions along a wide front of production problems." This implies that growth was a relatively balanced process, not dependent on breakthroughs brought about by "overwhelming, singular innovations" narrowly affecting only one or a few "leading sectors."

How universally this applies is a separate question. I have concluded earlier that there was indeed a handful of key innovations and leading sectors of growth at the heart of the Industrial Revolution in England, and will presently argue that our current phase of growth is heavily dependent on a few technological innovations like semiconductors and the computer industry they led to. But Fogel's underlying point, to look with considerable suspicion at a "hero theory of history applied to things rather than persons," is clearly an important and valid note of caution in assessing the role of any innovation: "Under competition. . . . Alternatives that could perform the same functions at somewhat greater cost are discarded and escape public attention. The absence from view of slightly less efficient processes creates the illusion that no alternatives exist."[30]

The Advent of Mass Production

At least up until World War II, the refinement of techniques of mass production was probably the most significant American contribution to industrial technology. It was not an unalloyed new discovery, of course; major technological breakthroughs seldom are. Fundamental to it were all of the slowly and dearly won lessons of "armory practice" that eventually gave substance to the American system of manufacture. Then those lessons had to be re-adapted or learned anew by companies needing to demonstrate their competitiveness in order to obtain access to capital markets, as well as to meet the test of the market for their products. But with due regard for its lengthy evolutionary origins, mass production was still a development that was immediately hailed as revolutionary upon its first, full appearance at the Ford Motor Company. And since then it has been consistently and universally recognized as one of the main pillars of effective industrial practice.

Only a wide-ranging economic, social, and cutural history could do justice to the way mass production has continued to unfold over the last ninety years or so. Moreover, it is only part—although a major part—of the dramatic expansions of scale and reductions of cost clearing the way for the flood of new consumer products that have, over roughly the same period of time, found mass markets and led to the creation of whole industries. A suitably balanced—let alone full—account of these developments would detract from this book's principal concern with the major characteristics of technological advance, and would in any case quickly outweigh the longer-term themes on which it concentrates. From this point forward, therefore, attention will be limited to only a few of the most salient features of technological advance and the industrial products or processes embodying them. But by any reckoning, the automobile belongs at or near the head of the list.

In terms of the ensemble of fabricating and assembling methods involved in its manufacture, the automobile was in most respects a linear descendant of approaches developed for other, newly devised consumer durables for which there proved to be heavy and increasing demand. The line of that descent, modified along the way to take account of the vastly greater complexity of the product, is the overly narrow technological theme on which considerations of space force me to concentrate. But the automobile was also the triggering mechanism for a revolutionary economic and social transformation. Virtually every pattern of American life has come to be built around the automobile as a mode of indefinitely flexible, individualized transportation—and, indeed, of self-expression. It would be difficult to overestimate the number of subsidiary, component, and service industries that have the automobile at or near their center.

The automobile has been accurately described as European by invention and American by adoption. Its early development, beginning in the 1890s, was directed exclusively at a small elite market. The key initial advances in small internal engines were made to meet this very limited demand, taking advantage of better-developed European roads. But beginning little more than a decade later in the United States, and taking advantage of its existing infrastructure of industrial technology better able to sustain mass production of complex machinery, the automobile became the material manifestation of an entirely different vision of economical mass transportation.

James Vance has identified what he describes as a series of "great ratchets" in American automobile ownership, linking them to phases of improvement and extension in American city streets and rural highways. Licensed automobile ownership rose from seventy-eight thousand in 1905 to one million in 1912, ten million in 1921, and thirty million in 1937. The earliest phase saw only limited improvement of facilities; the concerted effort on roads began later. But it did also see the introduction of easily serviced, reliable, reasonably powerful, and above all economical vehicles designed to operate even on an unimproved road system. Paramount among them was the Ford Model T, which unquestionably provided the primary stimulus leading to the development of the entire interdependent automobile system.[31] To understand the full magnitude of the transformation it effected, we need to return to the gradually expanding implementation of the American system of manufacture that led the way to it.

Cyrus McCormick, the dominant figure, has a largely undeserved repu-tation for having taken early leadership in introducing the American sys-tem, particularly for substituting special-purpose machinery for hand labor and for producing truly interchangeable parts, into the manufacture of reapers and other agricultural equipment. McCormick farm equipment had a solid reputation for basic design and quality, buttressed by success in competitive tests. But as already noted, the company's success was largely based for many years on its prowess in marketing. Only in the early 1880s, thirty years after he is commonly reputed to have done so, did McCormick finally hire personnel who could bring his mode of manufacturing up to the standard of the national armories where mass production methods had first been demonstrated.[32]

Treadle-powered sewing machines, introduced in the early 1850s, made an earlier and more substantial, but nevertheless still qualified, technologi-cal contribution in terms of interchangeability of parts. Once again, mar-keting prowess had much to do with their vast popular success. Almost at the outset of their appearance, the applicability of armory practice—the most fully developed form of the American system—was quickly demon-strated by a few manufacturers. On the other hand, the Singer Manufactur-ing Company, before long coming to dominate the industry, continued to rely for many years on "what contemporaries called the European approach to manufacture—skilled workers using general-purpose machine tools and extensive hand labor." Finally, as annual production approached a half-million units in the early 1880s, more or less contemporaneously with Mc-Cormick, the company recognized the necessity of adopting the production of fully interchangeable parts as a goal. But as David Hounshell, the lead-ing student of this transition, goes on to record in detail, "to the dismay of many, Singer's production experts found that such a goal was not easy to achieve. Indeed, manuscript records end before there is clear evidence that Singer realized its goal" (pp. 116–23, 328).

Bicycle manufacture proved to be a more significant intervening step along the uphill climb to the successful mass production of automobiles. Armory practice was introduced and perfected for a rapidly growing mar-ket, in what was viewed at the time as its first widespread adoption. Then midwestern producers during the 1880s and 1890s pioneered in the stamp-ing of bicycle parts out of sheet steel, a technique initially regarded as "cheap and sorry" by contemporaries although later it proved to be vital in automobile manufacture. And valuable experience was gained in the pro-duction and utilization of ball bearings and other high-precision parts where ruggedness and durability were essential.[33]

The introduction of Ford's Model T in 1908 represented a new synthesis of manufacturing practices. Well-established lessons of armory practice and pressed-steel construction played their part. Now, however, they were embedded in a more comprehensive approach that extended from radical improvements in manufacturing technology to vehicle design, and even to a heightened sensitivity to marketing requirements.

In manufacturing, there were innovations like multiple machining and new assembly methods, with conveyor systems becoming ubiquitous as labor-savers. These and other new concepts

> remained the industry standard for decades. . . . The spark advance and the torque tube, for example, marked a mating of technologies—electrical control concepts to the thermodynamics of engine design, and the dynamics of torque in motored propulsion to chassis design—that advanced and displaced the conventions of carriage makers. . . . The market implications of the innovations of this period are well illustrated in the case of . . . vanadium steel alloy. . . . [Its application] in engines and chassis components was an important part of the development of a lightweight vehicle. When Ford and his engineers found an alloy that afforded three times the design strength of traditional materials . . . they were released from many old constraints that had been adopted from carriage technologies. . . .[34]

As Hounshell notes, Ford production men, given free rein to experiment, were responsible for "a surprising rate of scrapping processes and machine tools," but more importantly "added fresh thinking about work sequences, tool design, and controlling the pace of work." "At the time, as in hindsight, it seemed that the Ford Motor Company did not want to make money as much as it wanted to build cars."[35]

The automobile industry provides an instructive illustration of an internal dynamic in the production of complex machinery. Technological improvements in one area can stimulate and interact with others, thus tending to preserve a kind of self-generating momentum. While these findings of Nathan Rosenberg seem to implicate technology as a kind of "prime mover," it clearly is so only in a very circumscribed sense and setting. Machine tool features and techniques, for example, not only were developed in response to automobile production requirements but began to borrow and embody innovations introduced for the automobiles themselves.[36]

The introduction of the moving assembly line, developed initially for the assembly of magnetos, was perhaps Ford's decisive innovation. An accompanying measure, successfully compensating for the increased stress the line placed upon the labor force, was the unprecedented incentive of the five-dollar day. In early 1913, first adopting the objective of reaching a mass market, assembly of a Model T required twelve hours and twenty-eight minutes; little more than a year later this had been reduced to an hour and a half. By 1920, in consequence, half of the automobiles in all the world were Model T's. In its heyday, steeply reduced purchase costs as well as escalating sales mark the success of what has been rightly described as a lever to move the world:[37]

Model T Costs and Sales		
1908	$850	5,986
1916	360	577,036
1923	290	1.8 million (+ 200,000 trucks)

Ford's reliance on time and motion studies to analyze work flow patterns and establish standardized work routines clearly was integral to the radical economies that were introduced into the production process. In these respects, the approach closely resembles the doctrines of "Taylorism," made widely influential at the time by Frederick W. Taylor. But there were also fundamental differences, and it is the differences that help to explain why mass production is now worldwide in scope while Taylorism has essentially disappeared from view:

> The Ford approach was to eliminate labor by machinery, not, as the Taylorites customarily did, to take a given production process and improve the efficiency of the workers through time and motion study and a differential piece-rate system of payment (or some such work incentive). Taylor took production as a given and sought revisions in labor processes and organization of work; Ford engineers mechanized work processes and found workers to feed and tend their machines.[38]

All of the most essential principles of mass production are exemplified by the installation of the Ford assembly line. Little would be gained by describing its later diffusion and refinements. With the onset of the 1929 depression, in any case, the mass consumer purchasing power needed to sustain high and growing levels of mass production suffered serious and sustained losses. Assembly lines crept back to life again beginning in the later 1930s. But they did not once again fully express the levels of output of complex machinery of which mass production methods were capable, or the speed with which those levels could be reached, until the country was faced with urgent military necessities as a consequence of World War II.

The earliest full development of the principles of mass production is probably the most salient U.S. contribution to industrialization. As noted, principles of standardization of labor inputs and components, emphasis on the employment of special-purpose machinery, and the maintenance of a high and uniform rate of throughput had their origins well back in the nineteenth century. They were brought together in a remarkable way during the first few decades of the twentieth. Although modified in some important ways that will be discussed in the following chapter, they remain generally intact and viable today.

However, this should not be taken to imply that mass production principles were, or ever could be, applicable in every setting. Large-scale, mass production enterprises form only one end of a very diverse continuum, and this was especially true in the earlier decades of the twentieth century. Nor were all of the enterprises distributed closer to the other end of the continuum weak and struggling. In fact, some of them made significant contributions of their own to advancing technological frontiers. A wide array of products and types of markets simply continued to be best fitted for approaches that are in many senses an antithesis of mass production. Among these were batch work, specialty production involving continuing adaptations and excellence of design, and products requiring a highly skilled, versatile, and motivated work force ready to respond to time-critical, diverse orders with primarily general-purpose machinery.

Locomotives, ships, turbines and generators, machine tools, paper and woodworking machinery, mining equipment, castings, fashion apparel, jewelry, and furniture all are good examples, and "all helped define American technical prowess." Baldwin paid its locomotive workers far higher than European wages, for example, but the worker-years required for building its products were only half those of its European competitors.

Philip Scranton, who has devoted careful study to this industrial niche, does not minimize the harshly uncertain environments within which such firms had to operate: "Unlike their throughput counterparts, batch firms could never sustain the illusion that they could control their market environments, manage technological change, and use formal rationality to make decisions. Theirs was a world of decision making under uncertainty, of extravagantly imperfect information, and of contingencies that pulled them beyond the boundaries of the firm." Yet for a considerable time they contributed significantly to the historical process by which American industry achieved world dominance. Many of the firms occupying this difficult niche not only survived but grew until the Great Depression of the 1930s. Subsequently many of them, or at any rate their functions, were absorbed into more broadly constituted corporate entities. But whether under one consolidated corporate roof or many independent ones, the diversity of manufacturing approaches they practiced remains alive and well today.[39]

The Age of the Independent Inventor

The foregoing account of the increasing scale and productivity of industrial operations is important not only for the impressive duration and cumulative character it documents but for the systemic breakthrough to mass production that was a part of it. These features clearly were major contributors to

America's climb to unchallenged industrial ascendancy from the Civil War era to the Great Depression and the buildup to World War II. But they were not the only contributors. Of comparable importance, at least in the earlier part of the period, were an impressive series of major inventions that permanently displaced the image of the United States as largely an adopter rather than innovator, and that associated American technology instead with a special kind of creative luster.

Emerging from obscurity to occupy positions of great social esteem in the late nineteenth-century was a substantial group of remarkably prolific, characteristically independent, inventors who were largely responsible for these breakthroughs. It was an era when scientific research was only beginning to become identified with universities, and when industrial research capabilities also were still rudimentary. The relative weakness of these other avenues of innovation no doubt helps to account for the public adulation that the group achieved. But as Thomas Hughes's penetrating discussion of them illustrates, it cannot wholly explain either their many similarities in personality and approach or the extraordinary fecundity of their work.[40]

Thomas A. Edison (1847–1931), "the Wizard of Menlo Park," was by all odds the dominant figure as well as prototype of the group. With virtually no formal schooling, already as a telegraph operator in his youth he began to display what became a lifelong penchant for study and, especially, experiment. ("No experiments are useless," he once replied to a critical business associate.) A tirelessly committed, hunt-and-try pragmatist, he dismissed formal theory and was content to rely upon trusted subordinates for mathematical rigor as well as the construction of working models. But the record of his accomplishments, both before and after he established and led a team of co-workers in what became renowned as something close to a manufacturing production line for inventions, was by any measure extraordinary.

The numbers are impressive enough—he received over a thousand patents in sixty years. His expectation for the forthcoming schedule at Menlo Park, "a minor invention every ten days and a big thing every six months or so," was a realistic projection of the pace he had maintained during the early years of his inventive career. But no less impressive was the radical, rather than merely incremental, character of many of his and his staff's contributions. Beginning with an electrical vote recorder, a stock ticker, and automatic telegraph systems, he went on to an electric pen that presently developed into the mimeograph; the carbon transmitter that was fundamental to the introduction of the telephone; the phonograph; an early variant of the incandescent lamp; the "Edison effect" covering his observation of a flow of electric current from filament to plate in incandescent lamp bulbs—the principle at the heart of the later development of radio tubes and hence of an entire industry that would presently follow; and many improvements in dynamos and electric propulsion and distribution systems. Later, in the 1890s and beyond, his attention would be directed still more widely, to magnetic ore extraction, storage batteries, a "kinetescopic camera," and a variety of chemical processes.

Almost mythical as this parade of interests and achievements seems, it is striking that there were others who were nearly as prolific. Elihu Thomson (1853–1937), concentrating on arc and incandescent lighting, alternating current transformers, motors, generators, and controls, electric meters, and high-frequency apparatus, also preferred to work with a small subordinate team isolated from other concerns, and held almost seven hundred patents. Francis H. Richards, Edward Weston, Alexander Graham Bell, and Nikolai Tesla are among many who exemplify roughly into the same pattern. Lee de Forest, Edwin H. Armstrong, and Wilbur and Orville Wright are further, somewhat later exemplars. Not all shared Edison's antagonism to formal science. Some had even received advanced European academic training and hence were familiar with James Clerk Maxwell's and Heinrich Rudolph Hertz's fundamental work on the propagation of radiomagnetic waves. Some, like Tesla, had a taste for overdramatic flamboyance that occasionally carried them outside the bounds of science. But they shared a burning concern to invent. Applying to the group as a whole is Thomas Hughes's remark on the Wright brothers: "Enthusiasm and a vague longing for fame and fortune explain more about their choice of inventing an airplane than do rational economic considerations."[41]

Elmer A. Sperry (1860–1930), holder of over 350 patents, is a variant from the pattern principally in that he became a successful manufacturer at the age of nineteen. Responsible for a succession of impressive achievements in the design and production of arc lights and search lights, he also introduced commercial electrolytic processes and designed and manufactured electric mining machinery and electric motors and transmissions for streetcars. Sperry's most enduring interests and contributions, however, lay in the direction of high-precision gyroscopic control devices. These found application in his introduction of the marine gyrocompass, automatic gyropilots for steering ships and stabilizing airplanes, gyrostabilizers for ships, and during World War I (in connection with his membership on the Naval Consulting Board) gyrocontrols for the guidance of submarine and aerial torpedos.

All in all, the half-century or so after the Civil War saw in the United States an era of unparalleled inventive achievement. With independent inventors in the leadership, there was a tendency

to concentrate on radical inventions for reasons both obvious and obscure. As noted, they were not constrained in their problem choices by mission-oriented organizations with high inertia. They prudently avoided choosing problems that would also be chosen by teams of researchers and developers working in company engineering departments or industrial laboratories. Psychologically they had an outsider's mentality; they also sought the thrill of a major technological transformation.[42]

Carried along with this transformative spirit, and with accompanying industrial growth, was a cultural outlook of sweeping technological optimism and even omnipotence; technology rather than science, for the role accorded to science and scientists was generally a subordinate one. Yet rising mainly from obscure origins and with access to rigorous scientific training only in a few cases, this highly unusual clustering of inventors made an extraordinary contribution to the amenities of modern civilization that we now take for granted.

Foundations for a Research and Development Complex

Technological advance in America changed in character as it approached the modern era. At least in some sectors, of which the manufacture of farm equipment is one, there is evidence of a transitional stage of organized product improvement before the full development of corporate research laboratories. An "inventive profession" was recruited into corporate ranks, composed not of academically trained scientists and engineers but of craftsmen with practical experience in farm-machine shops and a track record of some earlier success as small-scale entrepreneur-patentees. The trend may have begun slowly as early as the 1850s, but from 1870 onward most patents concerned with some of the technologically more complex machine requirements were received by hired professionals.[43]

From the mid-1870s onward, Rosenberg and Birdzell rightly speak of the most promising and active sector of the technological frontier turning away from exploitation of the long-familiar, visible world of mechanical objects and machine components. Beginning to receive attention instead was a hitherto invisible world of electromagnetic forces and microscopic or submicroscopic particles. This latter world offered few immediate attractions for the technological innovator. In order to enter it at all, new levels of organizational support needed to be assured not only for long-term investigative efforts but for expensive instrumentation and laboratory facilities.

Specialization was also becoming more necessary as the knowledge base in science and technology advanced. To an increasing degree, new inventions and innovations came to depend on the collaborative efforts of specialists. Individual inventors played a smaller and progressively declining part; even Thomas Edison stood at the head of a substantial staff and laboratory devoted to the business of invention. Before long a fundamental division of labor would appear between "work done by basic scientists, functioning in what amounted to an autonomous sector of their own, pursuing knowledge for its own sake, and funded by grants and subsidies not directly linked to economic values, on the one hand, and work done by industrial scientists, functioning in the economic sector, and funded on the basis of the economic value of their work, on the other."[44]

To be sure, the autonomy spoken of here was primarily a matter of institutional support structures: universities on the one hand and industrial corporations on the other. Individuals located in the two settings, as well as the lines of investigation they were pursuing, inevitably maintained many lines of communication. The challenge, in fact, was to build a continuous process of information exchange and cooperation across the institutional interfaces. Research and Development (R & D) is the phrase that presently became accepted as comprising the linkages contained in this process and to symbolize its unity.

In tracing the course of this important change, it is useful to return once more to the metaphor of convergent paths. It describes the Industrial Revolution not as a local development in eighteenth-century Britain but a mighty river of transcendent importance. The metaphor is no less appropriate to the history of American technological development in the late nineteenth and early twentieth centuries, especially to the merging evolutionary trajectories of the American scientific and engineering communities.

Civil engineering had crossed the threshhold of census recognition as an identifiable profession by the time of the 1850 decennial census. With the U.S. Military Academy at its origin, presently supplemented by the on-the-job efforts of American railroads, engineering easily became identified with heavy construction, work-force management, and the aspects of engineering most closely associated with architectural design. The near equivalency of engineering and civil engineering remained largely in effect through the Civil War and beyond, while the numbers in the new profession grew from two thousand in 1850 to seven thousand in 1880. The American Society of Civil Engineers was, for example, the only specialized subgroup of engineers to form at a national level during this period.

The following forty years constituted a kind of golden age for the profession's growth. Total numbers reached 136,000 by 1920, largely as a result of rapidly expanding industrial demand. That also accounted for a proliferation of new technical specialties in mechanical, chemical, electrical, metallurgical, and mining fields, each establishing its own professional engineering identification and working to enhance its standards and status. Slower growth continued thereafter, the rate declining most abruptly during the decade of the 1930s. There were 240,000 engineers in 1940.

Perhaps with the additional stimulus of the G.I. Bill following World War II, the number first exceeded a half-million in 1950.

Academic training at first played no more than a very minor part in this expansion. As much or more attention continued to be devoted to craft skills as to scientific subjects, and the advisability of including calculus in the engineering curriculum remained a matter of serious debate as late as 1920. There was an impressive, fivefold expansion in the number of engineering colleges in the last three decades of the nineteenth century. "Only with the twentieth century, however, did the college diploma become the normal means of admission to engineering practice."[45]

As a recognizably free-standing profession of engineering took shape, it tended to codify its own distinctive outlook. Its concern was not with an increasingly accurate portrayal and understanding of the natural world but with a world of its own creation—"with what ought to be, not with what is." Necessarily less discipline-bound in their associations than scientists, engineers saw themselves as enrolled in communities of practitioners whose membership was goal-defined rather than narrowly specified by function. Sufficient flexibility thus was maintained to include within such communities whatever specialties were necessarily engaged in a common class of undertakings or worked together in meeting the objectives of a particular firm.[46]

Among other features, the early engineering canon also placed greater emphasis on measurable entities than on more fundamental but less directly observable ones that were thought to present needless complications. While warmly accepting an experimental approach identified with the sciences, engineers for a long time tended to rely on approximations. Correspondingly little value was attached, at least initially, to rigorous, analytic solutions that would have required more advanced mathematical training. Science itself was not so much rejected as redefined in ways that brought it closer to engineering practice, using

> a sort of "uncertainty principle" to distinguish between "pure science" and engineering. In "pure science," it was stated at the time, . . . a single "positive instance" might suffice. But engineering deals with complicated situations in which the effects "are the joint production of many natural causes and are influenced by a variety of circumstances," so that engineers must generalize "not from single facts but from 'groups of facts.'" In addition, engineering must be concerned with scale effects and conditions of practice.[47]

As time went on, of course, the mathematical sophistication of the engineering profession developed at an accelerating pace. Simultaneously, from the other end of what was increasingly a continuum, the range of practical problems grew for which scientists had discovered elegant analytical solutions. Thus a point was reached where the approach of the basic sciences became a fundamental input in engineering education if not practice.[48]

Engineering, in short, tended to become a science of applications. What were gradually becoming identified as invariant laws of nature were brought into practical harmony with a knowledge of the properties of materials, and with tolerances that the wide latitude of experience indicated were often necessary. Intimate linkages in the form of feedback processes were increasingly recognized between engineering products and insights on the one hand, and the conduct and findings of every form of empirical scientific activity on the other.

Yet there are two serious problems with thus describing what was unquestionably a process of convergence between engineering and the sciences. The first is that it considerably understates the difficulties, irregularities, and duration of the effort that was required. To begin with, engineers increasingly were being employed in hierarchical industrial organizations in which senior management not only lacked scientific training but was slow to realize its potential relevance to their businesses. University-trained physicists, chemists, and mathematicians, by contrast, were at the same time finding that

> there was no significant industrial demand for practitioners in any of the three fields. In the burgeoning electrical industry, the technological innovator of the day was, of course, Thomas Edison, the self-taught genius whose spectacular success was generally taken as proof that in business, college training was not only unnecessary but a liability. Andrew Carnegie recalled that in the early 1870s chemistry was "almost an unknown agent in connection with the manufacture of pig iron." The blast furnace manager was usually "a rude bully, who was supposed to diagnose the condition of the furnace by instinct."[49]

It is also significant that, as a 1924 survey reported, most engineering students were drawn from "the poorer and less well-educated segments of the middle class," with only 40 percent of the students' fathers having finished high school. Yet consistently, their opportunities for advancement led away from further improvements in their technical mastery of their subject matter and in the direction of management instead. Engineering societies, pulled in one direction by the desire to raise their educational, experience, and other standards and thus their status, generally were pulled even more strongly in the opposite direction by the desire to recruit business superiors into membership even in the absence of technical training. Thus it was only as business itself began to discover the utility of more advanced scientific input, in the early decades of the twentieth century, that the leading, business-oriented elements of the engineering professions found it easy to argue for the same shift in orientation within their professions.[50]

But this process never moved to a complete merging of identities. Science may borrow from and communicate freely with engineering, but retains its central emphasis on fundamental entities and processes of change. The focus of engineering—to borrow Herbert Simon's fine title—is on *The Sciences of the Artificial* (1981). The very essence of engineering, as Layton insists, is that "design is everything. It represents the purposive adaptation of means to reach a preconceived end."[51]

Engineering, while converging with science in some respects, thus retained—and continues to retain—much of its autonomy of outlook. That is no less true of what may well be the characteristic engineering methodologies. Parameter variation, in particular, involves "repeatedly determining the performance of some material, process, or device while systematically varying the parameters that define the object of interest or its conditions of operation," the utility of the procedure stemming from "its ability to provide solid results where no useful quantitative theory exists."[52]

By contemporary standards, cadres of recognized professionals were slower to establish themselves in the major scientific disciplines than in engineering. Little time, and less opportunity for laboratory work, was allotted to these subjects in undergraduate college curricula of around 1870. Good graduate schools had not yet come into existence in the United States, so that students desiring to pursue careers in physics, mathematics, or chemistry generally went on to study abroad. Although twenty-five institutions formally awarded the Ph.D. by the mid-1870s, substantially less than a hundred individuals can be identified as seriously active in all three of these fields together before 1880. Virtually the only positions that were open to them lay in college teaching, with little time and less support or facilities set aside for research.[53]

At the same time, universities were extraordinarily quick to recognize that their existing programs of training in academic disciplines and laboratories fell short of meeting the emergent requirements of some of the new industries. The Massachusetts Institute of Technology introduced its first course in electrical engineering in 1882, the same year in which the inauguration of Edison's Pearl Street Station in New York City is conventionally described as having launched the electrical industry itself. Cornell followed only a year later, awarding the first doctorate in the subject in 1885, and by the 1890s a recognized profession had emerged that was supplied with graduates by a handful of major institutions. A similarly accelerated university response led to the emergence of chemical engineering as a recognized profession.[54]

For those with backgrounds only in academic science, public sector employment opportunities, for the most part, materialized somewhat earlier than private ones. Agricultural experiment stations, government agencies like the the Department of Agriculture, the U.S. Geological Survey, and the

U.S. Weather Service, and public health programs at the state and local levels all began to hire small numbers of trained scientific personnel in the 1880s. A few industries in the energy, pharmaceutical, and metallurgical sectors had begun to employ analytical chemists. But for several decades after the Civil War academic institutions continued as the "overwhelming" proportion of opportunities for employment, and "research remained unheard of in industry."[55]

By 1890, the situation had begun to change more rapidly. For physicists, and more especially mathematicians, nonacademic employment opportunities still were not numerous. A number of universities, on the other hand, had succeeded in establishing relationships with local industries that as yet lacked research facilities of their own. Records probably do not exist from which to plot the pattern with any accuracy, but Nathan Rosenberg and Richard Nelson have recently offered their general impression that "a century ago a large share of American university research was very much 'hands-on' problem-solving."[56]

Simultaneously, universities were evolving in the direction of offering some support for their faculties' less immediately utilitarian, basic research, and being recognized as an advantageous setting for it. Moreover, a new situation had begun to present itself in the rapidly expanding electric power industry. Despite the continuing public adulation of Edison and his own tenacious efforts, he was demonstrably losing his battle to prevent the victory of alternating current generation over the direct current networks he had introduced. With this defeat went a loss of opportunities for many self-taught technicians or even empirically trained engineers to retain their former leadership against academically better-trained counterparts.

The prodigously growing utilization of steel provides perhaps the earliest example of industry's discovery of the need for growing numbers of trained scientists. Bessemer converters of the 1850s had opened the door to relatively cheap and plentiful steel production. "One part fuel in the Bessemer process equalled 6 or 7 parts in the old method of steelmaking, and comparison of labor and machinery requirements would yield similar results."[57] But while effective in removing carbon and silicon as impurities in molten iron, phosphorus in any concentration was left as an unresolved problem. Moreover, the process proceeded so quickly, with so little opportunity for intervention or control by the operators, that it was difficult to specify the properties of the resultant product within commercially acceptable limits. A need quickly appeared, therefore, for quantitative chemical analysis.

Low-cost steel became a resource of enormous importance in practically every industrial field. Iron rails of Civil War vintage would support only eight-ton rolling stock before wearing out in a couple of years. The steel rails of 1905 carried seventy-ton cars and lasted ten years. In transportation, in high-performance machinery of all kinds, in steel girders for bridges and the new multistory buildings, in agricultural equipment, and presently in the fabricating of automobiles and airplanes, reliable steel components were a critical constituent. The transition from the Bessemer process to open-hearth furnaces contributed to this end, permitting the use of a wider range of iron ores and facilitating the widening use of alloys that met increasingly demanding specifications. But the essential point is that already by the early years of this century chemical analysis, and hence trained chemists, became an essential part of the process.[58]

Before the turn of the century General Electric had already organized a semi-autonomous consulting-engineering department, essentially an R & D laboratory—under the renowned Charles P. Steinmetz. Here ways were found to prepare tungsten lamp filaments that were ductile rather than brittle, and to achieve important results in the design of radio tubes and other electronic equipment. The American Bell Telephone Company (later AT&T) similarly began to strengthen its engineering department during the 1890s, and soon afterward, faced with eroding patent protection of some of its key products, took steps leading to the founding of the great Bell Laboratories.

From the limited available data, however, it appears that by the early years of the twentieth century the dominant area of industrial research activity lay in chemistry and its offshoots. Electrical machinery and instruments provided less than 10 percent of total industrial research employment in 1921, while by contrast slightly more than 40 percent—and an almost equivalent proportion of all the new industrial laboratories opened during the entire 1899–1946 period—was attached to the chemicals, glass, rubber, and petroleum industries. The industrial demand for organic chemists also swelled as new products and by-products proliferated. Drug, food-processing, and petroleum works identified a new need for organic chemists. The gross proportions are reported to have remained "remarkably stable," modified only by an increasing research intensity on the part of the automobile industry, even while there was a tenfold growth in total employment at industrial laboratories between 1921 and 1940.[59] Already by the turn of the century, the president of the American Chemical Society could claim that "many of the more important [industrial] works have corps of chemists numbering from 10 to 50, while very many more have smaller numbers."[60]

Du Pont's consistent policy of investing heavily in R & D laboratories, employing numbers of professional chemists that were a full order of magnitude higher than even the larger of those suggested earlier, followed by the time of World War I. Out of this complex presently came the firm's entry into the production of tetraethyl lead, cellophane films, freon refrigerant, and rayon. Later still, Du Pont was responsible for the major commercial successes of neoprene and nylon, although the bulk of Du Pont's other innovations during the interwar period were acquired rather than originating in its own research laboratories.[61]

The inauguration of the National Bureau of Standards in 1901 opened yet another new chapter. In an important sense, it followed in a pattern previously established by the Bureau of Mines and the Bureau of the Census, a pattern that would again be maintained in the Pure Food and Drug Act of 1906 and the National Advisory Committee for Aeronautics that was established in 1915. Part of its responsibilities could be characterized as basic research, or at least consisted of measuring physical and chemical constants, testing instruments, accurately setting standards, and otherwise lending support to the infrastructure of basic research. But at the same time, it was expected to answer the mundane queries of other government agencies and to deal with the needs of industry. It was, in fact, a manifestation of a new readiness of the U.S. government to seek "answers to those problems that industry needed to have solved but was unable or unwilling to answer for itself. With industrial research still in its infancy, the new bureau performed its functions for the building trades and mineral and petroleum production in the same way the Department of Agriculture served the farmer."[62]

Thus far in this section I have concentrated on changes in the different, relatively independent institutions and groups in which scientists and engineers worked, and around which growing numbers of them assumed their professional identities. However, we must not lose sight of the larger theme of technological development. To what extent, during the late nineteenth and early twentieth centuries, did university-based scientists contribute to the introduction of new industrial products and processes? It is not easy to answer this question. At least until recently, most technological advances have tended to draw on long-established pools of generalized scientific understanding, rather than following up quickly and directly on new findings. But apart from providing an increasing proportion of the trained scientific investigators that industry began to recruit, it must be said that a science contribution originating with U.S. investigators who were identified with academic disciplines was not of substantial importance.

Major technological breakthroughs occurring in this country, responsible for spawning whole new industries, are familiar to all: the telephone, the automobile, the airplane, electrical power and lighting equipment and appliances, radio and television. Important economies of process and scale also were achieved, many of them representing impressively innovative achievements in their own right. But the contribution of American basic science to these accomplishments remained quite limited until the enormous acceleration of R & D activity that began in industry as well as in the universities at about the time of World War II.

Agriculture Industrializes

Unquestionably by the time of World War II, American agriculture was well on the way to becoming an industrialized pursuit. Eighty years earlier, as another war was beginning to loom, the earliest steps in that transformation were already in evidence. Taken as a whole, however, mid-nineteenth century agriculture still contained a considerable domain that aimed primarily at self-sufficiency and was at least semi-autonomous in values and outlook.

Farms have always needed to be, in an important sense, locally specialized on the basis of highly variable resource endowments—facilities, experience, labor force, capital, and of course all those of the natural environment. But in the eighty-year period of time we are dealing with in this chapter, farm inputs that were no longer self-generated—seed, breeding stock, fertilizer, agricultural equipment, motor transport, energy, and an increasing proportion of the non–farm-related goods and services that farm families consumed—became increasingly indispensable while growing enormously in cost and complexity. The same is true of farmers' dependence on meeting the demands of increasingly time-sensitive, differentiated, and quality-discriminating markets.

As this occurred, the whole chain of processes by which agricultural output was directed to consumers came to represent a large (and today still-increasing) fraction of the total cost of farm produce. Necessarily, therefore, it becomes progressively more misleading to consider the subject of agricultural technology as limited to the farm workplace. Admittedly, it is not easy to establish a more inclusive definition. Moving onward from the immediate scene of crop and livestock production into the almost endless series of more or less intimately associated processes that lead ultimately to consumers makes it difficult to draw any boundaries at all around the subject. But only a more comprehensive approach along these lines can do justice to the profound changes that were underway in which technology was heavily implicated.[63]

It is only reasonable to include within a single, comprehensive rubric, for example, an innovation like commodity exchanges that, beginning with the New York Produce Exchange in 1862, introduced "a general stability in market prices" to the benefit of the farmer. No less relevant was the introduction of refrigerated railroad cars in the 1860s, and the rapid technical advances that began still earlier in the canning industry. Helping once again to enlarge as well as stabilize the market for agricultural commodities were the million cases of canned food that were already being produced by 1860, not to speak of the further, ninetyfold increase in this number by the time of World War I. Continuing further along the same line, how can we

TABLE 6-5
American Agricultural Exports, 1867–1901

	Wheat (bu.)	Corn (bu.)	Beef/ Products (lbs.)	Pork/ Products (lbs.)
1867–1871	35,032,000	9,924,000	128,249,000	902,410,000
1897–1901	197,427,000	192,531,000	1,528,139,000	3,477,910,000

Source: Schlebecker 1975: table 13.2.

fail to note the innovation of grocery chains (specifically, the A & P by the time of the Civil War, and of the principle of self-service in such establishments by 1913)?[64] And how then can we fail to range even further—that is, into packaging innovations, advertising and brand-name promotion, and the regulation of food quality standards?

With the growing scale and productivity of American agriculture, its entry into world markets was still another contextual feature that became central to its prosperity. The declining cost and increasing dependability of rail and ship transport were, of course, an essential part of this development. But it is sufficient for present purposes to record the scale of the change during the last part of the nineteenth century as a simple contrast. As table 6–5 suggests, the enormous growth that distinguished American agriculture must be seen in its rapidly developing, entirely unprecedented dependence at this time on overseas markets. "The increase in the size of the cities of Europe accounted for much of the prosperity of the American farmer."[65]

After 1900 or so, however, a new factor came into play—the rapid growth of American cities as a result of the high tide of European immigration. Agricultural production continued to rise in all commodities, but European markets were supplied from other sources while American consumption substituted for these earlier exports. Newly erected tariff barriers played a part, but the resulting diversion was in any case dramatic. Wheat and flour exports declined from a high of 235 million bushels to 146 million, for example, and fresh beef from 352 million pounds to a mere 6 million pounds. U.S. population in 1860 had been 31.5 million and 20 percent urban; by 1914 it had risen to over 99 million and was 48 percent urban. Herein lay an enormous new market.[66]

Given the unsurpassed diversity and complexity that came to characterize modern American agriculture, it is not possible to detail the full range of technological innovations that became indispensable to its operations. Even without taking into account the growth of increasingly elaborate transportation and distributional networks, there were few farm operations in which specialized labor-saving machinery did not at least begin to play a central part. Three technological features will be briefly described as illustrative of the wider pattern—tractors, fertilizer, and hybrid seed.

Attempts to move beyond animal traction to some form of mechanical motive force were almost contemporary with the clustering of other inventive efforts devoted to agricultural machinery in the 1850s. However, when limited to steam power alone, versatility proved very elusive and size remained an important drawback with regard to most of the routine requirements of farming. A steam plow was introduced by Obed Hussey in 1855, and after a series of incremental improvements steam tractors were introduced in the 1870s. Finding a very modest niche of acceptance, only three thousand or so were being built annually as late as 1890. Threshers were another story, for here large transient crews had already begun to move northward with the harvest, the impressive scale of their machines suggested by the need for as many as forty draft animals. Steam combines of the 1890s and early 1900s, some of them weighing forty-five thousand pounds, developed more than 120 horsepower.

In the meantime, John Froelich had probably made the first gasoline tractor that was an operational success in 1892. However, tractors with gasoline engines powerful enough to replace these enormous steam combines did not come into production until after World War I.[67] Gradual improvements in tractors involved not merely the power train and more broadly adaptable traction systems but a host of mechanisms for hydraulic lifting, power offtake for ancillary equipment, and specialized applications like seeding, mowing, corn-picking and -husking, and hay-baling.[68]

The replacement of animal energy by mechanical energy was not immediate. Mechanical horsepower did rise dramatically between 1880 and 1920, from 668,000 to 21,443,000. But the number of work animals in rural and urban use continued to rise during the same period, almost doubling from 11,580,000 to 22,430,000. Only after 1920 did the the reduction in the numbers of animals employed for traction really begin.[69]

Therefore, as late as the early 1930s the truly dramatic labor-saving that the use of the tractor implies had not yet been realized. The time requirements to raise one hundred bushels of wheat or corn were still seventy hours and 127 hours respectively. By a quarter-century later those requirements would shrink by more than three-fourths. There were still over nineteen million horses and mules in use on farms in 1930. By 1959 these numbers had grown so small that they were no longer included in the Census of Agriculture.[70]

The advent of the use of fertilizers as an integral, large-scale part of virtually all agricultural operations was different in detail—yet substantially similar in its increasingly industrial character and cumulative impact. The temporal and spatial horizons of the farmers who first opened new lands to cultivation in the wake of the moving frontier were, as noted earlier, brief and transitory. Tending to have only rudimentary equipment and limited capital reserves, they exploited the first flush of natural productivity and then moved on. Within a decade or two, they or their successors inevitably began to encounter declining yields as a result of their exploitative depletion of soil fertility.

Appropriate countermeasures were well understood by the early nineteenth century, not merely from English experience but from practices that had been adopted in older settlements along the Atlantic seaboard. Closer to the frontier, however, there was a different calculus. Manuring required a more plentiful supply of animals than was generally available. With the cheapness of land and the prevalence of other uncertainties, it made more sense to enlarge the cultivated area or abandon it than to invest in it by manuring and crop rotation.[71] Only when the supply of new and productive lands began to dwindle, and its cost to rise, was the way cleared for more long-term, capital-intensive measures to renew soil fertility.

One short-term solution to this problem, the importation of Peruvian guano during the 1840s and 1850s, primarily benefited the Atlantic seaboard because of high transport costs. A comparable, extractive approach followed somewhat later on the western plains, utilizing the bones of buffalo killed by professional hunters. Total supplies from both sources were obviously limited and quickly dwindled. By the 1870s eastern farmers had begun to make widespread use of a new mechanical technology in the form of patented endgate manure spreaders.

Manure thus became a valuable resource rather than a noxious waste product, with dairying, stimulated by the rapid growth of urban markets, as a major source of supply. Further technological developments were in turn called forth by this new, increasingly specialized focus of demand. Milking machines, specialized transport vehicles, and bottling facilities all followed in time, but among the first to appear were silos. Now a common feature on almost any agricultural landscape, they were initially a crucial innovation for dairies as a means for curing as well as storing winter feed for cattle.

For the West, with fewer and smaller cities, fewer animals, and large acreages, only chemical fertilizers could supply a fully effective answer after the disappearance of the buffalo. Chemically mixed fertilizers were first manufactured in 1849, and within a decade seven plants were in production. As early as 1870 output had already reached 321,000 tons.[72] Since that time specialized products and applications have proliferated. They can perhaps be exemplified by the introduction of anhydrous ammonia in the 1930s and 1940s. Having to be inserted into moist soil if it was not to dissipate ineffectively into the atmosphere, use of this gas required the invention of entirely unprecedented technologies for mass distribution and application. Nothing could better illustrate the increasingly intimate dependence of farming on industrial suppliers and their networks of local agents, and the increasingly industrialized aspect of farming practices themselves.

The story of hybrid corn is once again different in detail and yet generically similar. Charles Darwin himself had produced plant hybrids and noted that they were distinctive for their vigor. By 1880, within a very few years of Darwin's own published work on the subject, experiments at Michigan State College had already extended the principle to corn and shown that it could be productive of superior hybrid seed.[73] But thereafter the process of development slowed and became regionally compartmentalized.

Even in the Midwest, where experimentation had begun and the demand was best articulated, the practical availability of superior hybrids was delayed until the middle-to-late 1930s. This was inherent in the regionally differentiated nature of the environments in which agriculture was conducted, distinguishing it from industrialization prototypes applying to fertilizers and farm machinery: "Hybrid corn was not a once-and-for-all innovation that could be adopted everywhere, rather it was an invention of a new method of innovating, a method of developing superior strains of corn for specific localities. The actual process of developing superior hybrids had to be carried out separately for each locality."[74]

The process of innovation can be securely linked, not only in this case but also in the employment of tractors, to local demand and consequently also to speed of adoption. In effect, the economics of the spread of innovations accentuated rather than smoothed local disparities. New techniques, it would appear, tended to be supplied first to "good" areas that were oriented toward growth, at the same time often enhancing their superiority in levels of income and other respects.[75]

What is most important about hybrid corn for our purposes is that its superiority is not absolute but highly contingent, depending crucially on an adequate supply of moisture and nutrients: "Thus, the more extensive root system and the aggressive feeding characteristics of the hybrids enabled them, when first introduced, to extract fertility which was inaccessible to open-pollinated varieties. While the technical changes which contributed to the acceleration of fertility removal are not to be deprecated, such advancements relied on a continued supply of fertility, and became powerless once the stock of fertility was spent."[76]

In other words, to realize the improvements in productivity that hybrids promised, further technological enhancements were necessary to assure the replenishment of fertility. Without the simultaneous development of the use of anhydrous ammonia as a replenishing source of nitrogen in the early 1940s, hybrids might well have proved to be only another transitory "improvement" that merely extracted the natural fertility of the soil more rapidly and then left it more exhausted than ever.[77]

Tractors, fertilizer, and hybrid seed must serve to illustrate the general character of the process of agricultural industrialization. Like mass production, it has naturally continued to diversify and grow in complexity, often in unexpected directions, but this is not the place for a fuller account of those details. The examples are broadly suggestive of the pattern that has continued to be followed. It is enough to indicate that the trends toward growth in the scale of farm units, toward a deepening of investments in productive equipment, and toward proliferating linkages with industrial and corporate life all have continued almost inexorably. With regard to all three, the present condition is one that probably would have been utterly unforeseeable three generations ago.

The scale of farm units and the concentration of ownership is still increasing. By the 1980s, twenty-five thousand superfarms (with annual sales of $500,000 or more) accounted for almost two-thirds of total net farm income. Along lines suggested earlier, farming itself had become a relatively minor part of the agricultural mega-system, with on-farm production representing only 10 percent of value added for food. Less than 3 percent of the U.S. labor force found employment in farming, although nearly 23 percent worked in the food production and distribution system as a whole.

The family farm, while still occasionally invoked as an ideal-typical social unit that merges residence and production, is a species imminently threatened with extinction. Some 86 percent of agricultural workers, and 32 percent of all farm managers as well, no longer reside where they work. By 1991 the proportion of the population resident on farms had slipped below 2 percent, and in 1994 the Census Bureau abandoned its enumeration of them.[78] "In little more than a century, we could move from a world where industry depends on agriculture to one where agriculture totally depends on industry."[79]

It was stressed already in chapter 5 that American farmers have never been passive traditionalists with values centering on autarkic self-sufficiency. Readiness to experiment and innovate, and to accept the additional uncertainties of the market in order to enjoy the fruits of technology, have been enduring characteristics ever since the removal of the constraints of frontier life. Advanced scientific or engineering inputs have increasingly become a requisite in more recent times, but for much of the nineteenth century a large proportion of the successful agricultural inventions and major innovations (to say nothing of the innumerable ones that in the end proved impractical) originated at the hands of practicing farmers. Repeatedly made evident is a quality of reverence for good practice and self-improvement. A special status clearly was accorded to originators of new advances, with the expectation that they would be immediately and widely disseminated rather than concealed from one's neighbors for individual advantage.

> Since farmers were too numerous to recognize mutual interdependence in their production decisions it was all the easier for them to recognize their social interdependence through which the knowledge of production techniques was spread. . . .
>
> Toward the end of the nineteenth century a shift occurred toward government-sponsored research and experimentation. It is not at all certain that the Experiment Station system would have had an important effect half a century earlier. So long as farmers would experiment and diffuse rapidly results of their experiments, and so long as scientific methods were purely inductive, there was an advantage to the system of numberless private experiments, none bearing the official seal of Organized Science. . . . To be fruitful, large-scale research required sophisticated statistical tests of empirical results and close attention to theoretical questions of genetics and nutrition. Once these began to be studied, the way was cleared for the great twentieth-century rises in productivity. . . . [B]y some curious social process that guides the destinies of institutions, the Experiment Stations came into full use just as fundamental knowledge developed to the point at which it could be effectively employed.[80]

Government-sponsored research at agricultural experiment stations intensified during the immediately ensuing decades. It led to the successful identification of disease vectors, the introduction and free distribution of vaccines, and the eradication of pleuropneumonia in cattle. This gradual evolution of formalized research programs, carried on in close and supportive communication with farmers in the region of each experiment station, was outstanding not only in the importance of the resultant discoveries but in the rapidity with which they were accepted and put to broad use.[81] One inevitable effect, of course, was to add yet another element of great and growing importance to the network of dependencies that tied agriculture ever more closely to a wider world of external inputs as well as markets.

Science and Technology in World War I and Thereafter

This chapter is framed and punctuated by wars that, directly or indirectly, had a diffuse but apparently massive influence on technology. The Civil War at its outset, as we have seen, broke little new ground in a technological sense. Yet it clearly helped to define a direction of economic, industrial, and consequently also technical growth for most of the half-century that followed. By the time of World War I American science was for the first time coming into its own. The scientific as well as technological contribution to the national militarization efforts was consequently better organized and more effective. Once again, although for a shorter period since the interval of peace was only on the order of two decades rather than five, the newly emerging pattern cast its shadow forward in time. While the technological ramifications of World War II ultimately turned out to be considerably greater, at least the preparatory steps toward war leave the impression of having merely taken up afresh themes that had been temporarily suspended.

The outbreak of European hostilities in 1914 led only by slow and cautious steps to a concern for the part that science and technology might play in increasing the effectiveness of American military forces. The theaters of conflict were remote, and national interests at first did not seem immediately threatened. A neutralist stance was additionally supported by accounts of the unparalleled intensity of the fighting and the heavy accompanying casualties. Considerable popular sympathy for the Allied position was reinforced by the loss of American lives in the sinking of the *Lusitania* in June 1915, but formal neutrality remained largely unquestioned as a political stance through the 1916 campaigning that concluded with President Woodrow Wilson's re-election. Not until the German resumption of unrestricted underseas warfare in February 1917 did the issue of preparedness for impending conflict move to the top of the national agenda.

In the meantime, however, some preliminary but significant measures to mobilize scientific and engineering talent were undertaken. First off the mark, in June 1915, was the creation of a Naval Consulting Board, responding to an initiative of Thomas Edison and with him as its chairman. Like Edison himself, it became for the most part a citadel of old-line, independent inventors who thought of themselves as practical men with little tolerance for scientific theory or method.

Some innovative advances of lasting significance had their beginnings in the Board's work, among them improvements in aerial gunsights and bombsights, naval torpedos, depth charges, and aerial "torpedos" that were envisioned as pilotless, piston-engine-driven cruise missiles anticipating the German V-1s of World War II. But unfortunately of greater public salience was what proved to be a fiasco—the Board's call for new ideas or inventions from the grass roots that it was prepared to evaluate. "Of the 110,000 'inventions' sent to the board, only 110 were deemed worthy of development by overworked board committees, and only one of these reached production before the war ended."[82]

While deliberately and outspokenly shunned by Edison, the National Academy of Sciences was by this time highly regarded as an institution, both for the breadth and depth of scientific talent it enrolled and for the cordial ties some of its leaders had developed with government and industry. Sensing the approaching, if still unacknowledged, change of course and at the initiative of George Ellery Hale, its Foreign Secretary, the Academy moved more slowly in a broadly similar direction. It delicately proposed to President Wilson in April 1916 that "in the event of a break in diplomatic relations with any other country," it desired "to place itself at the disposal of the Government for any service within its scope."

As finally approved by Wilson in July (and made permanent by his Executive Order in May 1918), this led to the formation of the National Research Council under the auspices of the Academy. Included first in a very broad listing of its responsibilities was the charge "in general, to stimulate research in the mathematical, physical and biological sciences, and in the application of these sciences to engineering, agriculture, medicine and the other useful arts, with the object of increasing knowledge, of strengthening the national defense, and of contributing in other ways to the public welfare."[83]

Able, committed, and highly respected leadership for the new and very ambitious undertaking was provided by Hale, Robert A. Millikan, and Arthur A. Noyes. Especially noteworthy among its broad objectives was the explicit inclusion of basic research. As Hale wrote elsewhere, "true preparedness would best result from the encouragement of every form of investigation, whether for military and industrial application or for the advancement of knowledge without regard to its immediate practical bearing." Alexander Bache, could he have been on hand, would have been highly gratified that what he had hoped for in establishing the Academy itself in 1863 seemed so close to final realization.

Matters did not turn out that way. Through the intervention of the President's Council of National Defense, also newly formed and more heavily representative of engineering and business interests, the whole emphasis of the National Research Council had to be directed to defense-related work.[84] A pattern of practical advice to the government, as distinguished from the direct support of research, was laid down that has remained largely in place until the present.

Considered in strictly organizational and technological terms, American performance during the nineteen months of U.S. participation in World War I was a substantial achievement. Bringing an expeditionary force of two million men into action required a vast, quickly mobilized industrial and organizational effort. And although much military R & D did not reach fruition until after the conclusion of the armistice in November 1918, substantial progress was made in many directions that would remain important for years to come.

Some of these directions, such as poison gas, we now regard as inhuman even by the almost indefinitely elastic standards of modern warfare, and in any case had no significant later role. Submarine location devices and countermeasures immediately became of great importance again in World War II. But for many other areas of investigation in which the NRC played some role, a broad spectrum of nonmilitary as well as military uses subsequently developed: aerial photography, infrared signaling devices, large-scale production of helium and nitric acid through ammonia conversion, high-quality optical glass, mass psychological testing for illiterate as well as literate subjects, and improvements in the management of nutrition and infectious diseases. Large-scale scientific research, conducted in industrial as well as university laboratories (often on a cooperative basis) and linked to production, established its first firm beachhead under these circumstances.[85]

As was the case also with the Civil War, there are suggestions here that the experience of mobilizing for World War I had longer lasting, if also more diffuse, impacts upon science and technology that were much more significant than any listing of discrete innovations and advances. They involve transformative changes not only in the size and research commitments of major university science departments but in numbers of industrial researchers. Apart from a close coincidence in time, the linkages between wartime experience and what followed during the interwar period are only indirect and suggestive rather than determinate. But a closer look at the effects on physics as an example indicates that, at the very least, the participation of scientists in the war effort may somehow have powerfully accelerated not only the growth of science-in-industry but of basic science as well.

The average number of physicists who received Ph.D.'s from American universities in the years immediately before the war ranged from fifteen to twenty annually. Immediately thereafter it began a consistent, exponential climb, immune even to the Depression, to a level about ten times as high by 1940. A noteworthy feature of the pattern was its highly concentrated character. With striking consistency through the 1920s and 1930s (with membership in the American Physical Society defining the universe of professionalism), the twenty leading departments employed 40 percent of all physics teachers, produced three-fourths of the papers published in the leading journal, and trained nine-tenths of the new professionals. But concentrated or not, the growth was clearly extraordinary.

Not much less extraordinary was the growing number of physicists employed in industry. The proportion was about one tenth in 1913, but by 1920—while the APS membership was itself doubling—the proportion climbed to 25 percent. As was often claimed at the time, this tends to confirm that "the relations between physics and industry in America were revolutionized around the time of World War I. . . . there does seem to have emerged—in America before anywhere else—a new breed of scientist, the combination of physicist and industrial engineer."[86]

To a lesser but still significant extent, this growth was also cumulative. American Telephone and Telegraph and General Electric were the largest industrial employers, together accounting for about 40 percent of the industrial physicists who were APS members. AT&T, the largest, was indeed forced to reduce its R&D budget by over one-fourth during the worst years of the Depression. But increases soon resumed again, and by 1939 the 1930 level had been exceeded by 10 percent.

Making this general pattern of continuity of growth all the more remarkable was a pronounced shift in its sources of funding. No optimistic outcome seemed possible at first. The end of the war was rapidly followed by a removal of subsidies for defense-related research, and soon thereafter by inflation and a postwar recession. Efforts to convert the National Research Council to a federally supported research funding agency were unavailing, and American science had to weather a brief but difficult transition. By 1923, however, philanthropic instrumentalities of the giant Rockefeller and Carnegie fortunes had begun to step into the breach. This substantial private support brings us, in effect, to the birth of the research university as a new and enduring institutional form. Research universities not only have come to occupy a major position in American life today, but have begun to take part in a progressive blurring of the distinction between industry and the academic enterprise.

One component of the massive foundation intervention that initiated this process was directed toward promoting graduate training and research in the social sciences. Under the leadership of Beardsley Ruml, the Laura Spelman Rockefeller Memorial devoted some $20 million to this purpose in the mid-1920s, mostly in direct or indirect support of work at a handful of major private universities. More relevant to the present study were the activities of Wickliffe Rose, head of the General Education and International Education Boards at about the same period. Concentrating on the advancement of basic science, by the time of his retirement in 1929 he had directed the disbursement of $30 million to what were emerging as the preeminent research universities. "Both Ruml and Rose sought to advance knowledge by strengthening institutions, and for that reason they concentrated their support upon the best existing science programs—'to make the peaks higher' was the apt phrase associated with Rose's approach. . . . The overall effect, then, was to enhance the stratification of American research universities."[87]

There is little doubt that the foundation effort played a major part in raising the stature of American science. To the metaphor of raising the "peaks" in a few leading universities may be added another: a rising tide raises all boats. Among the basic sciences, advances in astronomy and nuclear physics were particularly noteworthy. Aided by the recruitment of a number of leading European theorists that began even before the influx of refugees from Nazi Germany, broad improvements in these areas brought the United States up to world standards already during the late 1920s and early 1930s. Unsurprisingly, the proportion of papers crediting external sources of support climbed substantially. With external support, American physicists were enabled "to go on a spree of cyclotron-building in the later 1930s, constructing two or three times as many as the rest of the world put together."[88]

Parallel with this development, and apparently complementary rather than competitive in its claims upon new resources, was an institutionalization of academic research in a number of key fields of engineering. Problem-solving for local industries continued to play a part in this, and recently has even been described as remaining the "hallmark of American university research" up until World War II. More generally, however, the movement maintained its independence from industrial research as well as from the more basic scientific disciplines. Electrical, chemical, and aeronautical engineering were among these "sciences of the artificial," all of them destined to play a decisive part in the hostilities that were shortly to come.[89]

Less directly evident at the time, however, were some more questionable structural and attitudinal changes that the new funding pattern tended to intensify:

> Academic science was largely directed by a tacit oligarchy of eminent scientists who shared a number of ideological convictions: university research should be supported by society because of the unforeseen benefits that basic discoveries would bring; funding should be directed to the best scientists, who would produce the most fruitful results; only scientists of established reputation could determine who the best scientists might be; and, private support was preferable to that from government in order to preclude the taint of politics in these delicate decisions. During the Ruml-Rose era, it suited the purposes of the foundations to operate in a manner consistent with these values. . . . But afterward, it proved difficult to expand support for research under those conditions. . . . By the end of the 1930s it was becoming evident that the privately funded university research economy was not generating adequate resources for the expanding capacity of universities to perform research.[90]

In 1940, on the eve of the major buildup of defense funding in the United States, total federal funding of R & D amounted to less than $75 million. With respect to technology, only in the aeronautical field were outstandingly effective research programs carried forward under largely government auspices.[91] Agriculture was still the major recipient, followed by defense, and total funding of basic research in universities (excluding agricultural experiment stations) amounted to little more than $30 million.[92] After 1940, of course, a great deal was to begin to happen very quickly.

7

The Competitive Global System

AMERICAN INDUSTRIAL technology in our own era differs sharply from its predecessors in a number of major ways. Most apparent, leaving aside its obviously unprecedented scale and complexity, is the rapidity with which it changes, and with which new innovations find additional, unanticipated applications. Beyond the most elementary processing of raw materials, very little of what we produce or how we produce it has remained untouched by radical change over the last half-century.

Of comparable importance is the ascendancy of research, or less restrictively Research and Development (R & D), as a fundamental, ongoing component, no longer merely an adjunct, of an enormous array of manufacturing processes. Harvey Brooks has rightly insisted that "the organized generation of technology and the creation of institutions to manufacture technology" is "perhaps the biggest revolution of all that has occurred in the last generation."[1]

Accompanying the new centrality of R & D is the heightened intercommunication linking many areas of science and technology. Although far from complete, this involves an increasing convergence of institutions as well as individuals—around common problems of broad societal relevance; around increasingly common standards and content in graduate and postgraduate training; around cooperative agreements between research universities and industrial firms; and around substantive synergies that have been found to interconnect sciences applying to the natural world and those that have been developed for the world of the designed or artificial.

Still another major difference is the appearance of a greatly enlarged, actively initiating, sometimes even overtly managerial, federal role. Originating in exigencies of World War II that will shortly be described, this focuses primarily on anticipated military requirements, but nonetheless stimulates an important two-way flow of ideas and expertise with the private sector.

Finally, the new international competitive environment poses many difficult challenges, especially in high-tech areas. Even for large, well-entrenched firms this has become an era of pervasive risk and uncertainty, of painful downsizing and plant closings, and of the threatened loss of formerly secure markets. But it is also an era when firms are rediscovering the art of nimble adaptations to such challenges, and of learning from and improving upon their competitors.

All these themes are inextricably intertwined with one another. None of them has yet run its full course, nor followed a linear historical sequence. This chapter will not so much trace their orderly development as seek to understand where and how, singly and in combination, they leave the state of American technology today.

Pools of Knowledge, Breakthroughs, Linkages, Feedbacks

Ongoing increases in technological scale and complexity, only made manageable by increasing intercommunication and science-technology convergence, are a major feature of the modern era. Formerly, the line of demarcation between science and technology was relatively much clearer and more stationary—a pattern still largely reflected in formal organizational affiliations. But with intensifying intercommunication the boundaries have blurred. The present overlap in activities and outlooks between individuals identified as engineers and scientists is extensive and growing. Much of science and technology are no longer readily distinguishable unless one moves outward toward the opposing ends of what has become a single continuum.

The distinguishing characteristic of those tending toward a technological identification continues to be an element of practical concern or intent as they engage in the deliberate manipulation and alteration of the material world. Those who think of themselves primarily in association with the scientific enterprise tend to be largely occupied instead with the longer-range elucidation of fundamental principles of natural order. But individual activities and allegiances can shift or even alternate frequently, each supplying necessary reinforcement to the other. Like the two strands of DNA, "which can exist independently, but cannot be truly functional until they are paired," science and technology complement and blend into one another.[2]

In the technological realm, immediate objectives largely determine the substances, tools, and methods used, normally with only modest departures from proven, customary practice. Since technological innovation is tied to heavily capitalized productive processes and marketing uncertainties, it tends to be slower, more cautiously incremental than scientific discovery. Even in the highly competitive information industry, as IBM's then chief scientist observed some years ago, technological development tends to be "much more evolutionary and much less revolutionary or breakthrough-oriented." Regularity and precision are also less uniformly characteristic of modern industry than of science. Once again at IBM, we are told that "things are sufficiently complex that much is done by rule of thumb and not by precise knowledge. Many factors enter in; some of them are even cultural." More generally, engineering designs of successful products commonly are holistic syntheses of elements that function well together, but represent perfection in no single dimension.[3]

Again there is obviously a continuum, but technological change tends to be relatively slower, more dispersed, anonymous, and unobtrusive. While there are significant exceptions, scientific advances are more likely to be driven forward by powerful and comprehensive new theories or syntheses. This leads many observers, especially those whose central concern is science, to deny a thought-component to technology—to hold that scientists generate new knowledge that technologists then merely apply.[4] Such was the position taken, for example, in the still highly influential report by Vannevar Bush to President Roosevelt in 1945:

> Basic research . . . creates the fund from which the practical applications of knowledge must be drawn. . . . They are founded on new principles and new conceptions, which in turn are painstakingly developed by research in the purest realms of science.
>
> Today, it is truer than ever that basic research is the pacemaker of technological progress. . . . *A nation which depends upon others for its new basic scientific knowledge will be slow in its industrial progress and weak in its competitive position in world trade, regardless of its mechanical skill.*[5]

Bush's views continue to be quoted occasionally with approval, providing a welcome rationalization for government support of basic research in primarily academic settings. Moreover, there continue to be instances of technological dependence on prior scientific advance to which they apply very well. But over the last half-century of accelerating progress in technology as well as science their inaccuracy as a general characterization has become increasingly apparent. Misrepresenting the importance and role of engineering, the Bush model induces "an excessive preoccupation with technical originality and priority of conception as not only necessary but sufficient conditions for successful technological innovation." And it neglects many influences that have proceeded in the reverse direction.[6]

"Everyone knows that the linear model of innovation is dead," affirms Nathan Rosenberg, a leading historian of technology. "It was a model that, however flattering it may have been to the scientist and the academic, was economically naive and simplistic in the extreme. It has been accorded numerous decent burials." Richard Nelson puts the matter even more strongly, holding that it is at least as correct to see new technologies as having given rise to new sciences as the other way around. But where the overall balance is drawn is less important than the fact that the two have become inextricably intertwined.

Especially but not exclusively in newly emergent, high-technology fields, Rosenberg argues that industrial activity often provides "unique observational platforms from which to observe unusual classes of natural phe-nomena. In this respect, the industrial research lab may be said to have powerfully strengthened the feedback loop running from the world of eco-nomic activity back to the scientific community."[7] Aptly illustrating this is the impact of the introduction of the turbojet upon both science and the aircraft industry

> by progressively pushing upon the limits of scientific understanding and by iden-
> tifying the specific directions in which this understanding had to be enlarged
> before further technological improvements could occur. Thus the turbojet first
> led to the creation of a new supersonic aerodynamics, only to give way to aero-
> thermodynamics as increasingly powerful turbojets pushed aircraft to speeds at
> which the generation of heat on the surface of the aircraft became a major factor
> in airflow behavior. Eventually, turbojet-powered aircraft would reach speeds at
> which magnetothermodynamic considerations would become paramount: tem-
> peratures would become so great that air would dissociate into charged sub-
> molecular ions.[8]

This is not in any sense to deny technology's substantial, longer-term dependence on the pool of scientific knowledge (while also replenishing the pool with discoveries of its own). The pool is drawn upon in two com-plementary ways. First and perhaps most importantly, it enters by way of the formal disciplinary training (and continuing supply of information in disciplinary journals) that engineers and applied scientists receive. In addi-tion, much R & D conducted under corporate auspices consists of widening circles of more or less directed search, iteratively moving back and forth between a firm's existing capabilities and marketing potential on the one hand and the winnowing of further prospects for development on the other.[9]

Another crucial science-technology link, already touched upon repeat-edly in preceding chapters, involves instrumentation. While there are forms of feedback operating in both directions, the substantial technological con-tribution that new developments in instrumentation represent to the prog-ress of science is generally less recognized than scientific discoveries as inputs to technology. In a word, instruments have an orienting and stimu-lating as well as a merely enabling role in the whole course of scientific advance.

Sometimes, not unreasonably, instruments have been termed the capital goods of the research industry, part of the antecedent investment without which the productive work of the industry could not go forward. Increas-ingly over recent decades, the existence of a vigorous, highly competitive industrial sector devoted to the production of scientific instruments has provided an "unexcelled infrastructure" for U.S. basic research as a whole. Among the outputs of the research industry itself is not only the wider use and continuing improvement of existing types of instrumentation but the development of new theories and experiments that lead to the devising of entirely new instrumental capabilities.[10]

Instruments naturally vary in the potential range and power of their applications. As noted in chapter 1, the full extent of their utility and significance is seldom foreseen at the time they are first introduced. Nowhere is this better illustrated than with modern computers, whose versatility in modeling and in solving theoretical problems in many fields was originally unsuspected.[11] But computers also illustrate how the discovery of progressively widening ranges of application has feedbacks leading to fundamental immprovements in the original capabilities of the instrument in question.

Computers have come to be of such central importance in all of contemporary science and technology that their explosive rate of growth is a driving force in many other areas of rapid advance. Characteristically, the major developments in computer architecture and components have taken place in discontinuous jumps:

> Processing power increases rapidly after the introduction of a fundamentally new technology. The rate of growth eventually slows as the technology is exploited to its full potential. Meanwhile, new technologies are incubating, and one ultimately supersedes the others to become the new dominant technology as the cycle is repeated. Under the right conditions, the shift to a new technology can be accelerated, resulting in speed increases of a hundred- to a thousand-fold in only a few years.[12]

The transformative rate of advance of computer capabilities is absolutely without precedent: "supercomputers of today run almost a trillion times faster than the fastest computer of fifty years ago" (p. ix). No less striking is the way in which, as the costs of acquiring and using computer power have been relentlessly driven down by technological progress as well as competition, computers have not only become indispensable for monitoring, control, and information management but have transformed the complexity, sensitivity, and analytical power of all of these activities. Precisely because this effect is so sweeping, computers by themselves must account in no small part for the multiplicity of linkages and blurring of boundaries that today have become dominant characterizations of the relationships between science and technology.

Wartime Federal R & D Support and Its Outgrowths

World War II proved to be perhaps the most significant watershed of changing direction for U.S. science and technology. It was the time of origination of most of the trends just described, and the occasion as well for major, upward deflections along their rising slopes of growth.

A conjunction of factors led to this outcome. It was a war of great destructive intensity, fought on a geographically unprecedented scale. The comparative complexity as well as the enormous quantities of weapons supplied to the Allied forces, and the development of reliable logistics systems to deliver them to distant, overseas theaters, made a very substantial—perhaps even decisive—contribution to the ultimate victory. As the United States was already the most powerful industrial nation at the outset, and provided a continuing base of production wholly removed from the direct effects of hostilities, American capabilities to move quickly into the mass production of complex, technologically advanced equipment in the immediate postwar period obviously received an important stimulus.

It was also a war in which scientific capabilities to meet new and very difficult challenges proved for the first time to be of major importance. Highly successful, mission-oriented science, heavily supported by federal funding under conditions of great urgency, established an unprecedented new pattern that still remains in place in defense-related spending today. Even more important, this federal involvement provided an impetus that led, in the years immediately after the war, to programs of general federal support of basic, undirected research that were quickly institutionalized and have continued to grow.

Initially as a result of war-related needs and subsequently through veterans benefits programs that encouraged advanced training in colleges and universities, federal support of graduate education in certain areas became accepted as a principle. Ultimately, permanent fellowship programs came into existence under the administration of agencies like the National Science Foundation and the National Institutes of Health that concentrated on the training of scientists. Permanent, major increases in the supply of trained scientists and engineers capable of participating in greatly enhanced R & D efforts thus derive at least in part from the war effort.

Prominently included in the flow of refugees from Nazi oppression in Europe that began already before the war was a significant proportion of highly trained professionals. Many of those immigrating into the United States were scientists and engineers, among them some of the outstanding leaders in their fields. After playing a major part in wartime R & D efforts, their contributions in many areas of basic as well as applied research have continued to be of the greatest importance.

Finally, there were a number of specific, war-related inventions and discoveries whose further development in the following decades laid a foundation for decades of American scientific and technological leadership. Most noteworthy are those that laid the foundations for the microelectronics, computer, and nuclear industries. In tracing the origins of the last great technological spurt or impulse—in fact, a cluster of closely interconnected impulses—we thus need to look first to the World War II experience.

The larger organizational entities responsible for the conduct of wartime R & D need only brief description. Dominant in terms of gross expenditures were the Army and Navy, but these tended to focus on immediately prospective procurements and hence were of limited long-term, strategic importance. More directed toward the latter end was the National Defense Research Committee, a blue-ribbon group of eminent, unsalaried experts that was established in 1940. In the following year it was superseded by an Office of Scientific Research and Development (OSRD), better configured for the widening scope of anticipated research activities and for hastening the deployment of the new technologies resulting from them. Under the politically astute, driving leadership of Vannevar Bush, president of the Carnegie Institution of Washington, the quality as well as the extent of the buildup was impressive. Total war-related research had been funded at about the $100 million level in 1940 and by 1945 had climbed to an annual rate of $1.6 billion.[13] NDRC-OSRD's smaller but still-impressive share throughout the war amounted to some $454 million. Most went in large contracts to a small group of the principal research universities and to major industrial corporations that had long maintained research laboratories of their own.[14]

Government secrecy notwithstanding, it was widely recognized at the time that "in the way World War I had been a chemists' war, this was a war of physicists." Some 1,700 physicists were already engaged in defense work in advance of Pearl Harbor, and by early 1942 OSRD's contracted expenditures on physics were four times larger than those in chemistry. The demand for physicists had, in fact, quickly risen to three or four times the aggregate output of American universities.[15]

Two great enterprises commanded the services of physicists: the Manhattan Project, devoted to atomic bomb development, and MIT's Radiation Laboratory, where pioneering work was done on radar and many other wartime applications of electromagnetic radiation.

Little need be said in this context about the Manhattan Project. It was launched with considerable uncertainty over whether a controlled nuclear chain reaction was possible, an achievement first demonstrated at the University of Chicago in 1942. No less difficult, but primarily technological, questions then had to be faced: the extraction of adequate amounts of weapons-grade concentrates of the fissionable materials, and the design of suitable triggering mechanisms.

There were, in a sense, only two wartime products of the vast and exacting effort that was carried on in great secrecy: the uranium- and plutonium-based bombs dropped on Hiroshima and Nagasaki respectively in August 1945, followed very quickly by Japan's surrender. In a truer sense, however, it can be said that the conduct and outcome of the Cold War with the Soviet Union during much of the following half-century was at least a derived product of that extraordinary effort. So also was an abiding popular awe tinctured with deep concern, on a worldwide scale, over the immense and dangerous powers that science and technology had together proved themselves capable of unleashing.

Lacking a comparably enduring "signature" product, the Radiation Laboratory is much less well known today. In its own time and setting, however, it won wide respect as a place of something approaching scientific wizardry. The radar war was a battle of constantly shifting measures and countermeasures, fought with great secrecy and urgency and in constant recognition that any advantage gained was likely to be transitory. Still, credited primarily to the "Rad Lab" are a string of major achievements— among them the development of much less bulky microwave radars, airborne submarine detection units and aids to bombing, and long-range aids to aerial navigation.[16] A facility run and largely staffed by scientists changed the image of science itself by proving of inestimable military value.

The OSRD policy was a linear descendant of the Ruml-Rose position in funding science during the 1920s and 1930s that was discussed in chapter 6. Concentrating federal support on a small circle of research-intensive corporations and distinguished universities assured high performance on challenging assignments. A contrary, more populist position was strongly articulated in the Congress, however, according to which these were elitist policies that failed to take full advantage of the nation's productive resources. By the summer of 1944, with the end of the war coming into view, these differences came to focus instead on what direction federal policies for postwar science should follow. There was general agreement only that the war itself had made the case for massive additional subsidies. But should control over them be retained within an unelected scientific and managerial elite like the OSRD or made a more politically responsive process?[17]

This was the climate in which Bush was successful in securing an invitation from the president to prepare the report promoting his own views on postwar policies for science that has already been referred to:

Bush called for a program of federal aid to basic scientific research and training, along with a civilian-controlled effort in fundamental defense research. Unlike [Senator Kilgore], he proposed no remedy for what he acknowledged were "uncertainties" and "abuses" in the patent system, omitted the social sciences from consideration, and made no mention of distributing the funds in his program in accord with either a geographical scheme or the planning of socially purposeful research.[18]

The underlying issue of principle was whether the country would be well served in its newly resolved support of science to place control in the hands of a small, unelected group of "wise men" and the presumably compliant director they alone would have the power to choose. It was an issue precipitating President Truman's veto of the National Science Foundation legislation as it was initially passed by Congress in 1947, leading to a five-year delay in the NSF's establishment. Ultimately what emerged was a compromise: a National Science Board along lines that Bush might have envisioned, but a presidentially appointed, senatorially approved director. With forty years of hindsight, this pragmatic outcome that probably satisfied no one at the time nevertheless seems to have worked well: ". . . the best conclusion may be some sort of healthy tension between planning and pluralism that never gets fully resolved."[19]

In the meantime, the Navy had already quietly received legislative blessing for an Office of Naval Research (ONR) in August 1946. Soon it was amply funded to move well beyond the apparently restrictive implications of its title into the funding of basic as well as applied research. Astutely recognizing the needs of academic science and moving to solidify its support in that quarter, ONR "encouraged the open publication of research results, stressed its interest in basic research, chose the most flexible possible contract forms, pledged continuing support, and involved scientists in its research planning."[20]

This liberality could not be indefinitely maintained, and arguments for maintaining greater military relevance presently had some effect in altering ONR's project support criteria. But for the NSF, the controversy, the resulting delay, and what seemed to some a compromise of principle were initially costly. Over time, the issue faded: university research became progressively more and more decentralized, with the earlier dominance of a handful of private institutions virtually disappearing; the NSF Board itself became a more representative body; and the oversight and appropriating powers of the Congress gradually imposed an extension of the mandate of the NSF well beyond its original focus on the best of basic research as determined by an autonomous peer-review process.[21] But ONR's success also meant that the creation of the NSF initially received only divided, tepid support from the science community. Failing to receive significant appropriations until the perceived crisis that arose over the Soviet launching of the first Sputnik satellite in 1957, NSF remained for a considerable period only a "puny partner in an institutionally pluralist federal research establishment."

The essence of this pluralism was that "the dominant presence of the federal government in the postwar research economy produced a research system that was heavily skewed toward programmatic ends." The largely mission-oriented budget of the National Institutes of Health, for example, climbed from less than $3 million at the end of the war to more than $50 million by 1950. In aggregate, the federal share of total academic research support—which itself was growing substantially—probably increased from around a quarter in the mid-1930s to over 60 percent by 1960. A recent study finds it "quite plausible" that by 1965 more than twelve times as large a flow of real resources was funneling into academic research as had been the case thirty years earlier.[22]

The greatly enhanced flow of budgeted funds is only part of the story. Hundreds of millions of dollars worth of advanced instrumentation and other equipment was simply declared surplus at the end of the war and transferred to the civilian sector. War-stimulated increases in the production of high-performance components for military purposes greatly reduced the need for custom fabrication of specialized equipment in university laboratories. With these stimuli to "industrialization," and with common, recent experience in wartime projects of hitherto-unimagined size,

> it is strikingly clear that the war had trained academic physicists to think about their research on a new scale, invoking a new organizational model. Not only did physicists envision larger experiments than ever before, they now saw themselves as *entitled* to continue the contractual research that both they and the government had seen function successfully during the war. This way of thinking molded both the continuation of wartime projects and the planning for new accelerators and national laboratories.[23]

Somewhat similar in its effects on the scale with which it became acceptable to envision research was the inauguration of the Apollo Project and the National Aeronautics and Space Agency (NASA) as an outgrowth of the launching of Sputnik. This was, for a time, an important stimulant to university-based research: "As a newcomer agency, eager to build a network with academic science, it was in a position analogous to ONR in the late 1940s. It provided generous funding on liberal terms to selected groups of scientists at many institutions. By 1966 NASA was supplying almost 10 percent of federal funds for academic R & D. . . ."[24]

Important as was the stimulus of the Apollo Project, its temporary character and heavy engineering emphasis must be recognized. The effects of its termination in the early 1970s were further intensified by a short but sharp economic downturn as well as by moderate brakings or even reversals in the ascending slope of some categories of research funding. Relative to its 1967 peak, it has been estimated that academic R & D funding (with biomedical research as an exception) had declined about 15 percent in real terms by 1975. As has also been the case at other times, the resultant atmosphere of crisis led to calls for less emphasis on allocations of research funds designed to meet the needs and priorities of disciplines and greater responsiveness to such external criteria as military needs or social and environmental problems.

Conflicting considerations like these, each with its advocates and allies, lead to policies composed of fairly heterogeneous elements. Especially in the heated political atmosphere of annual budgetary deliberations, agreements on priorities tend to need endless renegotiating and to be anything but permanent. Illustrative is the potpourri of quite disparate factors that came into play with the economic recovery in the latter part of the decade of the 1970s:

> the energy crises of the 1970s, which were accompanied by a remarkable diversification and intensification of energy R & D, the discovery of recombinant DNA techniques and the dramatic growth of new industrial applications of biology, the revival of the Cold War and the accompanying military build-up of the second half of the Carter administration, and the growth of computer sciences and engineering in universities with the maturing of the computer industry.[25]

Rare and striking episodes like Sputnik could have broad and lasting effects. However, as this suggests, the R & D system was more generally characterized by great diversity and was simultaneously subject to a wide array of positive as well as negative stimuli of only transitory importance. Program shifts and expansions or contractions on the part of particular funding agencies thus tended to have only limited effects. Research performers were able to maintain considerable operational continuity with new sponsoring agencies and minor re-direction of activities. Thus it is to the long-term, overall trends that we must look for an understanding of how the enormous advances made in postwar science and technology were financed. They are outlined in the accompanying tables.

Table 7-1 follows the major R & D funding sources and performers since 1960 in constant 1987 dollars. There are two categories that either remained relatively stable across almost the entire period or grew at only a moderate rate: federally supported research conducted by various federal agencies; and Federally Funded Research & Development Centers (FFRDCs), including both those in industry and those attached (sometimes loosely) to universities. Given their relative stability, these categories need not concern us further. But the others deserve attention as genuinely providing the fuel for growth.

There was almost a sixfold (581%) increase in federal support of R & D in universities, for example, as well as more than an eightfold (810%) increase in industry support of university R & D and more than a fourteenfold (1,404%) increase in R & D that universities underwrote themselves. The total of these forms of support had reached almost $14 billion by 1993, pointing to a qualitative change of state within the academic sector. That sector's capacity to generate new discoveries and innovations clearly had

TABLE 7-1

U.S. R & D by Perfroming Sector and Fund Source, 1960–1993
(billions of constant 1987 dollars, amounts rounded)

Fund Source:	Federal				Industry		University
			FFRDCs				
Performer:	Federal	University	Individual	University	Industry	University	University
1960	6.6	1.6	1.8	1.4	17.0	0.15	0.24
1965	10.9	3.8	1.3	2.2	22.7	0.14	0.44
1970	11.8	4.8	1.3	2.1	29.2	0.18	0.70
1975	11.2	4.8	1.5	2.1	31.7	0.24	0.88
1980	10.8	5.8	1.8	3.2	42.5	0.33	1.19
1985	13.7	6.4	2.0	3.7	60.4	0.59	1.71
1990	14.2	8.5	2.4	4.3	65.3	1.01	2.68
1993 (est.)	13.5	9.3	2.2	4.3	65.7	1.22	3.37

Source: National Science Board 1993: appendix table 4-3.

been greatly enhanced over the long run in spite of periodic setbacks. Corporate investment in industrial R & D also grew, and in absolute terms the net amount of growth exceeded by several times the total amount of additional R & D support that universities had received. On the other hand, the percentage rate of increase (386%), while impressive, is considerably smaller.

The same basic set of changes can be usefully examined from other perspectives as well. Federal support, to begin with, obviously came through many agencies that maintained quite different programmatic objectives for investments in R & D. The most fundamental distinction (although it may be obscured by different definitions) has consistently been between the support of basic and applied research, and a partial breakdown along these lines is given in table 7-2.

Only two of the agencies listed, the National Science Foundation and the National Institutes of Health, are charged with a primary responsibility for the support of basic research. This role is reflected in the steep increases in their budgets for basic research across the entire period, amounting to 542 percent and 1,174 percent respectively. On the other hand, at least a modest trend in the same direction is evident in the other columns of the table. While figures are not available on the applied category prior to 1980, it is apparent that, at least for the latest part of the postwar period, programmatic funding for basic research is generally receiving increasing support.

We should also consider different rates of increase in the support of R & D by and within the industrial sector. This is given in table 7-3. The figures given for 1958 provide a kind of early baseline, long antecedent to

TABLE 7-2

Basic/Applied Research Funding by Selected Federal Agencies
(billions of constant 1987 dollars, amounts rounded)

	Dept. of Agriculture	Dept. of Defense	Dept. of Energy	NASA	NIH	NSF
1960	0.13/	0.64/	0.40/	NA	0.39/	0.26/
1965	0.32/	0.93/	0.91/	NA	1.07/	0.60/
1970	0.33/	0.71/	0.83/	1.03/	1.48/	0.71/
1975	0.32/	0.49/	0.66/	0.65/	1.90/	1.02/
1980	0.39/	0.76/	0.94/	1.45/	2.33/	1.15/
	9.81	2.44	1.07	1.49	1.62	0.01
1985	0.47/	0.91/	1.00/	1.39/	3.20/	1.34/
	8.82	2.45	1.27	1.09	1.49	—
1990	0.46/	0.84/	1.33/	1.41/	3.78/	1.41/
	9.27	2.29	0.946	1.26	1.84	—
1994	0.50/	0.99/	1.39/	1.63/	4.58/	1.63/
	10.87	2.39	1.32	1.67	2.45	—

Sources: National Science Board 1993: appendix table 4-10 for 1980 and later; 1965, 1970 and 1975 from NSB 1981: appendix table 3-8, and 1960 from NSB 1977: table 3-4, using 1993: appendix table 4-1 price deflator. "Basic" on upper line, "Applied" on lower (where available). Dept. of Energy 1965 and 1970 represent Atomic Energy Commission; 1975 represents Energy Research and Development Administration.

TABLE 7-3

R & D Expenditures by U.S. Manufacturers, 1958–1990
(billions of constant 1985 dollars, amounts rounded)

	1958	1973	1980	1990	
Hi-tech Total	5.73	30.50	35.73	55.30	+81%
Drugs, medicine	0.95	1.59	2.34	4.61	+190%
Office and computer	NA	3.96	5.22	10.20	+158%
Electric machines	1.61	4.19	4.01	1.10	−73%
Radio, TV, communications					
equipment	1.08	7.01	8.07	13.20	+88%
Aircraft	1.42	11.55	12.11	21.17	+83%
Scientific instruments	0.67	2.20	3.99	5.01	+128%
Non-hi-tech Total	9.47	16.43	20.47	24.64	+50%
Chemicals	1.88	3.24	3.76	5.44	+68%
Motor vehicles	2.38	5.50	6.52	9.52	+73%
Other manufacturing	5.21	7.69	10.18	9.67	+26%

Source: National Science Board 1993: appendix table 6-8. The 1958 column of figures, not included in the percentage calculations, is taken from Mowery and Rosenberg 1993: 44, table 2.5.

the more consistent trends of growth at a much higher level shown for 1973–1990. They clearly indicate the very large increases that have occurred in virtually all branches of industrial research during the postwar period. More significant for present purposes, however, are the considerable differentials in relatively recent and current rates of growth between the high-tech sectors and the remainder of American industry. Hence the table provides contrastive rates of growth only for the period since 1973. It is no surprise to find that increasing investments in R & D for office machines and computers (not even separately recorded as a category in 1958), pharmaceutical products, and scientific instruments have substantially outstripped those for more traditional industries.

An exception is the case of electric machines (motors, generators), where R & D suffered a very substantial decline. It is likely that many major firms diverted their R & D programs instead into the many new commercial applications that were being found in electronics and especially microelectronics. Apart from electric machines, aircraft and communications equipment are the only other high-tech categories in which R & D failed to double after 1973. Probably this reflects the military significance of these categories, with a large share of investment in R & D having been borne by the federal government. As this suggests, military procurement is a highly significant factor in the larger R & D process. We will consider later the extent to which this has made a positive contribution to civilian applications of new technology, but the presence of the Department of Defense as a major R & D customer has had substantial structural effects that deserve to be mentioned.

Apart from the direct stimulus of R & D funding by the Department of Defense, there are a number of more indirect military influences on industrial technology. In the field of computers and computation, as an illustration:

> The sheer size of the increasingly high-technology armed forces ensured corporate investment in military-related R & D projects. The development of the transistor—privately financed by Bell Laboratories, but with military markets its major rationale—is the best-known example. But there are others of equal importance. The DoD sponsored the development of integrated circuits in the 1950s and purchased the *entire* first-year output of the integrated circuit manufacturing industry. . . . Two major programming languages were products of standard-setting efforts initiated by the military to assure software compatability. . . .[26]

Still another indirect effect of the mere size of most military procurements, of "hardware" as well as R & D, is that they have tended to lower barriers to the entry of new firms that otherwise might have been excluded because they could not demonstrate prior marketing success. In this way a number of originally small firms like General Radio, Texas Instruments, and Transitron were enabled to grow very rapidly. Federal antitrust policies during the postwar period exercised a similar influence. Consent decrees reached with AT&T and IBM, in particular, assured liberal patent licensing terms and created new openings for the commercial production of micro-electronic devices. In contrast to the pattern prevailing in other industrialized countries, the high-tech field in the United States has accordingly been characterized heretofore by many new, and conspicuously successful, corporate entries. Whether this pattern will continue to prevail is a matter of some uncertainty.[27]

Enhancing the Supply of Human Resources for Science and Technology

It is widely claimed—by whom originally I cannot identify—that 90 percent of all the scientists who ever lived are still living today. Probably impossible either to confirm or refute, the estimate is not implausible. The enormous recent growth of science has been closely linked with the appearance of great research universities, government laboratories, and industrial research facilities. As we have seen for the United States, most of this development has been concentrated in the half-century or so since World War II. Ninety percent may be too high an estimate, but not by a large margin.

Engineering presents a somewhat different problem. The community of scientists, defined as those engaged in the disciplined pursuit of a generalizable understanding of natural phenomena, has a fairly high entry threshold that limits the numbers of those recognizable as members before recent times. At least traditionally, engineering qualifications were less stringent. Roman road- and aqueduct-builders (of whom there were many) are surely classifiable as civil engineers. Even today, there would be widespread agreement that the practice of science is largely in the hands of holders of advanced degrees while the term engineer may apply to some who have received no postsecondary education at all. But if the proportion of all engineers who are still alive is lower than for scientists, any reasonable estimate still would reflect large increases in their numbers since World War II.

The main point is that we can now look backward, from the vantage point of a half-century's experience, on the essentially successful outcome of a vast social and educational experiment. Probably far beyond the levels that were initially anticipated, the demand as well as the supply for trained scientific and engineering personnel has grown enormously—in industry, in government, and in the not-for-profit sector. Colleges and universities, the institutional means for meeting the demand, have themselves grown to be a major source of the demand in their role as the primary performers of basic research.

Federal support of graduate and postgraduate education is essentially coterminous with the post-World War II era. At its beginnings, already in wartime, it reflected recognition of the increasingly complex set of responsibilities that would face America in the postwar world. Encouraged by ensuing federal initiatives in later years, and by a rapid, contemporaneous growth in college enrollments, educational institutions and private foundations multiplied their own support of these programs. But it may reasonably be concluded that the federal initiatives generally were the decisive ones in racheting up the scale as well as the average quality of graduate education.

It must be recognized that all such programs, directed at enhancing the supply of human resources but with very little control over the demand, are blunt, imperfect instruments. Aggregate funding levels are subject to abruptly shifting perceptions of other federal priorities. Anticipations of future personnel needs in different science and engineering fields are notoriously difficult. Paths taken by many individual participants in their later careers may involve fundamental shifts in direction and are difficult to trace. Fluctuations in employment possibilities, even if only relatively temporary and limited to certain sectors (e.g., defense), can be of critical importance for individuals caught in the "pipeline." But with all these difficulties, American success in a vast educational effort that underpins our scientific and technical leadership is beyond dispute.

Only a few turning points stand out as particularly significant in what has been a process of cumulative but somewhat irregular growth. Educational benefits under the GI Bill of Rights, although conceived primarily as an award to World War II veterans based on their service, may be considered the first of these. Opening the prospect of a college education to very large numbers of individuals who would not otherwise have considered it, this program began a cumulative broadening of the appeal of higher education for the American population at large.

The Soviet launching of Sputnik provided a second significant turning point. In the ensuing national concern over falling behind in the military uses of space, and over whether the unexpected Soviet achievement portended other surprises in efforts to maintain scientific, industrial, and military superiority, the climate for federal investment in research and graduate training improved dramatically.

The National Defense Education Act of 1958 inaugurated regular federal support for the training of graduate students, including those engaged in area studies and the study of foreign languages. In science and engineering fields, the ensuing decade saw a tripling of the number of doctorate degrees awarded annually, to a level of about eighteen thousand. With minor fluctuations, the new level remained relatively static until 1981, then once again began to rise much more slowly. Over the next twelve years the increase was only about 37 percent, although contained within this aggregate was a doubling in the annual number of engineering doctorates. Also during the decade following Sputnik, a time of relatively stable prices, there was a sevenfold increase in the federal underwriting of basic research in universities.

The impact of these changes, partly related to the speed with which they occurred, is difficult to overestimate. From a distance of three or more decades, the 1960s are often looked back on by university faculties as a kind of golden age of relatively unrestricted opportunity. Graduate education and research programs—the two mostly closely coupled—proliferated in their institutions (as did the institutions themselves), and received unprecedented emphasis. Subtle but significant shifts in faculty responsibilities and university governance ratified the enhanced importance of research and the commitments of university administrations to support it. Capital funds were made available in unprecedented amounts for infrastructural needs and the expansion of educational as well as research facilities. Consummating the trend at a national level, a target of thirty to forty research universities was formally promulgated in a report of the President's Science Advisory Committee. In effect, a pattern was established which had no earlier counterpart. Continuing until the present, it distinguishes the United States among industrialized countries in the emphasis given to public support of advanced teaching and research in the universities.[28]

It was expected at the time that the expansion of basic research activities within the universities would have a considerable, if unpredictable and often considerably delayed, "downstream" payoff in terms of U.S. technological as well as scientific leadership. A more immediate and directly sought product was a greatly enhanced flow of trained, technically qualified personnel for industry as well as for government. Both expectations can be said to have been fully met. But receiving little attention at the time were other changes in the composition of the scientifically and technically trained labor force in the United States that can be seen in retrospect to have been stimulated by the same programmatic innovations. One has to do with the accelerating entry of women as full-time, career-oriented participants, and the other with a growing recruitment of, and even dependence on, a stream of foreign, degree-seeking undergraduates and graduate students.

Many factors no doubt contributed to the historic transition in the role of women that probably occurred most rapidly during the 1970s but has continued ever since. The climate of expectations was materially altered as the proportion of women seeking higher education slowly rose in the direction of parity after World War II. Rising divorce rates and delinquent support payments for minor children forced increasing numbers of women permanently into the labor market as single heads of households. Suburban flight generated pressures for two-earner incomes. And, it is commonly thought, a proliferation of new technological devices "liberated" women from household drudgery and so made remunerative employment an increasingly attractive option.

This popular view involves an oversimplification of the impacts of technology. What might be thought of as the industrialization of the household cannot be adequately understood as the result of new and unconstrained consumer preferences, activated by rising disposable incomes and expectations of appropriate living standards. That development no doubt occurred to some extent and led in turn to the introduction of new, labor-saving technologies. But many other social processes also were at work, over which householders had very little direct control. In effect, "most labor-saving devices, in effect, have facilitated a rescheduled and reorganized work program, without substantially reducing the hours of work that a household requires."[29]

Tables 7-4 and 7-5 follow the changing character of the trained scientific and technological labor force over several decades. Federal intervention in the support of graduate education had an abrupt beginning, but then leveled out and began to taper off. These tables indicate that the numbers of individuals annually receiving earned doctorates in natural science and engineering fields are still maintaining a moderate, now apparently self-sustaining, rate of increase. On the other hand, they also indicate that this favorable outcome is essentially dependent on impressive increases in the number of women and noncitizens within the pool.

As table 7-4 shows, proportional increases in the number of women in key science and engineering fields accelerated after 1975 and mostly multiplied from two to four or more times in a sixteen-year period. Even in engineering, the serious laggard, the recipient of one in twelve of the new doctorates is now a woman.

The growing role of women in science and technology is no doubt a complex process with contradictory pressures from many directions. While the consistency of some steady progress in all fields that is shown in table 7-4 is impressive, there are sectoral differences and constraints that cannot be adequately touched on here. Significant differences in salary levels remain, even among full-time workers with comparable education, experience, and job characteristics. As a recent overview also emphasizes, however, "we need to get beyond issues of numbers." Unquestionably, there are other aspects of differentiation that are less amenable to direct measurement—for example, institutional location, access to interaction and collaboration with colleagues and supervisors, publication productivity, and subtle differences in evaluative standards.[30]

TABLE 7-4

Earned Doctoral Degrees by Field and Percentage of Women, 1960–1991

	1960	1965		1975		1991	
	Total	*Total*	*% Women*	*Total*	*% Women*	*Total*	*% Women*
Physical Science	1,861	2,865	4	3,076	8.6	3,623	18.4
Environmental science	NA	NA		625	4.8	816	21.8
Biology and agriculture	1,660	2,539	10	4,402	19.3	5,713	34.4
Computer science	NA	NA		213	6.6	797	14.6
Engineering	794	2,073	<0.5	3,011	1.7	5,212	8.7

Sources: For 1975 and 1991, National Science Board 1993: appendix table 2-27; for 1965, NSB 1979: appendix table 5-23; for 1960, NSB 1983: appendix table 3-29.

TABLE 7-5

Percentage of U.S. Doctorates Earned by Noncitizens, 1977–1991

	1977	1991
Natural sciences	16.3	28.0
Mathematics/Computer sciences	10.5	44.5
Engineering	29.4	47.4

Source: National Science Board 1993: appendix table 2-28.

TABLE 7-6

Science and Engineering Masters Degrees, 1966 and 1991

	1966	1991
Physical science	4,725	3,777
Mathematics/computer science	5,010	12,956
Environmental science	690	1,499
Biology and agricultural science	5,865	7,406
Engineering	13,706	24,013

Source: National Science Board 1993. Practices differ by institution and field, but it may be assumed that most of the masters degree recipients here tabulated did not (or at least not immediately) continue their formal education in a doctoral program. This clearly is the case for engineering and computer science (given the disparity in the numbers involved), and probably also for physical science since the trends across time run in opposite directions. But there is probably some overlap, and in any case masters degree recipients are somewhat less likely to have remained permanently attached to science and engineering as a field of employment.

The proportion of doctorates in science and engineering fields earned by foreign students has also increased rapidly, as table 7-5 shows. No doubt this once again reflects the complex interaction of many factors. With the great expansion of U.S. research universities, they have become a mecca for aspiring professionals from all over the world. In fields like engineering, virtually half of the new doctorates go to foreign students. It should be recalled, of course, that only approximately half of these graduates reportedly remain in the United States and become part of its trained labor force.

In some ways even more relevant for technology than the numbers of doctorates are the numbers of masters degrees in science and engineering. Recipients of doctoral degrees may function exclusively in teaching and basic research programs, while the great majority of masters degree recipients seek employment in industry. The totals show broad similarities with those given for Ph.D.'s in table 7-4, but table 7-6 also reveals significant differences. To hazard a general evaluation of the differences, computer science has grown strikingly rapidly, but to a much greater extent as an industrial and technological field than as an academic one. In the physical and biological sciences the doubling in the number of doctorates while the numbers of masters recipients have stagnated either suggests the reverse or indicates that there is little industrial demand for individuals with less training than the doctorate represents. Engineering, its practical role unquestioned, appears to have strengthened itself academically.

Current Technological Priorities and Frontiers

As described in chapters 5 and 6, modest federal support for technologically relevant R & D may in a sense be said to have originated almost at the outset of the Republic with early patent legislation. More actively, the nineteenth century saw gradually growing levels of budgetary support for the protection and promotion of commerce like the Coast Survey, and for the encouragement of agricultural research through the establishment of land-grant colleges and research stations. More affirmative and substantial steps were taken soon after the end of the century, including the creation of the National Bureau of Standards, regulatory agencies concerned with threats to public health and well-being, and presently the National Advisory Committee on Aviation. Evident in retrospect, if not so clearly at the time, was the sense that growing national integration required a wider exercise of governmental responsibilities. Actions that were largely argued for and taken on an individual, ad hoc basis can collectively be seen as an opening wedge of commitment to sustain a favorable infrastructure of research as well as regulation in emergent areas of potential risk and economic importance.

Although the impact was relatively brief, World War I was a pioneering exercise in the superimposition on this still only loosely articulated structure of new federal agencies directly committed to mobilizing academic scientists as well as engineers in support of military R & D. As we have seen, that theme re-emerged with much greater strength and more lasting effect in World War II. Established in a decision process that began already during the war and was consummated only a few years afterward was the general shape of national priorities for science as well as technology that has persisted ever since.

The cluster of emphases is well recognized. Their impacts on technology can perhaps best be identified in the total basic and applied research budgets of the major mission agencies, as shown in table 7-2 at five-year intervals since 1960. A slightly different breakdown shown in table 7-7 designates levels of R & D activity specifically in academic institutions on behalf of these agencies. It provides a measure of the involvement of academic scientists in agency R & D functions and hence of tendencies toward the functional convergence of the scientific and engineering communities.

Setting aside the role of agencies like NIH and NSF with mandated responsibilities for the support of academic research, at the head of the list of priorities is a primary commitment of large-scale support for military R & D under the direct administration of the Department of Defense. For the last decade or more, the Department's share of total federal R & D has substantially exceeded 50 percent, falling in the range of $21 to $35 billion (constant 1987) dollars annually. The share set aside for technology development has consistently been about one fifth of the total, with $3.6 billion of the 1994 allocation designated specifically for advanced technology development.[31]

Less direct, but no less fundamental as at least an initial stimulus, was the importance of technological rivalry with military overtones in the launching of the Apollo Project and the National Aeronautics and Space Administration. With a monopoly on manned launching capability through the space shuttle, NASA missions include placing unclassified scientific instruments as well as military surveillance satellites in orbit. The Apollo mission to the moon was itself, of course, an extraordinary, unequaled achievement that must be seen primarily as a technological rather than scientific triumph.

Because it is focused, the high priority given to defense-related R & D, and more especially to the advancement of technology with perceived military applications, deserves further attention. To be sure, it amounts to only one-third or so of industrial R & D in the private sector, and to slightly less than half of that in the high-tech private sector. But the greater part of these larger amounts is devoted to the development of a very broad spectrum of

TABLE 7-7

Federal Obligations for Academic R & D by Agency, 1971–1993
(millions of constant 1987 dollars)

	NIH	NSF	DoD	NASA	DOE	DoA	Other
1971	1662	734	581	369	259	198	727
1975	2262	914	427	227	277	227	732
1980	2674	970	702	223	404	307	759
1985	3154	1062	997	252	379	311	568
1990	3820	1172	1076	418	444	309	873
1993 (est.)	4212	1494	1267	549	468	336	1235

Source: National Science Board 1993: appendix table 5-9.

consumer products under what is arguably a different set of cost constraints. If so, military R & D, beyond the acknowledged superiority of a
high proportion of advanced American weapons to which it has no doubt
contributed, may have had some more questionable side effects. The coherence and specificity of military requirements, in particular, tends to stand in
sharp contrast to the heterogeneity of civilian ones. Faced with an "under-
determination" of the technical solutions needed to meet the demands of
the market, it is the military ones that prevailingly have had the greater
impact.[32]

A case can be made for doubting that military R & D has had a preponderantly positive impact on the competitive standing of other U.S. high-
tech industries in world markets, or on our R & D capabilities more generally. Making this judgment with any accuracy would be very difficult since
many politico-military as well as economic considerations are involved,
but David Mowery and Nathan Rosenberg have thoughtfully outlined
some of at least its economic parameters. Well-financed federal R & D, by
increasing the demand for scientists and engineers in certain fields, has
raised their rates of remuneration (and of course also, but more slowly, the
supply) and hence the costs of private R & D as well. That effect is likely
to have been particularly large in increasingly competitive, high-technology fields like microelectronic equipment, instrumentation, and aircraft,
and to have led to a displacement of research activity away from fields such
as chemicals and petroleum that receive few federal funds. These distor-
tions inevitably involve costs, even if the costs remain largely hidden.

More tentatively, Mowery and Rosenberg touch on the issue of differences between federal and private cost constraints that was mentioned
earlier:

Were engineers who had worked on programs where small performance improvements were sought almost regardless of cost effective designers of prod-

ucts for civilian markets where cost considerations and sensitivity to nuances of consumer preferences were likely to be far more significant? Have large federally supported "crash" programs shaped the approach and influenced the (perhaps implicit) trade-offs of U.S. engineers and product designers in ways that are dysfunctional for highly competitive consumer markets, such as consumer electronics?[33]

These are complex, subtle questions, not permitting easy, unambiguous answers. They center on the likely performance of individuals under changed conditions, while the only available data are almost certain to be aggregated at the level of firms. But it is worth noting that, according to a recent survey, the defense industrial base has been found to be "substantially 'dual-use'." A "vast majority" of defense-involved firms simultaneously meet commercial customers' requirements with the same equipment and work force, and in so doing are faced with seemingly comparable competitive pressures in both.[34] The doubts of Mowery and Rosenberg on this score are rendered somewhat less plausible, although not entirely dispelled, by these findings.

Another of their critical arguments is based on comparisons with our principal industrial competitors. Ratios of civilian R & D to GNP have been substantially higher in Japan and Germany, with relatively much smaller military budgets, than in the United States. To Mowery and Rosenberg, this suggests that "the true opportunity costs to the U.S. economy of high levels of defense R & D have been very high." Such comparisons are extremely difficult, they concede, but at least on purely economic grounds their case clearly has some substance (p. 160).

Providing some additional support for this position is a fuller account of individual technologies that currently are advancing rapidly and are of critical importance to the United States for their economic significance and numerous applications. Twenty-two technological sectors have been termed "essential to satisfy such national needs as defense, economic competitiveness, public health, and energy independence" by the U.S. National Critical Technologies Panel, and described as representing "the lion's share of the future growth of the nation's economy." As a prescription for federal policy the listing perhaps has deficiencies, some of its details possibly reflecting corporate self-interest or protectionist sentiments in a volatile atmosphere of international competition. But it provides a useful basis for briefly scanning what are today the key areas of technological advance.[35]

Five of the twenty-two fall within a broader category concerned with the synthesis and processing of advanced materials, to which the dominance of defense-related considerations seems clear. The diversity and potential capabilities of many of them are most impressive. Polymer and metal matrix composites, reinforced with high-strength fibers or particles, can achieve several times greater strength and greater stiffness than traditional metals or even superalloys, with a 20 to 30 percent weight reduction as an additional advantage. "Intermetallics," involving many combinations of nickel, cobalt, iron, lithium, and titanium with aluminum, have been found to promise the special advantage of greater strength at high temperatures. Carbon-based and ceramic matrix composites tolerate higher temperatures than any metal alloys, making them especially relevant for turbine engines, rocket nozzles, and space re-entry vehicles. High-strength, reinforced ceramics, in spite of persistent tendencies toward brittleness, are even being developed as the major engine component in missiles, drones, and other short-life applications.

High cost is, however, a regular trade-off for high performance. Not only expensive in themselves, the new materials generally require new and expensive processing methods: additional refining for ultrapurity, "near net shape processing" of alloys and ceramics through "hot isostatic pressing"; laser surface hardening; superplastic forming (slow deformation under high temperature and pressure to reproduce exact, intricate shapes); and very rapid cooling and solidification of metals in order to reduce cracking and weakness along grain boundaries. Another broad category involves sophisticated, complex materials that are "sometimes engineered literally an atomic layer at a time"[36] through chemical vapor deposition or molecular beam epitaxy, and the integration of electronic and photonic materials to form single circuits.

The integrated circuits are an exception, but few of the other examples cited are likely to be of near-term commercial utility. Most of these materials have been developed for specialized uses involving the need to operate at very high temperatures in order to achieve high performance, as well as exceptionally high strength-to-weight ratios and resistance to wear or corrosion under almost any conditions. Cost has been at best a secondary factor in their introduction, and so long as military orders are the source of funding is not likely to become a primary consideration. Hence, a commercialization phase of further development seems more likely to occur in Japan, and in a number of cases is reported to have already begun there.

Aeronautics is a somewhat countervailing example, with commercial developments clearly the net beneficiaries of R & D funding provided by NASA as well as the Department of Defense. Cost constraints are obviously different in the private sector, but there are many areas of convergent interest where federal support can assure more rapid progress. These include improved, computer-aided airframe (and especially integrated engine/airframe) design, more powerful engines with better fuel economy, avionics that enhance safety and improve man/machine interfaces, reducing and monitoring metal fatigue and corrosion, as well as many others. Partly because of the military stimulus, the U.S. commercial lead is still substantial. But with major, heavily subsidized European entries it is shrinking.

A broader category of "critical technologies" involves advanced manufacturing processes. By definition, initiatives for it lie primarily within the private industrial sector. Included under this heading are diverse approaches to flexible computer-aided design, inspection, and assembly; intelligent processing equipment and robotics; devices for micro- and nanofabrication and nanolithography of integrated circuits; thin films and other forms of surface treatment; and systems information and management technologies. Japan is once again a major, in many cases dominant, competitor, devoting, as noted earlier, twice as large a proportion of its industrial R & D to manufacturing processes as the United States does. Meanwhile, leadership in the race for important robotics patents is vigorously contested between Japan, the United States, and Western Europe.[37] As the National Critical Technologies Panel observes, "the culture and management practices of many U.S. companies must change if the United States is to remain strong in today's manufacturing environment."[38]

The information industry comprises seven of the twenty-two "critical technologies," the largest and perhaps most significant grouping. The whole ensemble may well be undergoing such rapid and far-reaching changes as to justify speaking of a full-fledged Information Revolution, distinct from the Industrial Revolution but at least potentially comparable to the latter in importance. It has widely been observed, for example, that there has been well over a 20 percent annual decline in the price of a given amount of computing power over the last three decades, "a steady rate that eclipses any sustained price decline in recorded history."[39]

But impressive as this achievement is, it would misrepresent the advances made in different but complementary sectors of the industry— involving software as well as hardware, and concerned with the assembly, storage, transfer, and processing of unprecedented masses of data—to think of their emergence and consolidation as an internally energized process. "For many years, these breakthrough computers were the carefully nourished offspring of government encouragement."[40]

Software is the most dynamic frontier of the information industry, as the meteoric rise of Microsoft and relative decline of IBM and other corporate giants associated primarily with hardware attests.[41] The broad versatility of computers depends upon software, and software is at the same time the key source of vulnerability in computer-based operating systems:

> Increasingly, the development of advanced software is an important limiting factor in the introduction and reliability of new military and commercial systems. Software requirements, as well as development costs, expand at a dramatic pace as automated systems proliferate and increase in sophistication. Despite these

growing demands, the generation of advanced software programs remains largely a painstaking, labor-intensive task. As a result, the ability of U.S. industry to provide high-quality, reliable software is in jeopardy.[42]

Also included under the information industries rubric are a dazzling array of additional technologies. Integrated circuitry, requiring demanding techniques of microelectronics and optoelectronics, merges with the category of advanced materials and has been mentioned earlier. Lasers play a vital part in information storage and printing, but also have become essential tools in manufacturing, components in compact disc players, and medical instruments with applications in eye surgery and dentistry. Fiber optic technology is recognized as the only mode of transmission consistent with the national goal of information superhighways. But at the terminals it interconnects, high-rate data transmission, high-density data storage, rapid real-time signal processing, and high-definition displays all are indispensable adaptations to a geometric rate of increase in the information base involved in making complex decisions—and to the decreasing intervals of time available to make them. In all of these areas save high-performance computing and computer networks, it is reported that Japan has at present a considerable lead.

The commitment of federal support for R & D in the biomedical field has been second only to that for defense. It has, moreover, maintained a considerably more consistent record of growth. Clinical as well as basic research that is funded by the National Institutes of Health is conducted both in-house and at independent hospitals and universities, and the targeting of funding into recognizable areas of concern is generally credited with having solidified public and congressional support for the program. The effect, in any case, has been to provide an unusual degree of continuity for studies in a very broad spectrum of biological fields ranging well beyond immediate issues of human health.

Without such support at least the pace of the truly revolutionary advances that began with the Watson-Crick discovery of the double helix of DNA strands in 1953 surely would have slowed very substantially. With it, recombinant DNA and monoclonal antibody technologies and other new and sophisticated forms of bioprocessing have transformed the health sciences. New vaccines, human insulin and growth hormone, treatment of a number of inherited diseases, "transgenic" experimental animals that can be used to develop human genetic therapies, and new treatments for cancer, anemia, blood clots, and many other conditions all are under development. The mapping of the entire human genome that is currently underway constitutes an unfolding process of major scientific discovery, but from it will come a vast number of further contributions to human health that will gradually take on a more "technological" character, and that as yet cannot even be anticipated.

TABLE 7-8

1991 Federal and Nonfederal R & D Expenditures at Academic Institutions by Field (millions of dollars)

	Physical Science	Environ- mental Science	Com- puter Science	Bio- logical Science	Medical Science	Engineering	All Other
Federal	1379	704	366	1951	2831	1631	1359
Nonfederal	558	415	178	1106	1738	1262	2141

Source: National Science Board 1993: appendix table 5-6.

TABLE 7-9

Corporate R & D Expenditures for Drugs and Medicines (millions of constant 1987 dollars)

1981	*1986*	*1991*
2,653	3,774	5,177

Source: National Science Board 1993: appendix table 4-32.

Similar, major impacts on agricultural productivity, waste remediation, energy conservation, and industrial chemistry flow from the same technologies. Concurrent with this have been striking advances in other medical technologies, most of them linked to electronics and the computer sciences. Magnetic resonance imaging (MRI), computer-aided tomography (CAT), and positron emission tomography (PET) are relatively noninvasive tools of great power. Cardiac pacemakers and fiber optics and lasers that have expanded the use of angioplasty, arthroscopic and other surgical approaches are merely the most salient and widely employed examples of major new additions in a diverse and rapidly growing field.[43]

In no area has there been a more substantial, federally orchestrated synergy between scientific discoveries by academic scientists in academic laboratories and their commercialization. The unparalleled federal funding extended to the closely related fields of biological science and medical science, in comparison with other discipines, is shown in table 7-8. Here also is indicated the extent to which this federal stimulus has been matched by increases from nonfederal sources, while table 7-9 indicates that corporate funding for (mostly in-house) R & D for drugs and medicines is not only large but has been growing at an impressive rate.

This discussion has covered the major elements on the National Critical Technologies Panel's list of twenty-two "critical technologies" that will continue to be the major growth sectors in the U.S. economy. Merely for the sake of completeness, the remaining members of the list may be summarily mentioned: surface transportation technologies; energy sources, conservation, and renewable energy technologies; and pollution minimization, remediation, and waste management.

A Global Setting for Industrial Strategies

On a global scale, high-technology industrial activity has become increasingly interdependent but at the same time intensely competitive. Driven by dynamic economies of scale and scope that are subject to many feedbacks, it is increasingly concentrated in the hands of a relatively small number of large concerns whose oligopolistic strategies are directed at dominating global rather than merely national or regional markets. More than four-fifths of the world's R & D expenditures that have fueled this growth, and more than two-thirds of the world's total R & D personnel, still are to be found in just five of the most industrialized countries: the United States, Japan, Germany, France, and the United Kingdom.[44]

Perhaps the predominant characteristics of the present world economic environment are its turbulence and uncertainty. Maximum flexibility becomes a high corporate priority—the ability to shift scale of output, product mixes, and even manufacturing locations and niches of primary activity quickly and at minimal cost. In large part as a result of the increasing utilization of robotics in assembly lines, heightened flexibility as well as economies of scale in industry have become possible at lower output levels. This is hastening the obsolescence of an older generation of giant plants, and regrettably also of some of the great industrial cities that grew up around them. With the progressively wider dispersal of manufacturing, there is a corresponding "change in the nature of markets from 'places' to 'networks,'" so that "work increasingly becomes detached from place, operations from their central headquarters."[45]

With R & D investments continuing at high levels, the pace of innovation remains high. This is accompanied by the increasing effectiveness with which new means of communication can generate new consumer preferences while also encouraging great latitude of choice within them. Pressure increases to make produce cycles shorter, driving down product development time. Enormously speeded as well as simplified and rendered more accurate by computer-aided design and manufacturing robotics, and by closely controlled inventories, orders trigger "lean" or "just-in-time" production. Customized manufacture, more responsive than ever to individual consumer preferences, ceases to be inconsistent with a high, sustained volume of output.[46]

The competitive advantages of powerful global oligopolies are enhanced by management innovations that facilitate complex, multi-plant operations across international boundaries. Their strategies of dispersal weaken the bargaining positions of both national governments and organized labor. The newer, most advanced technologies, more dependent upon economies of scope and rapid flexibility than of scale, still further reinforce the advantages of corporate size and diversification.

The importance of a nation-state's ability to control its supply of critical natural resources (other than sources of energy) has steadily declined. With Japan as an outstanding (although not the only) example of how well this can be dispensed with, the openness of world markets, the declining costs of long-distance transport and communications, and the increasing availablity of acceptable synthetics have made national control of resources a relatively minor competitive advantage. For high-tech commodities in particular, the principal value-added elements are the products of lengthy, specialized R & D and an educated labor force.[47]

Fundamental criteria of profitability and competitive success are also shifting decisively. The political climate to which firms must respond includes rising pressure for public access to fuller information. With the deterioration of servicing industries, *reliability* and *user-friendliness (in maintenance as well as operation)* take on unprecedented importance. *Affordability* is of course not a new concern, but the context of choice is affected by the explosion of new consumer goods and widening awareness of international marketing networks. *Mobility* is also not new, but receives greater emphasis because of widespread life-style changes.[48] *Safety* also has grown in importance as a consideration, at least partly because of growing testing and dissemination of information by governmental and public-interest organizations. Finally, *potential environmental impacts* have become a major public concern, articulated with great effectiveness at a global level by proliferating nongovernmental organizations (NGOs), and of course also a subject of governmental action.

As all of these trends continue, there is a dispersal of performance criteria accompanied by an erosion of control by even the most powerful nation-states over international corporate activities.[49] Employment levels and security of employment are among the first and most common of national interests to suffer. As a result of the worldwide slowdown in growth, this is perhaps in any case unavoidable. But it has now extended even to highly successful firms in fields of great technological promise and rapid advance.

The "lean" or "flexible" approaches to manufacturing that are currently regarded as essential all tend to involve at least selective reductions in labor force size and security of employment. Parallel to a significantly reduced, stable core, a marginalized, for the most part involuntarily temporary or part-time work force is created with sharply reduced benefits and working conditions, and with little opportunity for further training or advancement.

While it can be argued that new, small-firm start-ups represent a natural form of rejuvenation that will ultimately be beneficial, even in a vigorous sector like the computer industry current job replacements are overmatched by short- and medium-term job losses.[50] The effect is the gradual creation of a two-tier labor force, and the progressive de-skilling of one of its major components.

Exacerbating this problem in the United States are deficiencies in its educational system and the low priority given to supplementary work-force training in most sectors of American industry. United States per capita expenditures on education rank rather low among those of industrialized countries (twelfth of fourteen in OECD rankings) if we consider pre-collegiate schooling only. There are also disturbingly large variations in levels of spending and availability of advanced classes and specialized equipment that tend to favor schools and school districts with a high proportion of college matriculants. Students from low-income and inner city neighborhoods, who constitute the major source of supply of the industrial work force, thus tend to be ill prepared to be selected for the upper tier of permanently retained employees when there is an industrial contraction.

Nor are steps taken subsequently to overcome these deficiencies. The 1990 report of the Commission on the Skills of the American Workforce, written well before the recent contraction had reached its present proportions, notes with concern that prevailing practices in industry tend to reinforce rather than correct the disparity:

> [B]ecause most American employers organize work in a way that does not require high skills, they report no shortage of people who have such skills and foresee no such shortage. With some exceptions the education and skill levels of American workers roughly match the demands of their jobs. . . . More than 70 percent of the jobs in America will not require a college education by the year 2000. . . . No nation has produced a highly qualified technical workforce without first providing its workers with a strong general education. But our children rank at the bottom on most international tests—behind children in Europe and East Asia, even in some newly industrialized countries.[51]

In the most advanced industrial sectors, the higher technical and organizational requirements of automation and lean production lead to an increasing dependence on trained scientific and engineering personnel. For those who qualify, the news is good, although table 7-10 also shows that the United States has lost much of the competitive advantage it held in this respect over Japan and Germany a generation ago. But in any case, this numerically much smaller trend can in no way compensate for the disruptive social impacts suffered by the work force at large.

Very high R & D costs have come to constitute a necessary entry fee and continuing requirement in high value-added, high-technology industries.

TABLE 7-10
Scientists and Engineers in R & D by Country,
1965–1989

	1965	1989
United States		
Total (thousands)	494.2	949.3
Per 10,000 in labor force	64.7	75.6
Japan		
Total (thousands)	117.6	461.6
Per 10,000 in labor force	24.6	73.6
West Germany		
Total (thousands)	61.0	176.4
Per 10,000 in labor force	22.6	59.3

Source: National Science Board 1993: appendix table 3-22.

On average, basic research is the smallest part—only around one-twelfth—of the composite, while applied research accounts for one-quarter and development for a full two-thirds of the total.[52] Automation is another heavy fixed cost, to a considerable degree directly replacing the variable labor component of total production costs. Variable costs, in short, are tending to give way to fixed costs. The consequence is that high-technology change provides an avenue of escape from the usual assumption that economic actions tend to engender negative feedbacks and quickly stabilize prices and market shares around a new equilibrium. Advances of this type require large initial investments but then lead to steeply falling unit costs and provide leverage for further breakthroughs and entry into new applications. The new methodological prescription is that "situations dominated by increasing returns should be modled not as static, deterministic problems but as dynamic processes based on random events and natural positive feedbacks, or nonlinearities."[53]

The fixed costs are, however, front-end investments. Deep and immediate market penetration is a prerequisite for these to be promptly defrayed. Continuous and sensitive attention to every aspect of consumer demand, and a readiness to adapt quickly to shifts in consumer preferences, thus are also absolutely essential. And domestic markets alone, even for major industrial countries, are frequently not large enough to sustain fully automated plants in complex fields. Hence successful marketing efforts on a global scale become a further essential.[54]

It is sometimes suggested that a kind of asymptomatic function may be in sight on a truly global scale, setting limits to our collective possibilities of further growth. The raising of such an unprecedented—and it should be stressed, at this juncture absolutely unproven—eventuality obviously deepens the climate of pessimism. Virtually universal aspirations for improved well-being, consistently maintained across a century or more to come and enormously advanced by a host of discoveries and innovations of a scientific-technological character, now seem open to doubt. Moreover, the political mechanisms by which to deal constructively with forebodings of irreconcilable conflict and growing long-range uncertainties are simply not in place at present. It is difficult to see how such a complex of grave and divisive challenges can be addressed without moving beyond a framework of overriding national self-interest and a purely competitive economic marketplace.[55]

Beyond this general sense of growing insecurity and uncertainty, what distinguishes the world context within which modern technology finds its primary applications? Perhaps most disturbing is the widening gap between the most- and least-developed parts of the world. Major countries formerly in the middle of this range—China, Brazil, Mexico, India, and Indonesia are representative examples—have of course moved decisively toward the developed end of the contiuum. But similar progress is not in evidence in most of Africa and in other parts of Asia and Latin America.

In effect substituting capital investment in automation for labor, high-tech industries in developed countries are having some success in driving down the direct labor content of their costs to as little as 10 percent or less. With such reductions, the transfer of technology and manufacturing facilities to less-developed countries can be slowed or even brought to a halt in many but not all fields. But the result is a zero-sum game, raising prospects of irreconcilable political conflicts in the future. Residual rates of un- and underemployment among less-than-highly-skilled workers in developed countries are resistant to significant reduction, while the poorest countries find great difficulties in attaining a position on even the lowermost rungs of an ascending industrial ladder.

Remaining to be discussed as an unfolding competitive strategy is the use of patents and patenting. As a first-order approximation, they may be the best available surrogate for data on R & D activity—data which, in any other form, are more difficult to obtain and often seriously biased or inaccurate. But as noted in earlier chapters, aggregate series of patents also are flawed as an index to significant inventive activity in every historic period. Individual patents were never necessarily introduced into use at all, or for the purposes originally specified for them in patent applications. Inherent in the administrative and legal processes through which patents are granted and defended are reifications that subtly distort the notion of "invention" itself. Like technologies themselves, they "depend upon one another and interact with one another in ways which are not apparent to the casual observer, and often not to the specialist."[56]

Especially in the conditions of rapid technological progress and intensified competition now obtaining, their significance is becoming more and more limited. Advances have been so rapid in some fields like biotechnology that the distinction between basic science and technological R & D has almost disappeared. New products are introduced, altered, and replaced before the formality of a patent can be secured. Having a significant headstart in producing a new and complex product largely displaces the need for either patent protection or any form of secrecy. In addition, patenting inevitably involves a trade-off. Necessary for the issuance of a patent is the disclosure of essential information on the product or process for which legal protection of rights to its exclusive use is sought. In very rapidly changing fields like semiconductors and microelectronics, this trade-off is not necessarily advantageous. Earlier and more complete market penetration may be gained in a race for what will be at best short-term superiority by failing to disclose (in fact, by seeking to obscure) research discoveries that have been incorporated in a product than by seeking formal legal protection for them. Spiraling costs of litigation are still a third factor. While there has been an apparent decline in patenting activity over the last two decades, a more complex picture emerges when this trend is disaggregated. Patents continue to be a widely employed source of protection in industries producing chemicals, plastics, synthetic fibers, and devices whose design would be relatively easy to duplicate. Particularly in industries whose products take the form of complex systems, however, the attitude is different:

> Our respondents from industries producing aircraft and guided missiles, canonical complex systems, reported that it would cost a competent imitator three-fourths or more of what the innovator invested to come up with something comparable, that considerable time would be involved as well, and that it did not matter much whether or not there were patents. Producing complex systems effectively requires that many components and details be got right, and this is difficult to learn to do even if one has a model to take apart, or a blueprint to follow. These industries, and others like semiconductors, also involve complex production processes with tooling and equipment often finely tuned to product design. Simply getting the production line in place and running right can yield the inventor a substantial lead over potential followers.[57]

The enormously active biomedical field has special patenting complexities of its own. Medical devices have attracted the interest of many small start-up firms, which have recognized opportunities to develop specialized applications of microprocessors in a relatively relaxed regulatory framework. Patent protection is largely unavailing since many alternatives usually can serve the same therapeutic end, but there is the compensating advantage that a particular approach can undergo continuing improvements during the course of testing and subsequent production. The rate of product obsolescence is high, and firm failures greatly outnumber successes.[58]

The large, well-entrenched firms comprising the pharmaceutical industry, by contrast, are accustomed to lengthy time horizons and very large R & D commitments. While the international environment is certainly competitive, the usual role of consumer preferences as the ultimate arbiter of market mechanisms is sharply reduced and somewhat distorted. This is certainly the case if patients are considered as the consumers, most of whom lack knowledge of the efficacy, risks, and alternatives to their choices, and in fact leave the choices themselves in the hands of their attending physicians. The role of clinician thus often blurs the usual distinction between developers of and customers for new technologies.

From another direction, public and private third-party payers are more often the actual purchasers than the consuming public. Commensurate with their often very large purchasing power, they are increasingly exercising price leverage upon medical technology suppliers. The regulatory environment, by extending the duration of testing and heightening development costs, also places a premium on rapid, worldwide penetration of mass markets in order to assure profit and cost recovery before generic products become available. Little incentive is provided to direct R & D toward therapies for rare illnesses, or for illnesses found mainly in less-developed countries without hard currencies.[59]

Change with regard to patent protection, in short, has been complex, differentiated by sector and even micro-sector, and closely tied to marketing and financing conditions. The roles of science and technology are not easily distinguishable from one another, and both intercommunicate closely with corporate interests and strategies. Public policy and regulatory interventions, under the pressure of many interest groups and often subject to unanticipated second-order effects, are seldom comprehensive and wholly effective. And no effective means is yet in sight to hold in check the unprecedented share of GNP devoted to the gigantic health-care industry, among the many components of which technlogy may well be the one growing most rapidly.

All in all, the world context of industrial strategies is a volatile, highly competitive, correspondingly unsettling one. There is no apparent lack of new opportunities to be exploited, but the risks—including risks of unforeseen second-order consequences—have also grown enormously. Immediate and long-term clashes of interest are increasingly severe and difficult to contend with, and the real power and initiative in deploying new technologies has moved largely into the hands of corporate boards. Both as a unit of analysis for a study like this one and as a master of its own technological household, the nation-state, even the United States as the industrially most powerful nation-state, becomes more and more deeply embedded in, and difficult to hold separate from, its wider, international context.

Encountering the Japanese System of Production

Japan, having emerged as our most successful—some would say most dangerous—technological competitor, has nearly doubled its share of both world output and exports of high-tech manufactured products within little more than a decade. It displaced the United States as the leading high-tech exporter in the mid-1980s and today occupies the dominant position in fields in which the United States long thought itself securely pre-eminent.[60] No longer merely successful at commercializing foreign technologies, Japan has increasingly demonstrated its capability to operate at the technological frontier in key fields like fiber optics, advanced and composite materials, fermentation processes, computer peripherals, memory chips, and computer-numerically controlled machine tools.[61]

While the strength of Japan's position in this rivalry must be recognized, we should also take note that this strength is essentially confined to a fairly narrow sector of its industrial economy. Only some 13 percent of Japan's working population is employed in its extremely successful, high-tech, hardware-exporting industries. Many primary and secondary industries, as well as the entire service sector employing 56 percent of the population, meet only relatively low standards of competitiveness.[62] Moreover, the resilience of an emergent U.S. competitive response is evident in many scattered corporate reports and should not be underestimated. Under the new conditions of corporate interdependence and intensified competition, there are few impediments to the borrowing and adaptation by U.S. firms of the most advantageous features of Japanese industrial practice. Hence, any attempt to predict the overall long-term outcome of the rivalry would be unjustifiably speculative.

Our primary interest, however, is concentrated precisely on the advancing front of technology. Whether or not the United States is successful in borrowing and even improving elements of the Japanese approach, its initial, systemic features deserve to be considered. The coherence of Japanese industrial strategy in the high-tech fields sustains comparison with the American system of production that first emerged as an international presence in the 1850s—which, after all, was for a long time even narrower in its impact.

In the earlier case, U.S. industry had been able to pioneer its new and innovative approaches to mass production within a much more isolated milieu. It was powerfully assisted by the rationale of meeting military exigencies that the Congress was prepared to accept as overriding considerations of cost, and by what proved to be an atmosphere of public tolerance with regard to limitations of consumer choice that had no European counterparts. The long production runs of standardized products that became the distinguishing characteristic of American mass production now are giving way, as we have seen. Customized production preserves many of the same cost advantages as a result of greatly improved, more flexible strategies for automation, product innovation, inventory reduction, and the overall organization of production. These are all innovative strategies that are mostly Japanese in their origins, and that together constitute the essential manufacturing elements of the "Japanese System."[63]

Underlying these features, however, has been for many years an equally vital contributor to Japanese competitive performance in the form of a substantially higher rate of consumer savings. The outcome of a host of essentially cultural factors associated with lifestyles, intergenerational relationships, and much more, it has meant that Japanese capital costs are about one-third of what they are in the United States. As a result, capital investment per employee can be approximately double that of the United States in manufacturing, accounting in considerable part for significantly higher Japanese labor productivity in the economic sectors important for international competition. It also helps to explain Japan's tolerance for long time horizons for the recovery of investment capital, and hence for sustained investment in product development.[64] Perhaps it can be considered a modern counterpart of nineteenth-century American tolerance of cheaply finished, standardized products.

Still a third major element that contributes to the Japanese system has been a consistent emphasis on quality:

> quality, they say, is no longer simply the assurance of durability and reliability (the product works); that was the old way of thinking about it. Quality today is change, that is, ceaseless improvement, the continuing incorporation of new features that redefine the product and its uses and, so doing, make the consumer feel he wants it. Quality is the invention of needs.
>
> In such a game, speed means market share. Whereas in the automobile industry, for example, the lead time for new models was running four years and more, the Japanese reduced it to two. This kind of entrepreneurial advantage (in the Schumpeterian sense) can be translated into durable gains, and losers find themselves on a treadmill, running hard to stay in place.[65]

The relationship of an aggressive, nationally coordinated, long-term R & D program to all of these developments is obvious. As a percentage of gross national product, Japanese industrial R & D more than doubled between 1965 and 1986, increasing over this period from less than one-tenth to more than one-third of comparable U.S. expenditures.[66] Japan has long had the great advantage of being able to limit itself to a comparatively minor defense budget—on the order of 1 percent of gross national product.

Thus it has been able to devote a correspondingly much larger part of its R & D resources to the improvement of its manufacturing base as well as its consumer products.

But a fundamentally different valuation of the role of technology in planning for economic success is also apparent. Japanese firms, for example, are said to employ far more engineers on the factory floor than do their U.S. competitors, both to de-bug new production equipment and to improve manufacturing process know-how and extend its applications. More than two-thirds of Japanese firm-financed R & D is devoted to process research and improvement rather than new products, exactly the reverse of the proportions in the United States.[67] Japanese responses to a 1992 survey of leading technology-intensive firms, for example, indicated that more than 90 percent of their senior technical executives were members of their firms' boards of directors, as contrasted with less than one-quarter of their counterparts in the United States.[68]

Also testifying to a closer integration of R & D into management are constrastive Japanese and U.S. diversification strategies:

> [O]ver the last two decades, while some of the leading U.S. industrial corporations have looked to acquisitions to diversify their businesses and technologies, Japanese firms increasingly have made their R & D organizations the centers of diversification efforts. Firms in such mature industries as shipbuilding, steel, and textiles have exhibited an especially strong drive to technological diversification to provide opportunities for the growth their core businesses can afford no longer.[69]

With this exemplary emphasis on R & D has gone a corresponding Japanese readiness systematically to pursue the acquisition of foreign technology. Nearly three times the amount of technical information is purchased by Japan from the United States, for example, than by the United States from Japan.[70] Even more important is Japan's consistent de-emphasis on investing in fundamental or basic research, relying instead on open scientific literature, graduate training abroad, and continuing contacts with colleagues and programs in foreign (especially U.S.) laboratories. This sharp contrast with the priorities that have been followed unswervingly in the United States since World War II must be highlighted. It strongly suports the suggestion that international leadership in major areas of basic scientific research is not a *sufficient* condition for economic growth, although the advantage it sometimes confers in gaining time over competitors may be making it increasingly desirable or even *necessary*.[71]

This is not to imply that fundamental science and applied science or technology are seen in Japan as somehow opposed to one another. Quite the contrary, they are seen as constituents of a single community, but with different valuations of their respective contributions.

The Japanese cultural mindset of "learning by doing" as compared to the American mindset of "learning from principles" has proved to have a profound influence on technological innovation. It has led to the Japanese accepting, even enthusing about, incremental innovations which have a large cumulative effect. This is contrasted with an American predilection for home grown breakthroughs. It has also led to a Japanese willingness to undertake the manufacture of advanced materials for insignificant markets as a way of learning, with confidence that as learning proceeds and costs and quality improve, markets will develop.

A second profound cultural difference that permeates the mode of technological development is the strong American emphasis on individual achievement vs. the strong Japanese emphasis on group achievement. . . . This has been commented on by many observers, as has the difference in the Japanese system of lifetime employment and promotion based mostly on seniority as compared with the American system of contract employment, frequent job changes and promotion based on individual accomplishments.[72]

At first seemingly inconsistent with the importance Japan attaches to R & D, but on further reflection in underlying harmony with its favoring of pragmatic incrementalism over pursuing greater breakthroughs through basic research, is the disjunction between Japan's corporate sector and its universities. Technology-intensive firms seek out university graduates, but prefer to maintain control over their final, specialized training as well as the research in which they later engage. Ph.D.s are, therefore, produced in relatively small numbers.[73] As one of the country's most respected and senior engineering statesmen has described relations between Japanese universities and the corporate sector with regard to R & D, "the linkage is very tenuous, to the point of being negligible. Transfers from the private sector to the centers of higher education amounted to 0.55 percent of the private sector's total R & D spending for 1988, representing 3.3 percent of the corresponding budget of the educational institutions—the U.S. figures for the same year are 2.78 percent and 9.9 percent respectively."

Takashi Mukaibo, former president of Tokyo University and head of Japan's Academy of Engineering, goes on to characterize the universities as having been "singularly inactive in their quest to procure funds from the private sector," although more recently "some are timidly adopting steps in this direction."[74]

Technological Progress and Economic Growth

As recounted in this and the preceding chapter, the conditions under which technology originally contributed to the origins of mass production in the United States have undergone a profound transformation. At the outset, the frame within which action was enclosed was a largely internal one. A vigorous and growing population supplied both an urbanized industrial work force and a vast, relatively undiscriminating domestic market. Supplying it, as well as drawing upon abundant natural resources, required only an efficient internal transportation grid. The historic achievement of creating the latter helped to provide geographically dispersed models of management for the complex systems that would soon appear in other industrial sectors as well. Technological innovations were largely directed toward substituting machines for hand-processing to achieve uniformity as well as to lower costs, and economies of scale were the central objective of unprecedented industrial expansion. Partly as a result of having been insulated from the destructiveness of two world wars, the United States moved into a position not only of technological ascendancy but of dominance in many world markets for what were then high-technology products. Across-the-board U.S. support for free trade remained—although it was compromised in many particulars—a generally popular and sustainable position.

Today the position is different in several major respects. Laura Tyson, defining herself as a "cautious activist," has candidly and succinctly summarized them:

> During the last decade, new developments in trade theory have demonstrated that, under conditions of increasing returns, technological externalities, and imperfect competition, free trade is not necessarily and automatically the best policy. Promotional and protectionist policies by foreign governments can harm domestic economic welfare by shifting industries with high returns and beneficial externalities away from domestic producers and domestic production locations. Conversely, comparable policies at home can improve domestic economic welfare, sometimes at the expense of other nations.[75]

This alteration of our circumstances, and the changes in fundamental outlook that it may invite, have become a repeated focus of debate in newspaper and magazine columns and the subject of abrasive international negotiations. While our scientific leadership remains largely intact, the technological lead once thought to have stemmed directly from it has been substantially dissipated. Some role of trade restrictions in having brought this about is widely conceded, although the extent of the resultant losses and the details of its incidence on particular industrial sectors are tenaciously argued in internal as well as international fora. Trade restrictions of one kind or another are ubiquitous, and we are far short of the state of international understanding in which considerations of equity and the greater good of the greater number are likely to outweigh narrowly conceived national self-interest on any country's part.

A continuing series of less than wholly satisfactory interim outcomes is almost inevitable. Their details will take the form of complex trade-offs,

reflecting not only the bilateral or multilateral leveraging of overall economic power but the political strength of internal constituencies. The un-
deniable reality is that in many key, high-tech (other than military) fields,
the United States is repeatedly finding that its basic ability to remain competitive in international markets is in serious jeopardy. This is a problem
that must be grappled with primarily in its own terms, at home. It cannot be
effectively dealt with if it is viewed as a problem requiring only a readiness
on the part of U.S. competitors and trading partners to alter their policies in
its favor.

Advanced design, sensitivity to rapidly changing consumer preferences,
and superior manufacturing technology all have played a part in the successes of Japan and other U.S. competitors. But behind and alongside these
attributes are other strengths, in the domains of long-range corporate strategy and organization. Greater flexibility, and a readiness for more sustained efforts of greater vision, risk, and scope are characteristics that are
shared by our most effective industrial competitors—by no means all of
them Japanese. In relation to the U.S. chemical industry, for example, German and Swiss concerns have been remarkably more successful in their
diversification strategies. This effectiveness derives, in particular, from

> long-established organizational capabilities, close cooperation between firms—
> increasing the scope of their enterprises—and availability of capital. High rates
> of depreciation and a high degree of leveraging gave the German companies the
> cash flow that they needed to expand across a broad front.
>   . . . The great American chemical company might be disappearing, leaving the
> traditional field to the Germans, who rely on the scope of their varied chemical
> enterprises to give them long-enduring competitive advantages.[76]

Granting that conditions have changed in many essential respects, it is
impossible to reflect on this account without recalling its obvious parallel-
ism with the process, recounted at the end of chapter 4, by which the British
dye industry permanently gave way to German competitors in the 1870s
and 1880s.

However, there is a complex mix of secular changes affecting the position of technology in the U.S. industrial economy, on various time-scales,
and some are of a more positive and perhaps representative character than
this example suggests. Consistent with the very long time perspective that
is a principal feature of this study, one that is almost certainly of cumulative and fundamental importance has recently been identified by Moses
Abramovitz:

> In the nineteenth century, technological progress was heavily biased in a *physical*
> capital-using direction. . . . the bias of technological advance in the present cen-
> tury worked to reduce the relative marginal productivity of tangible capital com-
> pared with that of intangible capital, much of the latter embodied in labor.

Technological change tended to raise the relative marginal productivity of capital in the form of education and training of the labor force at all levels; in the form of practical knowledge acquired by deliberate investment of resources in research and development; and in other forms of intangible capital, such as the creation and support of corporate managerial structures and cultures and the development of product markets, which are the infrastructure of the economies of scale and scope.[77]

Of similarly long-term significance is the widespread introduction of corporate measures for the organized generation of new technology, a theme introduced at the outset of this chapter. More clearly than ever, technological change emerges as "a channeled historical process that exhibits strong aspects of inertia and continuity but that also is capable of generating abrupt changes in pace and direction."[78] Technological progress augments and is complementary with capital investment, accelerating the rate of advance of the most developed countries.[79]

While the distinction remains real and strong for many individuals and institutions, we have seen that there are broad areas of R & D in which science and technology have blurred into a near-continuum. Pronouncements from key agencies and legislators in favor of a shift toward greater funding of "strategic" research accelerate the trend.[80] Synergies multiply as computers, lasers, sensors, and many kinds of advanced materials find new uses other than those for which they were originally designed, and lead to further innovations and economies. Fixed R & D investments steadily mount, for manufacturing process improvements as well as new products.

The implementation of process innovations such as robotics drives down the cost of labor, creating the incentive of increasing returns if market share can be captured and assured. Other radical infusions into traditional manufacturing processes speed the introduction of new product and make possible the cost savings of mass production with greater flexibility to adjust to a wide variety of changing consumer preferences. Competitiveness in world markets becomes not merely a consuming corporate preoccupation but a national one. Even as this happens, many of the corporate entities involved tend to lose their national identity through mergers, buyouts, joint ventures, and cross-licensing.

But technology itself, as has always been the case, provides means and not ends. It is we as concerned citizenry, electing executives and legislators, appearing before regulatory commissions, conducting independent research, participating in nongovernmental organizations—who must set the goals and develop the mechanisms to assure that the benefits of well-being that can indeed be associated with technological progress will be wisely and equitably distributed.

8

New Paths: Technological Change in a Borderless World

THE GENERAL COURSE of technological advance, as it emerges from this study, has never been an orderly, predictable process. Technology itself is a multilevel phenomenon. Moving upward from a slowly evolving infrastructure of tools and techniques that has played little part in our account, one enters a more dynamic, vital field in which it is the more consciously designed battery of reproducible elements that plays the critical part in putting into effect every practical policy and purpose. Here, where technology is identified with systematic processes of analysis, choice, and implementation of plans and programs, it is inextricably a part of the larger social fabric. Here, too, it at least approaches Braudel's hyperbole—intersecting continuously with all the forces of change in history, if not quite covering as wide a domain as history.

Our concern has been primarily directed at successive impulses or surges of technological innovation, in historically and culturally varied settings. Their gathering momentum—the shortening, now almost disappearing, intervals between them; the progressively widened frontier of innovation and discovery; and the thickening network of interconnections and positive feedbacks—is most apparent in long historical retrospect and has not been adequately recognized. The waves or impulses form an accelerating, in some loose sense cumulative—but hardly inevitable or unidirectional—series.

At least from the standpoint of understanding the dominant forces that have converged to shape our present world, we have dealt with what appear to be some of the most decisive—permanently transformative—of those impulses. But it should not be forgotten that these are a small, selected sample. On a global scale, they reflect the historical experience of a geographically and culturally restricted segment of what could, and one day should, receive similar attention. Furthermore, having been chosen to demonstrate slowly accelerating, century- and even millennium-long cadences of change, they testify more adequately to the major avenues of earlier technological advance than to the simultaneous proliferation of fundamental changes in many fields to which we are now growing accustomed.

An underlying rationale in undertaking this study has been the need for a broadly contextual view of the predisposing conditions as well as the concomitants of technological advance. I believe that rationale has been amply justified. Limitations in the available sources of information, as well as the subsequent emergence of progressively greater institutional complexity, make comparison with pre-eighteenth-century episodes of rapid change difficult. But at least from the mid-eighteenth century forward until our own time, a convincing group of contextual features can be discerned to which the spurts of innovation seem to have been closely linked. Admittedly with few notable surprises, the list includes:

1. A generalized atmosphere of optimism, rising demand, rising expectations, openness and receptiveness to heterogeneity, to new external influences, and to change, experiment, and novelty. This coincides with an appreciable enlargement of the proportion of society (even if it was still largely limited to the upper strata) able to enjoy a significant margin of discretionary income.

2. Sufficiently broad-based literacy to sustain extensive networks of communication, embracing national as well as regionally based and local elites. This permitted a fairly rapid and effective circulation of news of shifting styles, standards, preferences, opportunities, and initiatives, in addition to the political, market, and administrative information that was their major preoccupation.

3. Few effective restrictions on the movement of personnel, including entrepreneurs and trained technical personnel as well as rank-and-file recruits for an industrial work force, into new activities and centers of opportunity. Status barriers may have discouraged, but did not foreclose, the possibility of upward social as well as economic mobility via these routes.

4. Not to be neglected is the role that the great cities like London and New York played in this as magnets for the adventurous and upwardly mobile as well as the dispossessed, as solvents of ascribed social distinctions, and as windows on a wider world of trade, discovery, and opportunity. I am mindful of V. Gordon Childe's old insight, mentioned in chapter 2, that an urban setting may have played an indispensable part in the inauguration of civilization itself. Many today view cities principally as sources of blight and insoluble social problems. But we should not forget that urban interchange and ferment may well have been a stimulus for some of the greatest cultural as well as material advances that humanity has ever made.

5. On the part of an elite capable of capital accumulation and investment, there was growing understanding of, and appetite and tolerance for, business enterprises with their attendant risks and uncertainties. A considerable segment of that elite was prepared, moreover, to devote time, energy, and enthusiasm to understanding the principles underlying new technologies, to participate in shaping technological choices and designs, and to take an active part in management.

6. An existing fund of organizational experience and administrative competence that could be adapted to the needs of new, unprecedented, technology-dependent enterprises.

7. Finally but of great importance, in the more profound of these impulses, and certainly in the epoch we have now entered, scientific theory and directed, experimental discovery played an immediate and crucial part. This must be understood as an intimate, two-way interaction, not a direction-setting flow of guiding ideas from theorists to practitioners. We have repeatedly been at pains to show that the role of technicians, and of the instruments for which they are often responsible, has been a fundamental, if usually badly neglected, influence in the progress of science itself. Ideas are to be found in technology as well as science. But my point is to insist that not simply practical discoveries but visions of new possibilities and interconnections form part of the context on the basis of which alone we can hope to understand technological advance.

Earlier scientific advances had a more delayed, diffuse relationship to technological progress, perhaps best described as a contributing or enabling role but not a sufficient, immediately precipitating one. The notion of an ordered, mechanical universe that found its supreme exemplar in the discoveries of Isaac Newton, for example, surely unleashed creative energies as it was broadly communicated and made popularly intelligible. We may never know precisely how that contributed to James Watt's insights nearly a century later that led to the transformation of the Newcomen engine into a much more efficient, more versatile source of steam power. But that there must have been a connection—that it helped to liberate human minds to think of transformations rather than merely marginal improvements—seems certain.

> Newtonian science, as embodied in applied mechanics, became the essential intellectual ingredient, the mental capital, of the Industrial Revolution. By the last quarter of the eighteenth century in Britain the same people who thought of themselves as enlightened, as teachers and appliers of Newtonian mechanics, were often the profit-seeking promoters of steam engines, canal companies, or factory-style manufacturing. In the pursuit of their interests they had spread the message of applied science more deeply and widely in Britain than in any other Western country.[1]

Similarly the explanatory, synthetic power of James Clerk Maxwell's electromagnetic equations opened vistas of new possibilities that at first were only intelligible to a handful of theorists like himself. By the latter part of the nineteenth century the lag for dissemination was shorter—a decade or two, rather than nearly a century. As understanding of the significance of his work spread, recognition of its potentialities for application surely made a contribution to the climate of increasing readiness to manipulate and experiment with sources and uses of electricity. Did this single-handedly unleash the flood of electrical innovations that followed? No, certainly not. Samuel Morse's telegraph pointed to the practical utility of electric currents carried over long distances decades before Maxwell wrote. But as with Newton, the comprehensive vision of a new world of lawful phenomena that were amenable to explanation and thence to previously unimaginable utilization also certainly had an important part to play.

But it must be emphasized that scientific discovery was not the only initiating or enabling agency behind waves of technological innovation, nor was it apparently a necessary one. That is certainly not the case, for example, for what some have termed the industrial revolution of the later Middle Ages. The new, diverse, ingenious, and astonishingly prolific uses of inanimate power in the later Middle Ages were inventive adaptations of working prototypes, and reflect no dependence whatever on the scholastic theorizing of the time. Moreover, while the precise lineage of the New-comen engine that was Watt's starting point remains obscure, it continues to seem most likely that it was an essentially pragmatic discovery. I believe it can best be visualized as having been patiently worked out by an inspired practitioner to fill an urgent social and commercial need, and as deriving little from scientific precedents or disinterested scientific curiosity.

Later still and on the other side of the Atlantic, the work at the Harpers Ferry and Springfield armories during the 1830s and 1840s that under-lay mass production with interchangeable parts was equally distant from contemporary scientific inquiry. It exemplifies a fundamentally different source of inspiration for technological advance—a breakthrough in manu-facturing processes that opened the way to new products with a previously unimagined array of new properties. In the same category are many innova-tions in applied chemistry—the search for cost-effective electrochemical refining methods that went on during the early twentieth century, for exam-ple, or the richly rewarded pursuit of whole new families of polymers of the years following World War II.

These narrowly focused illustrations of a systematic search for new in-novations set the stage for an approach that in more recent years has tended to become dominant. New institutions and incentives have been devised that explicitly systematize and stimulate the invention and adoption of new technologies. This was the central subject of chapter 7. The lags between even the most basic scientific discoveries and their adaptation to practical ends grow shorter, and may soon tend to disappear altogether. Instruments for manipulating and measuring, always a source of creative, two-way feedbacks between science and technology, play an ever more central part in opening new horizons of capability and understanding in both. As the trained scientific-engineering community grows exponentially, speciali-zation is carried to new extremes—but so also is the specialization of trans-disciplinary searches for ideas and discoveries that other specialists can learn from and incorporate, leading to new forms of integration and application.

What can be said in this expansive setting of the significance of the Information Age in which we now find ourselves? Beyond short-term, merely linear projections of present trends, most predictions are largely a delusion. But the power of the computer to transform our lives is so multifaceted and palpable that the temptation cannot be denied to see the fulfillment of its potential as the central axis of an approaching era of rapid and profound change fully comparable to the original Industrial Revolution itself.

Yet to be fully clarified are some of the major directions of change. The full extent of their probable impacts on society at large is still more difficult to discern. Many will surely find the uncertainty destabilizing and not a little threatening. Greatly accelerated and expanded, computer-dependent movements of financial and commercial proceedings, public and private databanks, and other information resources are already undermining international mechanisms for managing fluctuations in trade, credit, and currency rates. At a different level, they also pose grave threats of infringements to individual privacy. And a widely anticipated consequence of automation is that it will move us permanently closer to a two-tier society—indeed, a two-tier world—in which only a minority can find rewarding, productive employment.

But it would be myopic to see only the attendant risks and not the enormously widened opportunities that an Information Age will bring. As has been argued here consistently, the dangers lie not in the technology but in our failure to design and insist upon implementing measures that can control its new applications as they appear. The main thrust of computer-aided design and manufacturing ("CAD-CAM"), after all, is to widen markets and improve well-being by driving down costs and vastly increasing the range of choice and standards of quality of consumer products; meanwhile, human capabilities have been multiplied manyfold to master complex, nonlinear problems.

Some of these new capabilities open the way to fundamental shifts in our prevailing scientific paradigms. Computer simulations, for example, constitute a hitherto unimagined kind of laboratory for testing hypothetical behaviors and interactions in complex conditions affected by many feedbacks. Other challenges to which the new capabilities of an Information Age offer a response arise from growing "real-world" uncertainties. Unprecedented masses of information must be marshaled and analyzed if we are to grapple successfully with such issues as global climate change, new vectors of disease, and the economic volatility accompanying heightened world competitiveness and interdependence. Computer-aided education, including comunications networks to refine and diffuse improved educational materials and the new incentives to learning offered by virtual-reality techniques, may help to close the gap in employability that automation is opening. Even the sense of a decline of community that is widely complained of must take account of communities of an entirely new character that have sprung up through the internet.

Turning to the more remote past, anything recognizable as useful science becomes progressively more difficult to discern. A purely pragmatic source as well as course for technological advance seems all the more likely. But at least two instances suggest the need still to leave room for the possibility that large, enabling visions played a part. First, I have noted that a cluster of new pyrotechnological means for altering the properties of natural materials came to light fairly simultaneously late in the second millennium B.C. New ways of controlling the use of fire had been found to be the keys to the smelting, alloying, casting, and finishing of metals for ritual, ornamental, and utilitarian ends, and for the production of glazes, glass, faience, and frits. Iron's lower cost and superiority over bronze for many new uses may have furnished the basic incentive, but a sense of accomplishment and wonder at what human intelligence and skill could achieve with the aid of pyrotechnology was surely an important by-product.[2]

Second, it should be recalled that at least the remote origin of this study lay in the revolutionary quickening of pace accompanying the beginnings of cities, states, and civilization in fourth-millennium Mesopotamia. Here, too, there are suggestions of an organizational need and vision—the harmonious articulation of the work of different specialists supplied with centrally administered resources, and the invention of a written memory that could trace a pattern through time of how receipts were linked with disbursements.

Technology, as it emerges from this brief overview, provides few significant regularities in its inspiration or development. At certain places and times, it has moved ahead rapidly with a seemingly irresistible logic of its own. Most examples of this kind are recent, such as the almost unbelievable multiplication of computing power in a few short decades, or the transition from the first successful heavier-than-air craft to spacecraft traveling beyond the solar system within the span of less than a century. But that could be said also of the succession of innovations that transformed British textile production over a few short decades during the Industrial Revolution. It may even apply to the critical series of insights that opened the way to effective writing five millennia ago.

On the whole, then, we see technology as embedded in diverse larger systems while maintaining some dynamic qualities of its own that can be neither mandated nor accurately predicted. It probably has always been multicausal, path-dependent, subject to accident and opportunity, and in its most interesting and active dimensions far removed from those great, slow oceanic currents of which Fernand Braudel wrote. As we view it in the present, however, sweeping, mutually reinforcing changes that seem certain to continue into the indefinite future have become its overwhelmingly dominant characteristic.

Passive and Active Roles of Militarism and the State

The ultimate responsibility for promulgating and enforcing political and economic policies implicating technology continues to rest, of course, with the governments of nation-states. Across the long span of history described in the foregoing chapters, however, most states have played a relatively limited, largely passive role in this respect. Commonly included among their functions was little more than the setting of certain infrastructural or boundary conditions within which innovation and the advance of technology might, but not necessarily would, find encouragement and go forward. These included, for example, the provision of public education and monetary integration, measures providing for the assurance of property rights, and the maintenance of a degree of predictability and stability in taxation and the administration of justice. The times when states sought to seize and direct technological initiatives have been decidedly less frequent, although some of those interventions have been decisive.

Thus an impression is left that in most, although not all, circumstances technological advance has derived little direct support from the consciously promotional activities of states. Beyond the providing of some "externalities" that were no less necessary for other aspects of the social and political order, a fairly distant, noninvasive relationship between the two instead seems to have been the prevailing pattern. But does this somewhat surprising, perhaps even counterintuitive, conclusion withstand closer scrutiny?

For the earliest instance of a broad technological impulse that has been considered, the so-called urban revolution, it apparently does not. A persuasive case can be made that it was precisely the development of new, relatively powerful city-states, or the clusters of hierarchical, statelike institutions that emerged as rival cities grew and assumed hegemony over their surrounding dependencies, that provided a new and powerful technological impetus. Specialization aquired a new status in those cities. With increasingly coercive means, surpluses and reserves could be acquired not only to support crafts and administrative routines but greatly to extend and improve what those engaged in them were able to do. An inward flow of natural resources that were unobtainable locally, and that were needed as technological inputs, could be organized with state provisioning and protection. No doubt the relationship conferred benefits in both directions. New technological capabilities, while flourishing under state sponsorship, also contributed to the growing strength and effectiveness of the state itself. But at least it is clear that the state was not peripheral but central to developments in technology.

Not so in the classical world. By the mid-first millennium B.C., city-states had already been replaced in their initial position at the apex of the food chain (so to speak) for some two thousand years. While continuing to exist, and even thrive, at certain places and times, their relationship to new and larger, territorially more extensive units—empires—was often subordinate and always precarious.

What one can see of the Greek city-states under these conditions certainly does not suggest an active state role in the promotion of technology. Defense, active (not always successful) territorial aggrandizement, support of a navy that played some part in facilitating long-distance, commercial resource procurement, and meeting the other "boundary" or enabling conditions that were mentioned earlier all are in evidence. But technology and those actively concerned with it seem to have gone their own way largely unnoticed, and with a minimum of sympathetic involvement on the part of the sociopolitical elite.

The Roman Empire makes the same point in a different way. All the institutions of a state there certainly were, in great strength and complexity, and in territorially unprecedented extent. Immense efforts were devoted to consolidating the administrative structure and lines of communication of the empire, and to providing its cities with an appropriately secure and monumental ambience. But it is noteworthy that technology did not leap ahead under these seemingly most favorable conditions. The scale of technological applications grew enormously, but with a slavery-based economy the content remained remarkably static.

The next significant wave of advance, in the later Middle Ages, is equally instructive. A bristling, feudal countryside was dotted with fortified towns and the strongholds of a nobility habitually at odds with one another. Such resources as the crown could amass through capricious, predatory taxation and successful military adventures abroad were devoted to essentially nothing beyond the enhancement of the royal party's own political preeminence and the military prowess on which it rested. Yet on a highly decentralized basis there was no lack of vigorous investment and innovation. Supported almost entirely by local initiative, we have seen it widely expressed in the spreading utilization of inanimate power sources for many new purposes and in the introduction of new, craft-produced devices and manufacturing processes. Lords' manorial lands, ecclesiastical estates, and chartered towns all emerged as examples of effective, improvement-oriented management units. But of a favorable, let alone guiding, influence on these diverse processes by something recognizable as a state initiative there was essentially nothing.

Nor were conditions substantially different, at least within the territorial framework of Britain itself, in what has been seen of a long following span of centuries down to and through the Industrial Revolution. Royal power grew, but its resources continued to be devoted overwhelmingly to war. The aristocracy devoted little attention to commerce, and less to industry. Yet indirectly and with little evidence of conscious foresight, the pattern shifted in directions anticipating what was to come. Land enclosures, however narrowly acquisitive in their immediate objectives, precipitated an egress of rural villagers who would form the basis of an industrial labor force. The consolidation of large domains without customary restrictions then also encouraged capital investments in agricultural intensification, accounting for most of the gains in productivity that would help to meet the subsistence requirements of a growing, increasingly urbanized population.

It is important to note that the new centers of industrial growth were in formerly secondary cities and towns, as well removed geographically from the capital as they were from any royal stimulus. Like the new ports and arteries of communications, the factories that were their primary sources of wealth and employment were almost entirely private ventures. And those who led in the explosive growth that ensued were mostly nonconformist provincial entrepreneurs who had never previously achieved prominence.

Also illustrated by the English example, to be sure, are at least a few significant exceptions in which the state's military requirements hastened technological development. Naval strength, because it embodied the development of a vital stock of human capital and an articulation of many techno-economic subsystems that were of immediate civilian relevance, is an example of fundamental importance. Ship design and construction, transport logistics, navigation, seamanship and shipboard discipline, familiarization with foreign port facilities and routines and with the world's oceans—all were as vital for Britain's maritime commerce as for its naval supremacy. And a few other, considerably more circumscribed examples can be adduced as well.

> Cort's puddling-and-rolling technique was completed when its inventor was working on a contract for the Admiralty. Wilkinson's lathe, which bored the accurate cylinders needed for Watt's steam engines, was originally destined for cannon. The correct test for the net impact of military demand is, however, the question whether in the absence of military demand these innovations would have been substantially slower in coming. On that issue most scholars are wisely cautious.[3]

Similar generalizations apply as we cross the Atlantic and view at least the first century of the new American republic. There was an occasionally aggressive (if only briefly effective) imposition of import tariffs to support infant industries, or a program of sales or awards of bountiful western lands to further railroad construction, higher education, agricultural research, and, in general, a rapidly expanding frontier of settlement. Other than this, the federal intrusion into stimulating technology transfer and fostering industrial growth, by even the broadest definition, was very limited. The one major exception, again of military inspiration, was the development of elementary techniques of mass production involving replaceable parts in federal armories. It was indeed a significant exception, for it permitted the setting aside of considerations of cost in order to conduct the slow and difficult refinement of special-purpose tools and an articulated organization of work that underlay the eventual achievement of mass production. Hence we will need to return to it. But its significance lay in what it portended, not in the fairly modest annual expenditures of federal resources that were required along the way.

As recounted in chapter 6, the U.S. government's more direct and affirmative involvement in the support of science-and-technology (the two now beginning to cohere into this hyphenated form, with some recognizably common goals and interests) began with slow and tentative steps in the late nineteenth century. World War I powerfully accelerated the process, although only for a brief interlude. The federal impetus was withdrawn after the 1918 armistice almost as abruptly as it had been introduced, but the example remained fresh across the brief span of two ensuing decades until another world war commenced.

Building on this foundation and calling for a much greater, more demanding, and more prolonged effort on the part of the U.S. government than had been the case during World War I, World War II precipitated far more profound and permanent changes. There was a powerful demonstration effect through the government having directed the involvement of science-and-technology in a series of great, mission-oriented, and notably successful R & D enterprises. With the continuing need for heavy military expenditures associated with more than four decades of the Cold War that followed, a pattern of leadership became well established in which scientific, technological, military, and political elites opened and maintained effective channels of communication.

Lending weight to military influence were the greater urgency and specificity of military requirements than most scientific objectives, the less rigorous funding and review processes accompanying them, and the ambitious scale of effort that only military funding made possible. To be sure, the "vast bulk of civilian-oriented industrial R & D" continues to be funded on a decentralized basis "by the companies that expect to benefit from it."[4] But simultaneously and apparently irreversibly, a number of the most strategic, most rapidly advancing technological sectors were subjected to radically different mechanisms of coordination and control.

For the later stages of this long evolutionary process, in other words, there is no way in which the interest and support of the state can be regarded as other than a central element in at least certain key elements of advancing technology. Reinforcing this conclusion are the parallel trends toward closer and closer articulation of national and corporate strategies in the current era of intensified international competition. Japan may offer the most comprehensive example, with its Ministry of International Trade and Industry (MITI) adopting a flexible and unacknowledged combination of regulative, compulsive, and merely persuasive stances in coordinating long-term efforts to capture a larger share of world markets in selected, high-tech industries. Heavy government subsidies to Airbus Industries that have secured a substantial place for the European Community in the manufacture of commercial aviation are similar in principle, as is the European Strategic Program for Research and Development in Information Technology (ESPRIT).[5]

Nor is the United States fundamentally different. The joint establishment of the Semiconductor Manufacturing Technology Consortium (Sematech) by industry and the Defense Advanced Research Projects Agency was an early instance of the same strategy. High-density imaging and flat-screen displays are more recent examples in which defense needs have been offered as a rationale for government subsidies that would greatly improve an industry's international competitive standing. The political debate over "technology policy" and "industrial policy," essentially code words for differences in the degree and forcefulness of federal intervention, suggests, at a minimum, further movement in the United States toward a closer government-industry-university partnership. The pace of competition has become too intense and the stakes in terms of national well-being too high, for a relatively passive stance on the part of national governments to be a practical alternative any longer.[6]

Nothing better illustrates the crucial importance that state encouragement now has assumed than the rise to a dominant position in the U.S. economy of an enormous complex of aerospace industries. With military requirements as the primary rationale, sustained federal investments in R & D have consistently led the way to new horizons of possibility in every aspect of aircraft performance—range, speed, load, economy, and safety. In most respects, and with relatively little lag, these have proved as applicable to the needs of airlines as they moved into a position of unchallenged dominance in long-distance passenger transport as to the projected military needs for which they were originally subsidized.

There is no historic parallel for the ascent, in less than a century, from the first brief achievement of heavier-than-air flight to spacecraft venturing beyond the bounds of the solar system. Nor is there any question that the

United States has led the way throughout this process. And pioneering in virtually every successive technological breakthrough—in propulsion systems, airframe design, aerial navigation and avionics, and advanced, high-capacity production facilities—have been military applications. Government stimulus and subsidization have played a central part even where this was only more indirectly the case, as in airport construction and security, accident investigation, air traffic control, and the supervision of aircraft performance and personnel training standards. It is safe to say that the industry would be today only a small vestige of what it has become without these ubiquitous forms of state intervention.

As this example suggests, we must take fully into account the current extent of active state involvement in at least some domains of rapidly advancing technology. Military relevance, it would appear, is the key criterion. Recognizing this, any generalizations about the historically passive, only marginally supportive roles of the U.S. and British governments need to be carefully circumscribed in time and perhaps qualified in other ways.

In the British case, to begin with, subsumed under the Crown's preoccupation with militarism was the assurance of international trading pre-eminence with an unrivaled navy. Different views are sometimes expressed as to the magnitude of the benefits conferred on the mother country by the full extent of nineteenth-century expenditures as well as receipts from its imperial holdings. As noted in chapter 4, however, by any reckoning Britain found a virtually inexhaustible outlet for a preponderance of manufactured goods in the trade it assured for itself in overseas markets through its naval dominance. The textile and metalworking industries, constituting the very core of the British Industrial Revolution, thus were special, if indirect, beneficiaries of the most consistent and fundamental feature of the state's military policies. The rapidity of their growth in both scale of output and productivity unquestionably owes much to this conjunction of interest and policy.

Nineteenth-century America, on the other hand, is more truly exceptional. This was then a new country, expanding rapidly westward across a continent-wide area against very light resistance. Its great size, and the heavy flow of immigrants who arrived to help fill it, assured that there would be a large, uncontested market for its own manufacturers. After the success of its founding revolution its territorial boundaries, protected by two oceans, were secure against serious external rivals, and would in fact be repeatedly extended by peacefully negotiated purchases. Sectionalism could long remain a viable alternative to the vigorous growth of central state power under these circumstances. Hence the absence of much affirmative federal support for technology can be seen as just another consequence of long-maintained suspicion about all movement in the direction of centralism. In the United States the convergence of technical advance with decisive levels of governmental stimulus largely followed rather than preceded world industrial leadership.

The American and British examples apparently are, in any case, outliers toward one end of a wider distribution of states in which the advance of technology might have been considered. Had this account dealt instead with France or Germany, or with Renaissance Italy, the affirmative role of the state would have emerged earlier as a much more decisive, perhaps even fully determinative, force. Particularly in the field of railroad construction, reflecting the need for very large infusions of capital and for special authority in order to secure rights of way, state intervention uniformly played a far more direct and crucial part on the European continent than it did in the cases we have considered.[7] But at least this discussion does suggest that there is no single formula for technological advance. It has been a multicausal, multilinear process, following different paths in different circumstances. Adaptive to local particularities and criss-crossed by interactions that frequently produced positive feedbacks to redirect its course, the outcome in each instance appears also to be an expression of traditional differences in styles of governance.

And then, of course, there is the further, no less plausible possibility that the arrow of causality could sometimes fly in the opposite direction. Superior weapons, communications, logistics, and mobility—all products of technological advantage—may well, under some circumstances, encourage a state to rely on aggressive, militaristic policies. Broadly based industrial and economic superiority, again with vital technological roots, may serve not only to hold an empire in place but to give it "a new cutting edge overseas."[8] The state's powers, on this reading, may better be regarded as augmented by technological superiority than as a precondition for such superiority. But to test this proposition more fully would require a lengthy excursion into international political economy, well beyond the possibly already over-extended scope of this book.

Growth, Productivity, Equity, and Sustainability: Complementarities or Alternatives?

A regrettably common misperception of particularly some newer, less familiar, seemingly more menacing forms of technology is that they are an invasive, extra-human force. Apart from small cadres of scientists and irresponsible zealots, in this view, most of human society is unwittingly left at risk of a growth it cannot control. This book has partly sought to document the inaccuracy of this characterization, and to reinforce a different one. Technology is in every respect a human construct, a body of knowledge and practice that can only be designed, improved, disseminated, manipulated, and put to work at human direction. And the mechanisms by which it is directed, while diffuse and poorly understood, are no different from others that are firmly under the control of the society at large.

Technology itself is diffuse, existing at many levels and changing in response to many stimuli. Probably it can best be thought of in the plural, as a diverse, overlapping, interacting set of *sociotechnical systems*. Decisions as to technological ends and means may or may not be public, or even readily identifiable as decisions at the time they are made. But at least if they are at all nonroutine and difficult, they usually have to involve complex, consciously balancing processes or "trade-offs." It is unquestionably the case that most human decisions about how to deploy new technologies do not result in completely egalitarian distributions of their benefits. The overall outcome of technological advance throughout history may have been overwhelmingly positive, but that clearly does not apply to every individual episode or component. Trade-offs have all of the defects of the persons or processes deciding upon and imposing them.

Inherent in any trade-off, to begin with, is a recognition that there are costs as well as benefits. But who bears the costs, as often distinguished from who receives the benefits? What likelihood of result justifies what degree of risk or uncertainty? What mix of a pursuit of short-term or personal objectives is tolerable, alongside a concern for the common good or long-deferred advantage? What preference for clinging to the familiar as opposed to venturing in new and unknown directions? All such considerations as these naturally are mediated by social pressures, variable personal preference schedules, power hierarchies, culturally prescribed perceptions, and value systems. But the essential point is that it is nothing but a convenient, responsibility-shirking myth that technologies impose their own courses of action. The responsibility for positive or negative technological outcomes, and for the wisdom or equity with which they are allocated among the population at large, resides nowhere else than in the social order—in individuals, groups, and the institutions through which they organize their lives.

Erroneous as this common misperception may be, there is no denying that the creative power inherent in some advancing technologies is now confronting us with increasingly complex, urgent choices. In earlier times the apparent need to make such choices usually turned out instead to allow their almost indefinite postponement. Alternative solutions could be found that offered some additional satisfaction to every contending party. But many of the new generation of problems involve a painful, almost immediate balancing of costs and benefits in what the contending parties recognize has tended to become a zero-sum game.

Consider, for example, an emerging consequence within the United States of the intensified international competition in manufacturing that

was discussed in chapter 7. With automation, "lean production," and related, technological advances, manufacturing productivity has risen to a
point where renewed industrial employment for a significant proportion of
the formerly much larger work force is in serious jeopardy. As a proportion
of the total U.S. work force, manufacturing employment peaked at about
34 percent in 1960 and has declined ever since. Even as output rose sharply
during the prosperous decade of the 1980s, employment in manufacturing
dropped by 10.6 percent (and by 18.9 percent in larger firms employing
more than five hundred workers) according to the Bureau of the Census.
Bureau of Labor Statistics projections suggest that the value of manufactured goods will rise by an additional 41 percent in the next fifteen years
while the number of workers will fall at least 3 percent further.

The growth of the service sector, often visualized as a palliative, seems
unlikely to be able to compensate for such losses in value-added terms. An
involuntary transfer from the manufacturing to the service economy thus is
likely to be accompanied, for most of those without specialized skills or
advanced education, by a significant and permanent loss of income. Worse,
the services sector is itself threatened with the migration of a growing part
of its high-tech component to less costly offshore locations, a loss facilitated by technological advance in the form of high-speed computing and
improved communications.[9]

How do we muster the vision (and build the necessary political consensus) to meet a massive, long-term problem of this kind? It is not only illu-
sory but dysfunctional to think of technology as the source of the problem.
New technologies are merely elements in a larger battery of means employed to meet a different challenge. But as is more and more often the
case, challenges turn out to be linked and interdependent. Meeting one
exacerbates others.

The solution to quandaries of this kind can only lie in finding new ways
to deal comprehensively, within a larger, systemic framework, with policy
questions that in the past it has been politically more convenient to treat as
isolates. In this case, for example, the maintenance of an adequate level of
unskilled and semi-skilled employment opportunities, valuable enough in
its own right, also would contribute to overcoming a number of other social
ills. Somehow, then, a balance needs to be struck in which both the creation
of more job opportunities and the enhancement of industrial competitiveness find their appropriate places as national goals. And means as well as
ends—tax strategies, budgetary priorities, corporate planning, and investment incentives—all need to be weighed and employed in striking the right
balance.

The stresses that international economic competition creates in developed countries are only a symptom of tensions accompanying even broader
international problems. Virtually universally in the world community,

growth in GNP is viewed as a desirable goal or even an imperative. But at what point will parallel, self-interested pursuit of this goal by independent national states, beginning from different starting points and with different resource endowments, force a retreat on other goals of international comity and equity?

This is no longer a merely abstract question. Aspirations for growth and improved standards of living in the less-developed world already are described with increasing frequency in international meetings as threatened with postponement or foreclosure by the lending practices, investment decisions, and choices of consumption-oriented lifestyles of developed countries, and especially of the United States. More often emphasized by the developed countries is the counter-argument that the continuing high rates of population growth to be found in some countries in the less-developed world are the greater global danger and deterrent to well-being.[10]

Disputes like this all too easily can extend to ever-deeper levels of detail. The so-called "green revolution," itself a product of western biotechnology, unquestionably provided a margin of flexibility that for several decades has allowed the starkness of a Malthusian limit on population in some densely populated, less-developed countries to be held in check. Will similar advances continue indefinitely to serve the same end? Some authorities are doubtful, arguing that further productivity gains from conventional technologies are likely to come in progressively smaller increments and will not sustain aggregate demand beyond the middle of the next century. Others take a cautiously positive or even aggressively optimistic view. But complicating the picture, and magnifying the challenges to international relations that it poses, is an apparent deflection of advanced research away from world subsistence problems and toward providing unique food products for niche markets in developed countries instead.[11]

Nothing is to be gained, of course, by shifting blame in either direction. Behind the collisions of views are gathering forces that intersect to deepen a single, systemic problem: "The magnitude of the threat to the ecosystem is linked to human population size and resource use per person. Resource use, waste production and environmental degradation are accelerated by population growth. They are further exacerbated by consumption habits, certain technological developments, and particular patterns of social organization and resource management."[12]

As this suggests, it has become a matter of world concern that the environment and natural resources of the planet Earth may at some point impose their own limits of habitability, even if those limits still remain very ill-defined. There is no disagreement that present trends involving energy consumption, environmental pollution and degradation, and the like can only dangerously narrow the time interval available for readjustment before such a crisis might occur. How and at what point does the need to arrive at a more sustainable pattern for the whole of the planet impose limits on national rates and directions of growth?

To ask this question is to add an issue of intergenerational equity to the issue of equity for members of the present generation. Whether consciously recognized or not, the possibility of trade-offs can no longer be ignored between benefits sought by the present generation and the consequences as little as two or three generations hence. The relevance of technological advance to the larger issue of sustainability has thus become a real concern not merely of environmentalists and ethicists but of economists and planners.

Admittedly, sustainability is at best a somewhat elusive concept. Presumably it stands in opposition to unrestrained growth or emissions of pollutants, heedless of long-term problems of resource exhaustion or environmental degradation. But something more positive is implied than merely a strategy of moderation. The only acceptable position that I can identify is one that has been thoughtfully propounded by economist Robert Solow:

> If sustainability means anything more than a vague emotional commitment, it must be required that something be conserved for the very long run. It is very important to understand what that something is: I think it has to be a generalized capacity to produce economic well-being.
>
> It makes perfectly good sense to insist that certain unique and irreplaceable assets be preserved for their own sake. . . . But . . . it would be neither possible nor desirable to "leave the world as we found it" in every particular.[13]

Extending as well as affirming Solow's proposition, Harvey Brooks points out that

> We must take into account not only the resources we use up but also the resources and human capacities we leave behind, including especially newly created physical infrastructure, new knowledge, and new capacity for technological and social innovation. This includes the broad notion of "social capital" as a development of social organization, social norms, social networks, and social trust or sense of community that enhance the capacity of ever larger social groups for "coordinated action."[14]

Brooks accordingly suggests that "a sustainable policy for the selection and deployment of technologies is one that forecloses as few future options as possible." He frankly recognizes that any approach to sustainability is history- and culture-bound, and cannot aspire to universality or permanence. Scale, too, plays a part: what seems sustainable in isolation may not be so in a larger context, or vice versa (pp. 33, 35–36, 38, 42–43).

Technology's role in all this can rightly be thought of as paradoxical. On the one hand, the industrial economy is at least the immediate consumer of scarce resources and the primary source of much environmental disturbance or deterioration. On the other hand, it also provides the major means by which these and similar problems can be corrected. With regard to most scarce resources, improved prospecting and extraction methods have regularly led to a more rapid rate of growth of known reserves than of consumption. Similarly, the only way to address most serious pollution problems is through further advances in technology.[15]

Of course, these are merely abstract principles. The reality is more complex, difficult to attain, and occasionally contradictory. Take resource depletion, where measurement efforts are accompanied by formidable problems. So-called proven or economically recoverable reserves, not to speak of potential reserves, are moving targets, subject to upward revision with improving technologies. Prevailingly faulty statistics make quantifying the net volumes depleted on a worldwide scale at best very difficult and politically sensitive. While monetization of depletion losses seemingly provides an opportunity to reduce disparate trends to a single, easily recognized standard, there are many resources for which calculations of value in terms of imputed market prices are subject to grave disagreement.

In a more positive light, examples can be found illustrating a broad trend toward "dematerialization." In addition to an overall reduction in the weight of materials used and in the "embedded energy" content of many products, this typically involves a shortening of process chains to eliminate the production of intermediate products. Beyond increasing the efficiency of flows in the processing of materials, a long-term, economically advantageous strategy can be discerned of shifting "industrial metabolic technologies" in the direction of "reduced extraction of virgin materials, reduced loss of waste materials, and increased recycling of useful materials."

At least partly counterbalancing this, however, are diverse developments whose overall effect is negative. Photocopying and facsimile reproduction are among many ballooning uses of paper, even while they are at least partly duplicative of computers and electromagnetic disc storage. Extension of commuting paths as a result of suburban dispersal of settlement has a massive cost in energy as well as time. Many inherently dissipative uses of materials fall into the same negative category, including

> packaging materials, lubricants, solvents, flocculants, antifreezes, detergents, soaps, bleaches and cleaning agents, dyes, paints and pigments, most paper, cosmetics, pharmaceuticals, fertilizers, pesticides, herbicides, and germicides. Many of the current consumptive uses of toxic heavy metals such as arsenic, cadmium, chromium, copper, lead, mercury, silver, and zinc are dissipative in [a] strict sense. Other uses are dissipative in practice because of the difficulty in recycling such items as batteries and electronic devices. . . .
>
> It must be pointed out that short-term economic incentives do not necessarily point in the direction indicated. For example, market forces appear to favor product differentiation and specialization, but these trends increase the costs of repair

and recycling. In poor countries, such as India, there is virtually no such thing as "junk." Any manufactured product, no matter how old or obsolete, is likely to be repaired or rebuilt and retained in service as long as physically possible. When it can no longer be repaired, it will be disassembled and useful parts will be separated for further use. . . . By contrast, in advanced countries, manufactured products are becoming more and more complex and correspondingly difficult to repair. This is particularly true of electronic devices such as printed circuit-boards and cathode-ray tubes.

Moreover, as products are designed to be more reliable, so that repairs are no longer "normal" and expected, disassembly is becoming more difficult and in many cases is actively discouraged. In fact, for warranty reasons, critical subassemblies are often sealed and must be either returned to the factory or discarded. Finally, the complexity of products is often reflected by the increasing complexity of materials, which makes recycling inherently more difficult.[16]

The most serious as well as urgent challenges, and the correspondingly greatest economic liabilities, are found in the disposal of wastes and toxic effluents rather than in increasing the efficiency of use of materials and energy. As measuring instruments become more accurate, adequate removal of contaminants becomes a shifting target. The chemical evolution and dispersal of wastes does not cease at the moment of their disposal. Different contaminants react with one another in unforeseen ways. And "because causes, effects, and cures are still elusive in many large environmental problems and enormous challenges keep appearing, the condition that has developed is obviously one in which the legal profession can flourish."[17]

Technological Risks, Uncertainties, and Opportunities

Risk, as the term has come to be used by health and safety analysts and in the insurance industry, denotes not certainty but a measurable probability of loss or damage. Uncertainty implies no such threshhold of potentially quantifiable expectation. Actuarial needs fully justify the distinction. However, when used in another setting—as in trying to characterize complex technological risks and how they are perceived—a stress on accurate measurement can have the reductionistic effect of forcing an unnatural rearrangement of very divergent issues into a single linear scale of pecuniary valuations.

Technical and popular usages diverge so sharply that they "might be said to be speaking different languages." Only slightly to oversimplify, the former concentrates on calculating death rates, believing that deaths from any cause are (or should be) fungible units. From this vantage point, the technical view of risk is that its calculations produce "real" and "objective" data, in contradistinction to what the public merely "perceives" as risk. And unquestionably, the public view is more context-dependent and infused with not entirely consistent sets of values. Careful studies have elicited a number of broad clusterings of public perception that are independent of the technical "body count." Regarded as among the least "socially acceptable" are risks entailing sudden, heavy loss of life (e.g., from jumbo jet crashes); slow-acting, insidious poisons (e.g., from pesticides); involuntary exposure (e.g., atomic test fallout); and "dread" risks of truly catastrophic potential (nuclear war, great chemical disasters like Bhopal), as opposed to self-chosen risks (e.g., mountain climbing).[18]

Detectable behind many of these perceptions is a fairly coherent, widely shared value position. Among its features are a concern for distributive justice, the participatory rights of the governed, and the contractual autonomy of the individual. The exponentially higher fear or repugnance of "dread" risks or great disasters than what can be empirically demonstrated in past experience is noteworthy. Perhaps this reflects a valuation placed on the continuity and well-being of whole communities as greater than the sum of their individual parts. But comparing the far greater preoccupation with the undemonstrated but surely small likelihood of nuclear power disasters than with the substantial, known morbidity and mortality resulting from coal-fired generating plants (and mining the coal they consume), this may partly mean instead that the public reacts more strongly to disasters that can be sharply highlighted than to more diffuse conditions of risk.

To most of those holding them, their own value-influenced perceptions are no less valid, and considerably more important, than any unidimensional series of more "objective" findings. As Mary Douglas incisively comments,

> The wrong way to think of the social factors that influence risk perception is to treat them as smudges which blur a telescope lens and distort the true image. This metaphor justifies a negative approach. But the social point of view thus dismissed includes moral judgments about the kind of society in which we want to live. Why should they be summarily brushed aside? A better kind of analysis might treat such transformations of the image not as distortions but as improvements: the result of a sharper focus that assesses the society along with its assessments of risks.[19]

To be sure, not all divergences between "perceived" and "objective" criteria can be accounted for in terms of ethical or value positions. Essentially psychological explanations are apparent for some. The carnage on our highways is passively accepted as "accidents," with the imputation of a random mass of personal bad luck or misjudgment, while rare lapses in relatively much safer air transportation provoke exhaustive investigations and expensive countermeasures. Over-optimistic biases as to the consequences of personal exposure to risk are common. People are reportedly more concerned to avoid negative consequences than to secure positive ones of equal likelihood. Attitudes toward very improbable events tend to be speculative and undependable, sometimes dismissing them entirely and at other times elevating them to a supreme and undebatable level of importance. The complete elimination of a lesser hazard typically is overvalued, in comparison with a more significant but merely partial reduction of a greater one. But once again, public attitudes toward risk—conspicuously including risks thought to be associated with unfamiliar new technologies—are objective to those holding them.

A pessimistic but compelling case can be made that technological risks are inherently complex—in the last analysis neither "objectively" predictable nor avoidable. So many uncertainties, incommensurable factors, and nonlinear feedbacks are involved that neither their causes nor their effects can be predicted with much authority. "Normal accidents"—only superficially an oxymoron—is the term introduced by Charles Perrow to cover this aspect of unforeseeable inevitability: "Accidents are inevitable and happen all the time, serious ones are inevitable though infrequent; catastrophes are inevitable but extremely rare."

Technological risk probabilities are grossly underestimated, Perrow suggests, often by ascribing accidents to human or operator error that are in reality products of badly designed human-machine systems and interfaces. Typically given inadequate attention are new complexities introduced by alterations in the scale of operations, as well as critical transitional phases (start-up, shutdown) and the convergent effects of routine, minor errors that by themselves are easily correctable. Particularly dangerous are systems characterized by interactions that are complex as well as tightly coupled. Redundancy of safety features, often thought of as a countermeasure, may introduce new complexities and make the imminent failure of the system more difficult to recognize. Providing false reassurance, it can also lead those in charge merely "to run the system faster, or in worse weather, or with bigger explosives."[20]

This only begins to probe the difficulties in establishing "objective" evaluations of technological risk by any standards. Robert Kates has noted that while "scientific theories of comparative degrees of hazard are just being developed," it appears intrinsically more difficult to establish the probability of risk for some technologies than others. Low-level effects of toxic chemicals and radiation have proved particularly difficult to measure, and in many cases even to establish.

Other problems arise from the incapacity of courts and regulatory agencies to set consistent standards and enforce prompt compliance. Multiplying adversarial proceedings over the reputed toxic or carcinogenic effects of substances like agent orange, alar, asbestos, chlorofluorocarbons, fluoridated water, polychlorinated biphenyls (PCB's), and tetraethyl lead (merely to sample a long list) contribute to a generalized atmosphere of mutual suspicion and fear. There is an understandable tendency to respond to a particular accident by setting strict limits of tolerance or toxicity, neglecting the imbalances this creates with other dangers that may be even greater. Equally understandable, and unfortunate, is the tendency to concentrate on what appear to be the most obvious risks while neglecting to pay attention to the tails of the probability curves.

Still another problem is rooted precisely in the nonlinearity of popular responses to different kinds of accidents. As was the case at Three Mile Island, for example, very heavy damage compensation and cleanup costs may be imposed by public alarm even in the absence of immediate fatalities (and expectation of at most relatively few later cancer fatalities). Even heavier was the burden of nonacceptability that this accident may have placed on the nuclear power industry. Other second- and third-order consequences that are often not taken into account include adverse impacts on consumer products, insurance rates, property values, tourism, and even (as in the case of Three Mile Island) on the mental health of the adjacent population. In aggregate, their costs may vastly exceed those of the first-order consequences that are easiest to calculate.[21]

Technical expertise in risk assessment thus can fall far short of providing an adequate guide to the risks associated with new technologies, either directly associated with them or arising from how they are popularly perceived. There is a natural tendency to bring disciplinary specialization to the field, but this can have the adverse effect of impeding recognition of second- and third-order consequences that fall within the purview of other disciplines. Technological risk, in other words, is a subject calling for the interaction of specialists not only with the concerned public but with generalists prepared to maintain an organizationally independent, broad, and consistently unconventional—even provocative—perspective.

The historical evolution of approaches to improving the safety of public water supplies provides an excellent example of problems that have arisen from the tendency to work within a narrow disciplinary focus. Physicians were the first to turn their attention to the adverse effects of human contamination. Through their efforts the impacts of cholera were much reduced by removing sewage from populated areas. Raw sewage added to water bodies that were also the sources of drinking water then led, however, to a new generation of threats to public health. These were met, on the advice of a different set of specialists, by chlorination and the consequent, virtual elimination of typhoid. But now, once again, a new generation of problems has been created. At great cost, most cities have installed facilities that admirably met earlier design criteria but still failed to take a sufficiently broad, ecosystemic view. Besides wasting tremendous quantities of water and nutrients, most contemporary installations bypass expensive treatment facilities with uncontrolled storm drains. They also fail almost completely to check discharges of heavy metals and exotic chemicals, and have contributed to spreading eutrophication and contamination of increasingly valued wetlands and natural water bodies.[22]

Were it possible for scientific and expert opinion to be comprehensively brought to bear on issues of technological risk, the disjunction between "objective" evaluations and popular "perceptions" might be brought within reasonable limits if not overcome. Panels of individuals qualified to deal with all aspects of such questions, including generalists as well as specialists and balancing the representation of different scientific viewpoints, are regularly established under the auspices of the National Research Council to advise federal agencies on matters of general policy. But day-to-day regulatory and judicial processes proceed under different, essentially adversarial, legal principles. Hence the outcome is often starkly different:

> Too many organized interests have both the incentive and the ability to pattern as experts and to create accounts of risk that find plausible support in the record of available information. The ambiguities and uncertainties of policy-relevant science undoubtedly facilitate the production of such alternate readings. But equally important to the politics of risk in America is the relatively nonhierarchical organization of science as well as politics.[23]

The problem is less often one of direct subordination of scientists to political or economic interest groups than of the nontransferability of a consensual mode of fact- and theory confirmation to a fundamentally more disputatious realm:

> the validity of scientific "facts" and "theories" depends on tacitly negotiated, interpretive conventions that are shared among communities of researchers. Science, thus "socially constructed," retains its authoritative status as long as its underlying premises are not scrutinized too closely or with hostile intent. It is, however, always vulnerable to deconstruction, and particularly so in the policy arena, where pressure from competing interest groups and organizational cultures routinely militates against consensus-building or closure around any particular account of "reality."
>
> . . . Advantage can be gained, moreover, not only by advancing one's own account of the "facts," but also by weakening the accounts of others. Thus, procedures that allow for open criticism do indeed "bring new data to light, and challenge gaps in reasoning," as the U.S. courts foresaw, but the commitment to scientific rationality, coupled with the commitment to democratic openness, leads in the policy environment not to enlightened consensus but to a mounting cacophony.[24]

But in the end, what means are there for materially improving human life and overcoming its many adversities, other than scientific-technical and sociotechnical ones? Let us readily concede the adverse human and ecosystemic impacts of some developments that have been characterized as technological "progress." It is certain that, the frailties of human institutions and foresight being what they are, there will be more. Having granted this, what other means has humanity ever had than "extrasomatic adaptations"—in the broadest sense, technological ones—to enlarge our reservoir of useful knowledge and provide ourselves with a great and growing band of insulation from the ordinary risks and catastrophes that beset life? This would seem to provide better reason for hope, and perhaps even pride, than for fear and regret, if we adequately take cognizance of the immense technological progress of our own time.

Toward an Unbounded Field of Interaction

The strategic importance of technology for human well-being in the modern context is the imaginative flexibility with which an immense repertoire of knowledge and technique can now be routinely drawn upon to meet unprecedented, newly emerging needs and goals. Less and less tied to narrow prescriptions aiming only at a range of traditional solutions, it should be thought of as a primary provider of new combinations of means to reach consciously selected, socially constituted ends. Some of those ends may well involve ill-considered or positively harmful misapplications of technology. We should not confuse the further enlargement of our capabilities, however, with the choices we make over how to employ them.

Yet technology is also more than a battery of capabilities waiting passively to be employed for ends that are specified by an entirely separate social order. As we have repeatedly seen, especially through the example of the largely neglected contributions of instruments (and instrument-makers) to science, means-ends relationships involving technology may all be reciprocally interactive or dialectical. New or enhanced technological capabilities lead to new scientific discoveries. By so doing, they contribute to an indefinitely ongoing alteration of existing judgments and perceptions. Thus playing a part in exposing the limitations of earlier goals and outlooks, it cannot be denied that they also play at least some part in encouraging visions of new ones. Yet the larger quest within which this fits, as in this case the disciplined pursuit of scientific discovery, is one that has long been socially sanctioned and is not merely technologically generated.

There is yet a further technological contribution that actively shapes the social order rather than merely responding to its directions. Many new technologies (possibly all to some degree), with the computer perhaps the most salient example, turn out to be broadly versatile and unexpectedly open-ended in their applications. As new uses for them proliferate, previously undreamed-of pathways of analysis and communications are opened.

Uncovering important new unities, computers may be almost uniquely effective in breaking down the barriers between the increasingly specialized realms that compartmentalize our lives.

This book has not attempted to take into account all technology. It deals only tangentially with the continuities of everyday life, and not at all with those great, unconsciously slow-moving oceanic currents through history that Braudel identified. Industrial technology is, by contrast, driven forward at a more dramatic, still accelerating pace by irregular, focused waves of innovation. It is an historic emergent of a more systematically ordered, goal-directed character. Although some of its roots are ancient, it is largely synonymous with modernity.

This contextual view of impulses of industrial change has had to look outward from it in two primary directions, both increasingly closely interactive with technology to compose a single transformative force. One is toward science (or its precursors), the other toward the larger environment of business enterprise, government institutions, and public policies. Technology serves, in fact, as a vital nexus in which the three must come together. Their dynamic convergence, complex and indeterminate as we have seen its past (and prospective) outcomes to be, is admittedly responsible for a number of serious global challenges still waiting to be addressed. But more importantly—overwhelmingly—it provides us with the only effective means we have for finding a way beyond our present problems to a better life.

Technology can no longer be thought of as a bounded domain at all, if indeed it ever was. On the one hand it links the advancing scientific understanding of the natural world with the newly burgeoning world of the human-directed and artificial. On the other, it is wholly dependent for its growth and effectiveness on a world of human organizations that has learned to measure needs and preferences and respond to them; to formulate realistic plans and programs; to balance conflicting claims, rationales, and issues of equity; to explain and persuade or if necessary compel; to meet crises; and to build cautiously and cumulatively on a growing reservoir of knowledge and tradition. Technology is best thought of, within this framework, not as an entity at all but as a field of interaction—as a prospect that lies open before us for a growing, enormously fruitful, mutually enriching interchange of ideas and resources.

Notes

Preface

1. Marx 1992a: 407.
2. Hall 1978: 92–96; Braudel 1981, 1: 431.
3. Willey 1988: 186.

Chapter 1
Paths of Fire

1. Hughes 1994; Scranton 1994a: 150.
2. Von Hippel 1988: 25, 30, and related sections document the existence of whole classes of relatively recent product innovations for which, contrary to traditional expectations, users and not manufacturers have been responsible.
3. Nelson 1990: 194.
4. Jordanova 1993: 478.
5. Marx and Smith 1994: xii.
6. Bennett 1980: 204.
7. For an up-to-date group of studies that reflect both the present state and the promise of the field, see Gordon and Malone 1994.
8. Wajcman 1994: 193–94. One possible explanatory approach that is "central to the radical feminist analysis," Wajcman points out, "is a concept of reproduction as a natural process, inherent in women alone, and a theory of technology as intrinsically patriarchal. Technology, like science, is seen as an instrument of male domination of women and nature." Wajcman's cogently stated grounds for finding "fundamental problems" with this view deserve re-telling:

> The belief in the unchanging nature of women, and their association with procreation, nurturance, warmth, and creativity, lies at the very heart of traditional and oppressive conceptions of womanhood. Rather than asserting some inner essence of womanhood as an ahistorical category, we need to recognize the ways in which both "masculinity" and "femininity" are socially constructed and are in fact constantly under reconstruction.

9. Stanley 1993: xxxvii–xli; Macdonald 1992: 576.
10. Braudel 1981: 334.
11. Braudel 1973: ix.
12. e.g., Singer et al. 1954–58: 1, vii.
13. National Academy of Sciences 1969: 16.
14. Merrill 1968: 15, 585.
15. Brooks 1980: 65–66.
16. Bijker and Law 1992: 3.
17. Girifalco 1991: 10.

18. Lazonick 1991: 8, 24, and related sections.

19. Scherer 1984a: 8; Schumpeter 1934.

20. Kuznets 1979: 12–13; cited in Lazonick 1991: 322–23.

21. Dosi et al. 1992: 7. The "cumulative and self-propelling advance" characterization is taken from Landes 1969.

22. Polanyi et al. 1957.

23. Granovetter 1993.

24. Adams 1991; Larsen 1987.

25. DiMaggio 1990; Muldrew 1993: 169, 173.

26. Persson 1988: 12.

27. Dosi et al. 1992: 7.

28. Oral comment quoted in Waldrop 1992: 119.

29. Kline and Rosenberg 1986: 302.

30. Lee 1986: 28.

31. OECD Development Centre 1989: 23.

32. Abernathy and Clark 1985: 4.

33. Schumpeter 1939: 1, 102; cf. Freeman 1982: 207–11.

34. Thirtle and Ruttan 1987: 2.

35. Smith 1993a.

36. Schumpeter 1950: 82; Schumpeter 1939: 104; cf. Elster 1983: 113 and related sections; Rosenberg 1994: 49–57.

37. Girifalco 1991: 32, 88.

38. Arthur (in press). On the concept of punctuated equilibrium in biology see Eldredge and Gould 1972, and Gould and Eldredge 1993. Cf. Mayr 1988: 457–88.

39. Arthur (in press).

40. Girifalco 1991: 31–32, 81–86; the quote is from p. 32. Cf. Mensch 1979.

41. Berry 1991: 7–8.

42. Rosenberg and Frischtak 1994: 68–79.

43. Berry 1993: 3–4; cf. Hall and Preston 1988.

44. Berry 1991: 187.

45. Hughes 1987: 54.

46. Hughes 1983: 6, perhaps the most convincing and effective account of large technological systems yet published.

47. Constant 1987: 224–25, 228–29, 231–32. It should be noted that I have retained the single, overarching term *system* for the three, successively more embracing levels that Constant distinguishes.

48. Scranton 1994a: 151.

49. Landau and Rosenberg, eds. 1986: 302.

50. Hughes 1987: 52.

51. David 1991: 72. But while this is certainly a plausible surmise, and while the importance of Edison's personal influence is unquestioned, the fact that on this issue Edison was engaged in a dogged but ultimately unsuccessful effort lends the suggestion only moderate support at best.

52. McGuire et al. 1993: 216.

53. Hughes 1983: 79–80.

54. MacKenzie and Wajcman 1985a: 13.

55. Keniston 1990.

56. Ausubel 1989.
57. Merton 1973a.
58. Nelson 1990: 197.
59. Steward 1955: 182; cf. Thirtle and Ruttan 1987: 175–83; Rogers 1983.
60. Coombs et al. 1987: 121, 124–25; Brown 1981: 177.
61. Ausubel 1991: 12–17.
62. Fogel and Engerman, eds. 1971: 206.
63. Griliches 1957: 502.
64. David 1993: 235–36.
65. White 1978. But it should be noted that some of the specific instances he used in support of his position have received a sharp rebuttal. Cf. Hilton and Sawyer 1963: 90–100.
66. Rosenberg 1994: 15.
67. Schumpeter 1934; Nelson 1962a: 13; Solow 1957; Kuznets 1978; Brooks 1982; and the like.
68. Solow and Temin, 1978: 26.
69. Coombs et al. 1984: 94, citing Schmookler 1966.
70. Debate in Graubard, ed. 1980: 12.
71. Coombs et al. 1987: 103; Thirtle and Ruttan 1987: 6–11.
72. Rosenberg 1994: 16–17.
73. Mathias and Davis 1991a: 3.
74. Rosenberg 1994: 219–24.

Chapter 2
The Useful Arts in Western Antiquity

1. Childe 1950.
2. Frankfort 1939: 23.
3. Smith 1972: 499.
4. Nissen et al. 1993: 19, 21.
5. Nissen 1993: 128–29.
6. Nissen et al. 1993: 30, 47, 49, 134, 138.
7. A preliminary—although regrettably still far from comprehensive—overview of the extensive published and unpublished cuneiform sources on this subject under the Third Dynasty of Ur (during roughly the last century of the third millennium B.C.) is provided by Neumann 1987.
8. Waldbaum 1978.
9. Wertime and Wertime, eds. 1982.
10. Finley 1965: 29; cf. Ste. Croix 1981: 38, 645–47.
11. Reynolds 1984: 123.
12. Wikander 1984: 7, 9–10. 22. Strongly supporting his view (p. 916), in the context of a comprehensive and more positive evaluation of all of the evidence of classical and medieval evidence of agricultural technology, is Raepsaet 1995.
13. Landels 1978: 199–211; White 1984: 180–83, 185–87.
14. The compound pulley, essential for cranes and hence of great importance for monumental construction and the loading of ships, is a good case in point. Simple pulleys were already depicted much earlier, in Neo-Assyrian reliefs, but only in

Vitruvius do we find a detailed technical account of their application to the crane (White 1984: 14). Was this really so slow to develop, or is it merely a reflection of the inadequacy of our sources?

15. Lindberg 1992: 26–27.

16. Finley 1965: 32.

17. White 1984: 15.

18. There can be no doubt that the results of Archimedes and his successors such as Hero of Alexandria were applied to lifting devices, war machines, steelyards and so on. The difficulty with so many other problems in engineering is that their solution almost always demands the use of differential or integral calculus, which was not invented until the seventeenth century. Nor should we forget the difficulties of making calculations without the place-number and decimal systems (Hill 1984: 6).

19. White 1984: 33.

20. Finley 1973: 41–42.

21. Ibid.: 75–76.

22. Ibid.: 139; Randsborg 1991: 94–102.

23. Hopper 1979: 101–2, 104, 129–30.

24. Reynolds 1988: 47–48.

25. Vernant 1980: 11–12.

26. White 1972: 146–53.

27. Brenner 1985: 233, citing especially Searle 1974: 147, 174–75, 183–94, 267, 329. It should be noted that Postan and Hatcher (1985: 73, 77–78) are in some disagreement with this interpretation. Observing that acquisition of new lands was exalted above all other forms of capital investment by medieval landowners, they concede that this would have detracted attention from improvements in productivity. But they also maintain that the backwardness and stagnation of prevailing agricultural technology is "partly to be accounted for by the insufficient supply of new technological possibilities."

28. Postan 1975: 38.

29. Boserup 1981.

30. Pleket 1967: 19.

31. Whitney 1990: 2.

32. Lilley 1973: 214.

33. Whitney 1990: 3.

34. Reynolds 1984: 126–28; Gimpel 1988: 66–68.

35. Ovitt 1987: 200.

36. White 1978: 77–81; Lilley 1973: 188–90.

37. Vance 1990: 42–49, 83–86.

38. Lieberman 1993: 547.

39. Cipolla 1977: 56–57, 62–63.

40. Price 1959; cf. Landes 1983: 54–58.

41. Thompson 1967: 64; Mayr 1986: 6–9.

42. Maurice and Mayr 1980; Mayr 1986: 31, 41–42.

43. Mayr 1986: 116–17.

44. Hall 1959: 17.

45. Ibid.: 21; cf. Crombie 1959: 77–78.

46. Cipolla 1977: 33.

47. Ovitt 1987: 57.
48. Ibid.: 165.
49. Vickers 1984a: 9–10.
50. Pagel 1970.
51. Walker 1975: 200–202.
52. Westfall 1983: 86–87.
53. Westfall 1984: 315.
54. Cunningham and Williams 1993: 417, 421.
55. Kuhn 1969: 427; Landes 1966: 293; Burstyn 1979: 60–62.
56. Houghton 1941: 36.
57. Rosenberg 1972: 8–9.
58. Kuhn 1969: 428.
59. Layton 1974: 31.
60. Price 1984: 105–6.
61. Vicenti 1979: 743.
62. Merton 1937.
63. Cipolla 1993: 230.
64. Hall 1959: 18, 21.
65. Mayr 1986: 55–56, 124–25; Kuhn 1976: 17.
66. Cope and Jones 1959: 83.
67. Shapin 1989: 556, 561.
68. McKelvey 1985.
69. Price 1984: 110.
70. Ibid.: 18.
71. Van Helden 1983: 63–64.
72. Ibid.: 67–68.
73. Warner 1990: 83–86.
74. Musson and Robinson 1969: 14.
75. Hall 1959: 20.
76. Kuhn 1976: 14–15.
77. Webster 1975: xiv, 494.
78. Westfall 1983: 89–91.
79. Musson and Robinson 1969: 18–19.
80. Westfall 1983: 104, 106–7.
81. Wallace 1982: 23–24.
82. White 1978: 130.
83. Musson and Robinson 1969: 398–99.
84. Ibid.: 47–49.
85. Harris 1992a: 6–7, 10–11.
86. Wrigley 1987: 20, 65, 80; Wrigley 1988: 54.
87. Customs records indicate that the average number of bags of raw wool exported annually during the decade ending in 1370 was 28,302, and of woolen cloth equivalents 3,024. Already by the decade ending in 1500 these proportions were well on the way to being reversed, with English exports of bags of wool having declined by more than two-thirds to 8,149, while there had been a fourfold growth in the number of cloths to 13,891 (Cipolla 1993: 260).
88. With a characteristic turn of phrase, Landes notes that Adam Smith "does not

talk about putting-out; he is too taken with pins. He missed a rare opportunity there; but then, one must not expect even the most brilliant and practical minded theorist to give up the unexpected for the banal (the purloined letter syndrome) . . ." (Landes 1994: 651).

Chapter 3
Technology and the New European Society

1. Cardwell 1980: 453–54.
2. O'Brien 1993b: 135.
3. Lazonick 1991: 2–3.
4. Rosenberg and Birdzell 1986: viii.
5. Wallace 1982: 8.
6. Ibid.: 153.
7. Stone and Stone 1986: 289–90.
8. Merton 1970: 136; cf. Ben-David 1985: 218.
9. Inkster 1991: 36, 72–73.
10. Rosenberg 1982: 36–39.
11. Wrigley 1987: 133–34.
12. Ibid.: 138, 149.
13. Fisher 1935.
14. Jackman 1962: 212.
15. Ibid.: 212, 348.
16. Vance 1990: 163.
17. Jackman 1962: 208–9, 449.
18. Wrigley 1987: 80.
19. Smith 1976: 22.
20. Wrigley 1987: 152.
21. Campbell and Overton 1993: 100, 194.
22. Hayami and Ruttan 1985: 82; Wrigley 1988: 42; Deane 1965: 37, 39; Jones 1968: 59–60.
23. Jones 1981: 66, 75.
24. Deane 1965: 39, 41–43.
25. Crafts et al. 1992: 115; Wrigley 1985: 720.
26. Jones 1981: 77.
27. Jones 1968; Jones, ed. 1974; Jones 1981: 66–71.
28. Shammas 1993: 182; Wrigley 1987: 139, 144–45, 147.
29. Harris 1992: 145, 79.
30. Chapman 1987: 12.
31. Cole and Deane 1966: 7–9.
32. McKendrick 1982: 9–10.
33. Ibid.: 24.
34. Mokyr 1993a: 60–61; Weatherill 1988.
35. Smith 1976: 87–88.
36. de Vries 1993: 89.
37. Scranton 1994b: 478.
38. Levine 1987: 54.

39. Fine and Leopold 1990: 177. Two additional, supporting considerations are that "probate inventories for the late seventeenth and early eighteenth century show no simple association between socio-economic status and ownership of material goods" and that "upper- and middle-class consumerism, whilst adding to effective demand, may have been disproportionately directed at imports . . ." (Hudson 1992: 177, 181).

40. King 1936: 31.

41. Wrigley 1987: 234.

42. Shammas 1984: 254, 261.

43. Ibid.: 257, 264–67.

44. Shammas 1990: 200–201.

45. Deane 1965: 132.

46. O'Brien 1991: 11.

47. von Tunzelmann 1981: 146.

48. Griliches 1984: 1.

49. MacLeod 1988: 41.

50. Ibid.: 6–7.

51. Ibid.: 144–49.

52. Ibid.: 5; cf. Griffiths et al. 1992.

53. Landes 1969: 78.

54. Sullivan 1989: 424; Sullivan 1990: 358–59.

55. MacLeod 1988: 180.

56. Smith 1976: 14.

57. Chapman 1987: 13, 16.

58. The higher temperatures resulting from the coke-fired process increased silicon content, making the subsequent production of bar iron considerably costlier and discouraging the use of coke for this reason. But by making melted iron more fluid at a given temperature, the admixture of silicon made it possible for Darby to produce thinner-walled cast iron pots. "Using half the metal, he produced a superior pot that sold at a higher price." His casting methods were kept "a well-guarded secret long after the expiration of his patent" (Hyde 1977: 40–41).

59. Hyde 1977: 62.

60. Harris 1988: 30, 34, 36, 41, 46.

61. Relevant statistics are episodic in their coverage, and are further affected by intervals of international hostilities and other exigencies. An additional complication is that, until the time when coke iron took over, domestic production was supplemented by an almost equal tonnage of superior Swedish wrought iron that was imported through London. British ironmasters, having been among the highest-cost producers in Europe early in the eighteenth century, were only able to achieve cost parity, thus displacing imports, by around the end of the century (Harris 1988: 209).

62. Ibid.: 50–61, Appendix 2, p. 79.

63. Ibid.: 66–73.

64. Musson and Robinson 1969: 47–48; Cardwell 1991: 83–84.

65. Musson and Robinson 1969: op. cit., pp. 47–48.

66. Olson 1990: 328–29.

67. Vicenti 1990: 138–39. For fuller discussion see below, p. 196.

68. McKendrick 1973: 279–81.

69. Musson 1981: 32. Cf. the considerably earlier testimony of Gabriel Harvey (ca. 1545–1630) on the contributions of individual mathematical mechanicians, architects, navigators, gunners, chemists, "or any like cunning and subtle empiric, . . . or any sensible industrtious preacticioners, however unlectured in schools or unlettered in books" (Musson and Robinson 1969: 11).

70. Hills 1989: 22–23, 31.
71. Cardwell 1980: 455.
72. Robinson 1974: 98.
73. Olson 1990: 373.
74. Scherer 1984a: 13–16.
75. Mathias 1972: 40.
76. Rosenberg 1972: 7.
77. Gillispie 1957: 399.
78. Olson 1990: 323.
79. Stewart 1986: 188; Jacob 1988: 141–45.

Chapter 4
England as the Workshop of the World

1. Rose 1981: 255.
2. Thompson 1963: 197.
3. Coleman 1992: 3–4, 11. It should be noted that Friedrich Engels' work ([1845] 1968) was not translated and published in English until forty years after its original publication in German.
4. Thompson 1967: 82.
5. Berg 1994: 8–9.
6. Rosenberg and Birdzell 1986: 186 (italics added); Hudson 1992: 28.
7. Rose 1981: 267; McCloskey 1981: 109.
8. Wrigley 1987: 1–13.
9. Landes 1993: 148. Both Landes's chapter and Mokyr's own, long contribution as editor of this volume provide up-to-date discussions of the disputatious literature on this subject—and strong affirmations of the continuing indispensability of the term.
10. McCloskey 1981: 103.
11. von Tunzelmann 1978: 1–2.
12. Hills 1989: 75.
13. Musson and Robinson 1969: 60.
14. Chapman 1987: 18–19.
15. McCloskey 1981: 111.
16. Landes 1969: 84, 88.
17. Walton 1989: 63–67.
18. Berg 1994: 241–42.
19. Chapman 1987: 17, 21.
20. von Tunzelmann 1981: 145.
21. Chapman 1987: 22.
22. Cartwright's account of how he came to devote himself to this invention is

arresting. Apparently knowing nothing of handloom weaving until after he had taken out his first patent in 1785, some "Manchester gentlemen unanimously agreed" in conversation that his casual speculation about the possibility of mechanical weaving was "impracticable." Telling them that "there had lately been exhibited in London an automaton figure which played at chess," he apparently took up the intellectual challenge forthwith (Bythell 1969: 68).

23. Bythell 1969: 82; Chapman 1987: 23–24; Walton 1989: 67.
24. Bythell 1969: 270.
25. Landes 1969: 92.
26. Hyde 1977: 73.
27. Harris 1988: 39–40.
28. Hyde 1977: 119.
29. Ibid.: 56–61, 173.
30. Lilley 1973: 223–24.
31. Hoover and Hoover, eds. 1950: 156.
32. Vance 1990: 184–202.
33. Lilley 1973: 224–26.
34. Musson 1981: 26.
35. Ibid.: 35–36.
36. Ferguson 1981: 23n.
37. Ibid.: 38.
38. Mokyr 1993a: 35.
39. McCloskey 1985; Jackson 1992: 21.
40. Crafts 1985: 7; Crafts and Harley 1992: 721.
41. O'Brien 1993a: 15; Thomas 1993: xv–xvi.
42. Landes 1991: 13; cf. Hudson 1989: 8.
43. Berg and Hudson 1992: 29–31.
44. Schumpeter 1939: 134; O'Brien 1993a: 15–16.
45. Hudson 1992: 182.
46. Hohenberg and Lees 1985: 227; Rose 1981: 257.
47. Hudson 1989: 22–23.
48. Rowlands 1989: 124.
49. Inkster 1991: 43.
50. Thackray 1974: 679–81.
51. Ibid.: 686.
52. Ibid.: 674, 678.
53. Hudson 1992: 53–56.
54. Whether this deserves to be thought of as what Thompson termed a "political *counter*-revolution" may be questionable, but unquestionably there were consistent and concerted efforts by many state as well as private entities to maintain the inequities of the existing order. On the scope of relevant legislation, cf. a brief and useful summary by Ira Katznelson (1993: 248–52).
55. Thompson 1963: 194, 197.
56. Cf. the wide range of views offered by different contributors in Thompson, ed. (1993) and the monographic discussion by Patrick Joyce (1991), the latter one of E. P. Thompson's harsher critics. Insofar as political radicalism is seen as a

correlate of working-class consciousness, Duncan Bythell's views, offered in a monograph on the handloom weavers (by most accounts, among the casualties of the Industrial Revolution) are worth noting:

It is difficult to avoid concluding that, over the first half of the nineteenth century, political radicalism on the whole was no more than a kind of lowest common denominator which all workers might seize upon when some temporary setback seemed particularly severe and when the more usual forms of defence seemed inappropriate or had proved unsuccessful. It is always difficult to assess the numerical significance of any one section of the working class in movements which appear to involve all sections, but there seems no reason to suppose that radicalism was not the same vague, ill-thought-out last resort to the cotton weavers as it was to most other working men. (1969: 217)

57. Thompson 1963: 94. But Thompson is surely open to the (sympathetic) criticism:

where is the reciprocal analysis of social change—the structured material limits within which the English working class had to make itself? The absence of any systematic discussion of the ways in which English industrialization concretely affected the given patterns of class and community relations makes it appear that history did not in any way happen behind the backs of the English working class. In his zeal to correct the "objectivist" biases of economic historians and more orthodox Marxists, Thompson fleshed out only one side of the full dialectic of being and consciousness. (Ellen Kay Trimberger, quoted in Katznelson 1993: 75)

58. Pollard 1978: 161–64.
59. Thompson 1963: 249–50; Pollard 1978: 127.
60. T. S. Ashton's measured and authoritative views, while insufficiently sensitive to the un- and underemployment issue, deserve to be quoted in greater detail:

During the period 1790–1830 factory production increased rapidly. A greater proportion of the people came to benefit from it both as producers and as consumers. The fall in the price of textiles reduced the price of clothing. Government contracts for uniforms and army boots called into being new industries, and after the war the products of these found a market among the better-paid artisans. Boots began to take the place of clogs, and hats replaced shawls, at least for wear on Sundays. Miscellaneous commodities, ranging from clocks to pocket handkerchiefs, began to enter the scheme of expenditure, and after 1820 such things as tea and coffee and sugar fell in price substantially. The growth of trade unions, friendly societies, savings banks, popular newspapers and pamphlets, schools, and nonconformist chapels—all give evidence of the existence of a large class raised well above the level of mere subsistence.

There were, however, masses of unskilled or poorly skilled workers—seasonally employed agricultural workers and hand-loom weavers in particular—whose incomes were almost wholly absorbed in paying for the bare necessaries of life, the prices of which, as we have seen, remained high. My guess would be that the number of those who were able to share in the benefits of economic progress was larger than the number of those who were shut out from those benefits and that it was steadily growing. But the existence of two groups within the working class needs to be recognized. (Ashton 1954: 158–59)

61. Lindert and Williamson 1983: 7, 11–12, 21, 24.
62. Pollard 1978: 121–23.
63. Thompson 1963: 212.
64. Mokyr 1993: 122–23.
65. Pollard 1978: 162.
66. Chapman 1987: 46–47.
67. Thompson 1963: 313.
68. Pollard 1978: 126.
69. Thompson 1963: 197.
70. Huck 1995: 547.
71. Pollard 1978: 162; Nicholas and Nicholas 1992: 16.
72. Pollard 1978: 123.
73. Rosenberg and Birdzell 1986: 160, citing detailed studies of A. J. Taylor and B. R. Mitchell. Reflective of the general slipperiness of statistics in this area, they mention scattered handloom piecework rates and factory weekly wage rates and conclude that ". . . the question whether handloom weavers earned more or less, before the advent of power looms, than the workers who ran the early power looms, can be answered either way, depending on how much one allows for differences in skills and on one's choice of handloom weaver—the average worker or the Stakhanovite" (1986: 176).
74. Bythell 1993: 32–33.
75. Pollard 1978: 120.
76. Ibid.: 163.
77. Payne 1978: 182.
78. Ibid.: 188–89.
79. Mokyr 1985: 27–28.
80. MacLeod 1988: 277–78.
81. Landes 1969: 306.
82. Lazonick 1991: 140; Payne 1978: 189.
83. Lazonick 1991: 46, 192–93.
84. Payne 1978: 194–95.
85. Veblen 1939: 23, 32–37, 132, 142–43, 249–50 and related sections.
86. Pollard 1992: 45.
87. Lazonick 1986: 19; Mokyr 1990: 265–66.
88. Crafts et al. 1992: 123.
89. Elbaum and Lazonick 1986: 2, 6–7.
90. Lindert and Trace 1971: 263–64; cf. Freeman 1982: 28–30.
91. Inkster 1991: 19–20; Brock 1993: 276–93.
92. Brock 1993: 296–308.
93. Chandler 1993: 29–30.
94. Travis 1993: 190, 202, 237–39.
95. Landau and Rosenberg 1992: 77; Smith 1993b: 142.
96. Chirot 1991: 31–32; Mokyr 1990: 265–66.
97. Mokyr 1990: 268.

Chapter 5
Atlantic Crossing

1. Rosenberg 1981: 49.
2. Marx 1964: 147–49; Marx 1992: 462–64.
3. Lemon 1987: 143–44.
4. Smith 1976: 78–79.
5. "By the time of the American Revolution of 1776 a full third of the British merchant fleet was registered in American ports, and an unknown but presumably large number of ships were built in America for British customers. Perhaps as much as half the British merchant fleet was thus built of American resources" (Hugill 1993: 25–26).
6. Lemon 1987: 141.
7. "Britain's lead in transportation by canal in the late eighteenth and early nineteenth century was largely compensated for in America by small canals around rapids in rivers, as on the Susquehanna and the Mohawk, and the use of first the sailing shallop and then the river steamboat" (Cochran 1986: 257).
8. Ibid.: 255.
9. Lee and Passell 1979: 94.
10. Ibid.: 256–57.
11. Cochran 1986: 258.
12. The 1810 Census listed 2,016 in Pennsylvania alone, and credits them with an annual production of 74,538,640 board feet of ready-cut lumber (Ibid.: 259).
13. Ibid.: 260.
14. Chapman 1987: 41.
15. Jeremy 1977: 6.
16. Jeremy 1981: 82.
17. Clark 1929: 1, 247.
18. Niemi 1980: 41.
19. Lee and Passell 1979: 85, 88–89.
20. Jeremy and Stapleton 1991: 35.
21. Purcell 1969: 4.
22. Carter 1986: 213.
23. Ibid.: 220.
24. Purcell 1969: 72–74.
25. Temin 1966: 204.
26. Ibid.: 188, 196.
27. Purcell 1969: 81.
28. This diffusion apparently took place quite rapidly, since until 1812 steam engines reportedly could be manufactured only in New York and Philadelphia (Purcell 1969: 56).
29. Temin 1966: 190.
30. Jeremy 1981: 180.
31. Rosenberg 1981: 52–54.
32. Hoke 1990.
33. Jeremy 1973; Jeremy 1990: 37.
34. Uselding 1977: 168–69.

35. Mayr and Post 1981a: xvi.
36. Ferguson 1981: 23n (italics added).
37. Uselding 1981: 109–10.
38. Musson 1981: 27.
39. Stapleton 1987: 30.
40. Smith 1985: 44.
41. Ibid.: 63–64.
42. Burstyn 1979: 70.
43. Smith 1985: 63, 83–84, 86.
44. Rosenbloom 1993: 684.
45. Mayr and Post 1981a: xvii. Ferguson, in the same volume, strongly supports the separation of interchangeability from the core of the American system, adding that "not until the twentieth century did interchangeable parts become a generally viable component" (1981: 4–5).
46. Hoke 1990.
47. Rosenbloom 1993: 698.
48. Hounshell 1981: 128.
49. Sawyer 1954: 369–71.
50. Hindle 1981; Hindle and Lubar 1986.
51. Sawyer 1954: 371.
52. Ibid.: 375–77.
53. Habakkuk 1962.
54. Musson (1981: 40–41) has pointed out that "wages of the most skilled British machinemakers were not only relatively but even absolutely higher than those of comparable American workers. It is clear that from the late eighteenth century onward, British engineering employers had experienced a considerable scarcity of skilled workers, and that high wages and trade-union apprenticeship and machine-manning restrictions were a constant source of complaint. Consequently, contrary to Habakkuk's thesis, employers had constantly sought laborsaving innovations. . . ." But he goes on to agree that "[i]n many manufactures, other than machine-making, it would indeed appear that labor was more abundant and cheaper in England than in the United States, where standardized mass production therefore developed more rapidly in the second half of the nineteenth century."
55. Habakkuk 1970: 23–25.
56. Musson 1981: 41.
57. Mokyr 1990: 165–66; cf. von Tunzelmann 1981: 158.
58. Saul 1970: 17; Uselding 1977: 198.
59. Field 1985: 388, 398.
60. Chandler 1981: 153.
61. Rosenberg 1963: 435.
62. Ibid.: 418–20, 423.
63. Niemi 1980: 48–53, 69–71.
64. Lee and Passell 1979: 133.
65. Jeremy and Stapleton 1991: 39, 42.
66. "John Jervis developed the pivoted forward truck for locomotives, to allow them to negotiate sharp turns; Mattias Baldwin doubled the steam pressure, increasing tractive power; others developed lever systems that helped keep railroad cars on

the rails despite rough, uneven road-beds. The effect . . . was to adapt the railroad to American geographical and economic conditions" (Layton 1987: 153–54; cf. Vance 1990: 270–73).

67. Jeremy and Stapleton 1991: 43.

68. Vance 1990: 268–95.

69. Danhof 1969: 6–9.

70. Ibid.: 18.

71. Ibid.: 19.

72. Barron 1990: 7–8.

73. Jones 1968: 66–67.

74. Lee and Passell 1979: 137–38.

75. Danhof 1969: 50, 138–39, 82–83.

76. Ibid.:151–52.

77. Ibid.: 142–43. A horse produces about ten times the effort of a man, while an ox can produce half to three-quarters of this amount.

78. Rasmussen 1962: 579–80.

79. Danhof 1969: 189–92.

80. Schlebecker 1975: 174–75.

81. Rasmussen 1962: 580; Danhof 1969: 181, 218.

82. Rasmussen 1962: 580.

83. Ibid.: 5.

84. David 1971: 223.

85. Olmstead and Rhode 1995.

86. Danhof 1969: 182.

87. Feller 1962: 560–77, appendix table 1.

88. Parker 1971: 176–77.

89. Lee and Passell 1979: 137–38.

90. Dupree 1986: 21.

91. Ibid.: 16; Greene 1984: 35–36.

92. Layton 1987b: 158.

93. Layton 1986: 3.

94. Sokoloff and Khan 1990: 365, 367, 387.

95. Sokoloff 1988: 818; Dupree 1986: 46–47.

96. Dupree 1986: 50–63; Manning 1988: 1–2; Layton 1986: 28.

97. Dupree 1986: 64–65.

98. Cohen 1976: 374.

99. Molella 1976.

100. Molella and Reingold 1973: 333.

101. Dupree 1986: 48–49.

102. Miller 1972: 15–16.

103. Dupree 1986: 104.

104. Ibid.: 114.

Chapter 6
The United States Succeeds to Industrial Leadership

1. McPherson 1988: 451–52.

2. Cumulatively, about 11 percent of 2 billion acres of public land were thus

disposed of. As McPherson notes, "To some degree these laws functioned at cross purposes, for settlers, universities, and railroads competed for portions of the same land in subsequent years."

3. First ad hoc steps in this direction began as early as 1836, when the Commissioner of Patents informally initiated efforts to secure and distribute new plant materials. Modest appropriations to support this activity shortly followed, and then steadily grew. The introduction of a cabinet-level department, however, clearly envisioned programs on a larger and more comprehensive scale.

4. Danhof 1969: 66–72.

5. Dupree 1986: 151.

6. Hattaway and Jones 1983: 139.

7. Letwin 1989: 655.

8. Hattaway and Jones 1983: 139.

9. Dupree 1986: 120–26.

10. Ibid.: 135.

11. Cochrane 1978: 88–89.

12. Untested though he felt they were, Kuznets's seminal suggestions continue to be of great interest as the outline of a whole program of prospective research:

> Could the very rapid rate of growth of population and labor force . . . have restricted the rates of growth in per capita and per worker output? . . . Surely one cannot assume that the supply of natural resources had any limiting effects. . . . Could the limitation stem from difficulties in supplying adequate capital per worker, engendered by a rapidly growing labor force, despite the high long-term capital formation proportions . . . ? Or did the problems of adjustment and assimilation faced by immigrants lower productivity, despite the fact that most immigrants were in the prime labor ages and presumably endowed with strong economic incentives? Or, finally, did the very high level of per capita income induce a lower rate of growth by permitting the exchange of work for leisure, since there was no great pressure to "catch up"? (Kuznets 1971: 17–18, 21).

13. Wright 1990: 651.

14. Mathias 1991.

15. Rosenberg and Birdzell 1986: 213. The authors point out that, since the industrial use of waterpower also increased until at least 1880, no replacement of waterpower by steam power is involved in at least the early phases of this increase.

16. Examples of these changes, given as percentages of all manufacturing employment, are as follows (North 1966: 694):

	1870	1910
Leather and leather goods	9.56	4.72
Cotton and cotton goods	7.29	6.16
Woolen and worsted	5.00	2.90
Iron and steel products	7.58	15.19
Transportation equipment	5.63	8.88
Lumber and lumber products	8.00	12.00
Printing and publishing	1.50	6.00

17. Chandler 1990: 127–40.

18.

Percentages of Pig Iron Produced With Various Fuels

| | Bituminous | Anthracite | |
Year	Coal and Coke	Coal and Coke	Charcoal
1860	13	57	30
1870	31	50	20
1880	45	42	13
1890	69	24	7
1900	85	12	3
1910	96	2	1

Source: Temin 1964: 268–69.

19. Rosenberg and Birdzell 1986: 213.
20. Temin 1964: 101–2.
21. Chandler 1990: 136.
22. Ibid.: 25.
23. Lazonick 1991: 273–74.
24. Chandler 1990: 8.
25. Landau and Rosenberg 1992: 90–93; Mowery and Rosenberg 1989: 26–27; Rosenberg and Nelson 1993: 14.
26. Chandler 1990: 76, 89.
27. Chandler 1977: 120, 132, 204. Salsbury (1988) provides a fine, generalized account of the emergence of American railroads as a system.
28. Chandler 1977: 188.
29. Fishlow 1966: 425. No doubt the increasing use of steel for these purposes was accelerated by its declining price as noted above.
30. Fogel 1971: 202–3.
31. Vance 1990: 496–99.
32. Hounshell 1984: 5, 7.
33. Hounshell 1984: 328–29; Rosenberg 1963: 437.
34. Abernathy and Clark 1985: 9–10.
35. Hounshell 1984: 220.
36. Ibid.: 439–40. The specific examples he adduces are not included in the quoted passage since they are not directly relevant to this discussion of the Model T. But as revealing illustrations of his point, they are as follows:

[T]he transmission for the drive and feed mechanisms of machine tools was considerably improved when machine tool builders adopted the alloy steel sliding gears and integral keyshafts developed by automobile designers. Moreover, the introduction of antifriction bearings into key points of the machine tool resulted from the demonstration of their usefulness in automobiles. Finally, the whole approach to the lubrication of machine tools was radically revised as a result of the automobile.

37. Vance 1990: 500–501; Hounshell 1984: 220, 329.
38. Hounshell 1984: 252.
39. Scranton 1991: 29–41, 88–89.

40. Hughes 1989. The remainder of this section draws heavily on this fine account.

41. Ibid.: 56.

42. Hughes 1987: 59.

43. Nader 1994: 482.

44. Rosenberg and Birdzell 1986: 252.

45. Layton 1986: 3–4.

46. Layton 1988: 92; Constant 1987: 224.

47. Layton 1976: 691–93.

48. Layton 1971: 573; Hall 1978: 99.

49. Kevles 1979: 139.

50. Layton 1986: 29–40.

51. Layton 1976: 696.

52. Vicenti 1990: 139, 166–67. For a much earlier appearance of this principle, see p. 96.

53. Kevles 1979: 140–42. It should be noted, however, that at least a few professional chemists were already active in industrial research. The first research laboratories were reportedly established by chemists as early as 1836, and railroads concerned with problems like the longevity of rails opened chemical and testing laboratories by the mid-1870s (Rosenberg and Birdzell 1986: 246).

54. Rosenberg and Nelson 1993: 10–11.

55. Ibid.: 144–45.

56. Ibid.: 4.

57. Temin 1964: 131.

58. Mowery and Rosenberg 1989: 33–41.

59. Mowery and Rosenberg 1993: 33–34.

60. Ibid.: 150–51.

61. Ibid.: 62.

62. Dupree 1986: 272, 288.

63. Fitzgerald 1991: 120; Schlebecker 1975: 173.

64. Danhof 1969: 169–71.

65. Schlebecker 1975: 158.

66. Ibid.: 159–60.

67. Dieffenbach and Gray 1960: 26–30; Schlebecker 1975: 190; Danhof 1969: 194–95.

68. Bogue 1983: 4–5.

69. Rosenberg and Birdzell 1986: 187–88.

70. Durost and Bailey 1970: 2.

71. Danhof 1969: 253–55.

72. Rasmussen 1962: 580–81; Schlebecker 1975: 181.

73. Bogue 1983: 10.

74. Griliches 1971: 208.

75. Ibid.: 213.

76. Bray and Watkins 1964: 761.

77. Ibid.: 763.

78. *New York Times*, October 10, 1993.

79. Hansen et al. 1986: 29–30.

80. Parker 1971: 182–83.

81. Schlebecker 1975: 186–87.

82. Hughes 1989: 124–27.

83. Cochrane 1978: 644.

84. Ibid.: 208–9, 215, 223; Dupree 1986: 309.

85. Cochrane 1978: 229–32.

86. Weart 1979: 296–98, 302–3, 305.

87. Geiger 1990: 18.

88. Weart 1979: op. cit., p. 313.

89. Brooks 1993; Rosenberg and Nelson 1993: op. cit., p. 30.

90. Geiger 1990: 19.

91. Dupree 1986: 334.

92. Mowery and Rosenberg 1989: 92–93.

Chapter 7
The Competitive Global System

1. Brooks 1980: 10; cf. Deane 1965: 118.

2. Brooks 1994: 479.

3. Gomory 1983: 576, 579; Constant 1984.

4. Layton 1974: 31.

5. Bush 1945: 13–14 (italics original).

6. Brooks 1994: 478–79. As "the rarest, but therefore also most dramatic" examples of science having led directly to new technological ideas, he cites the discovery of uranium fission and, slightly more problematically, of the laser, X-rays, and nuclear magnetic resonance. More common are cases such as the discovery of the transistor in the Bell Laboratories in which "the exploration of a new field of science is deliberately undertaken with a general anticipation that it has a high likelihood of leading to useful applications, though there is no specific end-product in mind." For the turbojet as an example of the more fully reciprocal process, see p. 215.

7. Rosenberg 1991: 335, 338; Nelson, ed. 1993: 7.

8. Rosenberg 1992: 82–83.

9. Nelson and Winter 1982: 206–12; Mowery and Rosenberg 1989: 8.

10. Janich 1978; Abelson 1986; Rosenberg 1994: 251; Brooks 1994: 484.

11. Singer 1993: 4.

12. Kaufmann and Smarr 1993: ix, 25–26; Flamm 1988: 12–13.

13. Dupree 1986: 369–75.

14. Pursell 1979: 359–64.

15. Kevles 1978: 320.

16. Ibid.: 302–17.

17. Pursell 1979: 325.

18. Kevles 1978: 347.

19. Brooks 1993: 211.

20. Sapolsky 1979: 386.

21. Ibid.: 364.

22. Committee on Criteria for the Federal Support of Research and Development 1995: 42; Rosenberg and Nelson 1993: 25.

23. Galison 1988: 1: 77–79.

24. Geiger 1990: 24–26.

25. Brooks 1993: 207.

26. Edwards 1994: 261–62.

27. Mowery and Rosenberg 1993: 46–48.

28. Ibid.: 48; National Science Board 1994.

29. Cowan 1983: 196–97, 208–10.

30. Fox 1994: 214–15, 222–23.

31. National Science Board 1993: tables 4–10 and 4–19.

32. Mendelsohn et al. 1988: xxiv.

33. Mowery and Rosenberg 1989: 159–60.

34. Kelley and Watkins 1995: 525, 531.

35. National Critical Technologies Panel 1991: 1; National Critical Technologies Panel 1993: x. Branscomb (1993b: 62) notes that the second report was "published for the OSTP [Office of Science and Technology Policy] by the CTI [Critical Technologies Institute], but because of political sensitivities no institutional sponsor or publisher is indicated on the document." Branscomb also provides a succinct account of why he believes " 'criticality' is a poor way of identifying commercially relevant R&D projects that merit public funding."

36. National Critical Technologies Panel 1991: 9.

37. National Science Board 1993: 178–80.

38. National Critical Technologies Panel 1991: 49.

39. Flamm 1988: 1.

40. Kaufmann and Smarr 1993: 26. Fuller details are provided by Flamm 1988: 17–18, 26–27, and 78–79.

41. Ferguson and Morris 1993.

42. National Critical Technologies Panel 1991: 55.

43. Ibid.: 83–90.

44. Tyson 1992: 4, 18, 44.

45. Bell 1987: 12.

46. Womack et al. 1990: 13.

47. Mathias 1991: 19.

48. I am indebted to Thomas J. Murrin, Dean of Duquesne University's School of Business and former Deputy Secretary of Commerce and vice-president of manufacturing at Westinghouse, for an illuminating conversation on this issue during an intermission at a NASA Advisory Council meeting in April 1991. I had suggested that building in redundancy was a common and effective approach to heightened reliability. Characterizing this as mere "brute force," he cited reports of IBM's 90 percent reduction in the number of moving parts in its electric typewriter, in the face of Japanese competition, as a more effective alternative. (Ralph Gomory [1988: 13] has published a broadly similar account of IBM's re-design of a printer to improve manufacturability, effecting a more than a 50 percent reduction in the number of parts including the elimination of all screws.) Motorola, finding that the designed two-year mean time-to-failure of its pager was only half that of Japanese competitors, reportedly redesigned it completely to achieve a 180-year mean time.

49. Logan and Molotch 1987: 248–49, 252.

50. Harrison 1994: 39–40.

51. Commission on Skills of the American Workforce 1990: 3–4; cf. Adams 1991: 245. For a detailed, critical comparison of prevailing training policies followed by American firms with those of European and Japanese competitors, see Lynch 1994.

52. Rosenberg 1992.

53. Arthur (in press).

54. Ohmae 1985: 7.

55. Group of Lisbon 1995.

56. Griliches, ed. 1984: 11, 14, 16; Cooper 1991; Rosenberg 1979: 26–27.

57. Nelson 1992: 63–64.

58. Kahn 1991.

59. Halm and Gelijns 1991: 4; Grabowski 1991: 46–47.

60. White 1991: 52.

61. Mowery and Rosenberg 1989: 220. They cite such indices as the increasing ratio of new technological exports to imports, the steeply declining rate of growth of Japanese annual payments for imports of technology, increasing numbers of Japanese contributions to basic science, and qualitative by-field assessments of industrial products.

62. Ohmae 1985: 55.

63. Ohmae 1985: 3–4, 55–57; Hatsopoulos et al. 1988: 300; Cohen and Zysman 1988. An excellent overview of the essentials of the "Japanese System" is provided in Imai and Yamazaki 1994.

64. Rosenberg 1988: 5.

65. Landes 1992: 94–95.

66. Office of Japan Affairs 1990: 2.

67. Mowery and Rosenberg 1989: 231–32.

68. Branscomb 1993d: 16.

69. Office of Japan Affairs 1990: 7.

70. Gamota 1992: 23.

71. Rosenberg 1988: 2–3.

72. Kingery, ed. 1991: 283–84; Gamota 1992: 40.

73. Office of Japan Affairs 1990: 9–10.

74. Mukaibo 1991: 6–7.

75. Tyson 1992: 3, 254.

76. Smith 1993b: 154–55.

77. Abramovitz 1993: 224–25, 229.

78. David 1993: 218.

79. Boskin and Lau 1992: 50.

80. Mikulski 1994: 221–22.

Chapter 8
New Paths

1. Appleby et al. 1994: 23.

2. Without implying a necessary derivation, this attitude can be found richly resonating in the work of the elder Pliny more than a millennium later:

And now that we have described everything that depends upon Man's talents for making Art reproduce Nature, we cannot help marvelling that there is almost nothing that is not brought to a finished state by means of fire. Fire takes this or that sand, and melts it, according to the locality, into glass, silver, cinnabar, lead of one kind or another, pigments or drugs. It is fire that smelts ore into copper, fire that produces iron and also tempers it, fire that purifies gold, fire that burns the stone which causes the blocks in building to cohere. There are other substances that may be profitably burnt several times; and the same substance can produce something different after a first, a second or a third firing. Even charcoal itself begins to acquire its special property only after it has been fired and quenched: when we presume it to be dead it is growing in vitality. Fire is a vast, unruly element, and one which causes us to doubt whether it is more a destructive or a creative force. (*Natural History*, XXXVI, LXVIII [Loeb Classical Library ed., London: William Heinemann Ltd.])

3. Mokyr 1993a: 56.
4. Nelson 1990: 211.
5. Branscomb 1987: 254.
6. Branscomb 1993a.
7. Sylla and Toniolo 1991: 17.
8. Ward 1994: 62.
9. White 1993.
10. Jasanoff 1992: 174–75.
11. Ruttan 1994: 344–45; Holmes 1993: 1517; Waggoner 1994; Abelson 1994b: 1363.
12. Population Summit of the World's Scientific Academies 1994: 5.
13. Solow 1992: 14–15.
14. Brooks 1992: 31.
15. Ausubel et al. 1989: 3.
16. Herman et al. 1989: 26, 34.
17. Tschinkel 1989: 161.
18. Gould et al. 1988: 46–49; Teuber 1990: 237; Slovic 1987: 283.
19. Douglas 1985: 18.
20. Perrow 1991: 23, and related sections; cf. Sagan 1993.
21. Kates 1986: 211; Abelson 1994a: 591; Slovic 1987: 283–84.
22. Tschinkel 1989: 159–61.
23. Jasanoff 1990: 75–76.
24. Jasanoff 1992: 162, 164, 167.

References Cited

Abelson, Philip H. 1986. "Instrumentation and Computers." *American Scientist* 74: 182–92.

———. 1994a. "Reflections on the Environment." *Science* 263: 591.

———. 1994b. "Continuing Evolution of U.S. Agriculture." *Science* 264: 1363.

Abernathy, William J., and Kim B. Clark. 1985. "Innovation: Mapping the Winds of Creative Destruction." *Research Policy* 24: 3–32.

Abramovitz, Moses. 1993. "The Search for the Sources of Growth: Areas of Ignorance, Old and New." *Journal of Economic History* 53: 217–43.

Academy Industry Program. 1988. "Industrial R&D and U.S. Technological Leadership." Washington: National Academy Press.

Adams, Robert McC. [1974] 1991. "Anthropological Perspectives on Ancient Trade," 141–60. In Silverman, 1991.

———. 1991a. "Cultural and Sociotechnical Values," 26–38. In Sladovich 1991.

———. 1991b. Strains in the American Industry-Science-Triad," 243–48. In Kingery 1991.

Almond, Gabriel A., M. A. Chodorow, and R. H. Pearce, eds. 1982. *Progress and Its Discontents*. Berkeley: University of California Press.

Appleby, Joyce, Lynn Hunt, and Margaret Jacob. 1994. *Telling the Truth about History*. New York: W. W. Norton.

Arrison, Thomas S., C. Fred Bergsten, Edward M. Graham, and Martha C. Harris, eds. 1992. *Japan's Growing Technological Capability: Implications for the U.S. Economy*. Washington, D.C.: National Academy Press.

Arthur, W. Brian. In press. "On the Evolution of Complexity." In Cowan et al. in press.

Ashton, T. S. 1954. "The Standard of Life of the Workers in England, 1790–1830," 123–55. In von Hayek 1954.

Aston, T. H., and C.H.E. Philpin, eds. 1985. *The Brenner Debate: Agrarian Class Structure and Economic Development in Preindustrial Europe*. Cambridge: Cambridge University Press.

Ausubel, Jesse H. 1989. "Regularities in Technological Development: An Environmental View," 70–91. In Ausubel and Sladovich 1989.

———. 1991. "Rat-race Dynamics and Crazy Companies: The Diffusion of Technologies and Social Behavior," 1–17. In Nakicenovic and Gruebler 1991.

Ausubel, Jesse H., Robert A. Frosch, and Robert Herman. 1989. "Technology and Environment: An Overview," 1–20. In Ausubel and Sladovich 1989.

Ausubel, Jesse H., and Hedy E. Sladovich, eds. 1989. *Technology and Environment: An Overview*. Washington: National Academy Press.

Barlett, Peggy F., ed. 1980. *Agricultural Decision Making: Anthropological Contributions to Rural Development*. New York: Academic Press.

Barnett, George E., ed. 1936. *Two Tracts by Gregory King*. Baltimore: Johns Hopkins University Press.

Barron, Hal S. 1990. "Listening to the Silent Majority: Change and Continuity in the Rural North," 3–23. In Ferleger 1990.

Bell, Daniel. 1987. "The World and the United States in 2013." *Daedalus* 116:3, 1–31.

Ben-David, Joseph. 1985. "Puritanism and Modern Science: A Study in the Continuity and Coherence of Sociological Research," 207–23. In Cohen et al. 1985.

Bennett, John W. 1980. "Management Style: A Concept and Method for the Analysis of Family-operated Agricultural Enterprise," 203–37. In Barlett 1980.

Berg, Maxine. 1991. "Revisions and Revolutions: Technology and Productivity Change in Manufacture in Eighteenth-century England," 43–64. In Mathias and Davis 1991.

―――. 1994. *The Age of Manufactures 1700–1820.* 2d ed. London: Routledge.

Berg, Maxine, and Pat Hudson. 1992. "Rehabilitating the Industrial Revolution." *Economic History Review* 45: 24–50.

Berry, Brian J. L. 1991. *Long-wave Rhythms in Economic Development and Political Behavior.* Baltimore: Johns Hopkins University Press.

―――. 1993. "Long-term Economic Cycles and American Politics." *Occasional Paper*, Nelson A. Rockefeller Center for the Social Sciences at Dartmouth College.

Bijker, Wiebe E., Thomas P. Hughes, and Trevor J. Pinch, eds. 1987. *The Social Construction of Technological Systems: New Directions in the Sociology and History of Technology.* Cambridge: MIT Press.

Bijker, Wiebe E., and John Law. 1992. "General Introduction," 1–16. In Bijker and Law 1992.

―――, eds. 1992. *Shaping Technology/Building Society.* Cambridge: MIT Press.

Bogue, Allan G. 1983. "Changes in Mechanical and Plant Technology: The Corn Belt, 1910–1840." *Journal of Economic History* 43: 1–25.

Boserup, Ester 1981. *Population and Technological Change: A Study of Long-term Trends.* Chicago: University of Chicago Press.

Boskin, Michael J.; and Lawrence J. Lau. 1992. "Capital, Technology, and Economic Growth," 17–55. In Rosenberg et al. 1992.

Branscomb, Lewis M. 1993a. "The National Technology Policy Debate," 1–35. In Branscomb 1993c.

―――. 1993b. "Targeting Critical Technologies," 36–63. In Branscomb 1993c.

―――, ed. 1993c. *Empowering Technology: Implementing a U.S. Strategy.* Cambridge: MIT Press.

―――. 1993d. "Comment," 16. In Cutler 1993.

Braudel, Fernand. 1973. *Capitalism and Material Life 1400–1800.* New York: Harper Colophon.

―――. 1981. *The Structures of Everyday Life: Civilization and Capitalism 15th–18th Century.* Vol. 1. New York: Harper and Row.

Braund, David, ed. 1988. *The Administration of the Roman Empire 241* B.C.–A.D. 193. Exeter Studies in History, vol. 18. Exeter: University of Exeter.

Bray, James O., and Patricia Watkins. 1964. "Technical Change in Corn Production in the United States, 1870–1960." *Journal of Farm Economics* 46: 751–65.

Brenner, Robert. 1985. "The Agrarian Roots of European Capitalism," 213–327. In Aston and Philpin 1985.

Brewer, John. 1990. *The Sinews of Power: War, Money, and the English State, 1688–1783*. Cambridge: Harvard University Press.

Brewer, John, and Roy Porter, eds. 1993. *Consumption and the World of Goods*. London: Routledge.

Brock, William H. 1993. *The Norton History of Chemistry*. New York: W.W. Norton.

Brooks, Harvey. 1980. "Technology, Evolution, and Purpose," 65–81. In Graubard 1980.

———. 1982. "Can Technology Assure Unending Human Progress?" 281–300. In Almond et al. 1982.

———. 1992. "Sustainability and Technology," 29–60. In International Institute for Applied Systems Analysis 1992.

———. 1993. "Research Universities and the Social Contract for Science," 202–34. In Branscomb 1993c.

———. 1994. "The Relationship between Science and Technology." *Research Policy* 23: 477–86.

Brown, Lawrence A. 1981. *Innovation Diffusion: A New Perspective*. London: Methuen.

Bugliarello, George, and Dean B. Doner, eds. 1979. *The History and Philosophy of Technology*. Urbana: University of Illinois Press.

Burke, John G., ed. 1983. *The Uses of Science in the Age of Newton*. Berkeley: University of California Press.

Burstyn, Harold L. 1979. "What Can the History of Technology Contribute to Our Understanding?" 57–80. In Bugliarello and Doner 1979.

Bush, Vannevar. 1945. *Science, the Endless Frontier: A Report to the President*. Washington, D.C.: U.S. Government Printing Office.

Bythell, Duncan. 1969. *The Handloom Weavers: A Study in the English Cotton Industry During the Industrial Revolution*. Cambridge: Cambridge University Press.

———. 1993. "Women in the Workforce," 31–53. In O'Brien and Quinault 1993.

Campbell, Bruce M.S., and Mark Overton. 1993. "A New Perspective on Medieval and Early Modern Agriculture: Six Centuries of Norfolk Farming c. 1250–c. 1850." *Past and Present* 141: 38–105.

Cardwell, D.S.L. 1980. "Science, Technology, and Industry," 449–83. In Rousseau and Porter 1980.

———. 1991. *Turning Points in Western Technology*. Canton, Mass.: Science History Publications/USA.

Carter, Edward C. II. 1986. "Benjamin Henry Latrobe, 'Learned Engineer,' the American Philosophical Society, and the Promotion of Useful Knowledge and Works," 201–23. In Klein 1986.

Chandler, Alfred D., Jr. 1977. *The Visible Hand: The Managerial Revolution in American Business*. Cambridge: Belknap Press of Harvard University Press.

———. 1981. "The American System and Modern Management," 153–70. In Mayr and Post 1991.

Chandler, Alfred D., Jr. 1990. *Scale and Scope: Dynamics of Industrial Capitalism.* Cambridge: Belknap Press of Harvard University Press.

———. 1993. "Learning and Technological Change: The Perspective from Business History," 214–39. In Thomson 1993.

Chapman, S. D. 1987. *The Cotton Industry in the Industrial Revolution.* 2d ed. Houndmills, Hampshire: Macmillan Education Ltd.

Childe, V. Gordon. 1950. "The Urban Revolution." *Town Planning Review* 21: 3–17.

Chirot, Daniel. 1991. "After Socialism, What? The Global Implications of the Revolutions of 1989 in Eastern Europe." *Contention* 1: 29–49.

Cipolla, Carlo M., ed. 1972. *The Middle Ages.* Vol. 1 of *The Fontana Economic History of Europe.* London: Collins/Fontana.

———. ed. 1973. *The Industrial Revolution.* Vol. 3 of *The Fontana Economic History of Europe.* London: Collins/Fontana.

———. 1977. *Clocks and Culture 1300–1700.* New York: W.W. Norton.

———. 1993. *Before the Industrial Revolution: European Society and Economy 1000–1700.* London: 3d ed. Routledge.

Clagett, Marshall, ed. 1959. *Critical Problems in the History of Science.* Madison: University of Wisconsin Press.

Clark, Victor S. 1929. *History of Manufactures in the United States.* 3 vols. New York: McGraw-Hill for the Carnegie Institution of Washington.

Cochran, Thomas C. 1986. "Cotton Textiles and Industrialism," 251–69. In Klein 1986.

Cochrane, Rexmond C. 1978. *The National Academy of Sciences: The First Hundred Years 1864–1963.* Washington, D.C.: National Academy of of Sciences.

Cohen, Eric, Moshe Lissak, and Uro Almagor, eds. 1985. *Comparative Social Dynamics: Essays in Honor of S.N. Eisenstadt.* Boulder: Westview Press.

Cohen, I. Bernard. 1976. "Science and the Growth of the American Republic." *Review of Politics* 38: 359–98.

Cohen, Stephen S., and John Zysman. 1988. "Manufacturing Innovation and American Industrial Competitiveness." *Science* 239: 1110–15.

Cole, W. A., and Phyllis Deane. 1966. "The Growth in National Incomes," 1–55. In Habakkuk and Postan 1966.

Coleman, D. C. 1992. "Myth, History and the American Revolution," 1–42. In *Myth, History and the American Revolution,* edited by D. C. Coleman. London: Hambledon Press.

Commission on Skills of the American Workforce. 1990. *America's Choice: High Skills or Low Wages!* Rochester: National Center on Education and the Economy.

Committee on Criteria for the Federal Support of Research and Development. 1995. *Allocating Federal Funds for Science and Technology.* Washington, D.C. National Academy Press.

Committee on Technological Innovation in Medicine. 1991. *The Changing Economics of Medical Technology: Medical Innovation at the Crossroads.* Vol. 2. Institute of Medicine. Washington, D.C.: National Academy Press.

Constant, Edward W. II. 1984. "Communities and Hierarchies: Structure in the Practice of Science and Technology," 105–14. In Landau 1984.

————. 1987. "The Social Locus of Technological Practice: Community, System, or Organization?" 223–42. In Bijker et al., 1987.

Coombs, Rod, Paolo Saviotti, and Vivien Walsh. 1987. *Economics and Technological Change*. Totowa, N.J.: Rowman and Littlefield.

Cooper, Carolyn. 1991. *Shaping Invention: Thomas Blanchard's Machinery and Patent Management in Nineteenth-century America*. New York: Columbia University Press.

Cope, Jackson I., and Harold Whitmore Jones, eds. 1959. *History of the Royal Society, by Thomas Sprat*. St. Louis: Washington University Press.

Cowan, George, D. Pines, and D. Meltzer, eds. In press. *Integrative Themes*. Santa Fe Institute Studies in the Sciences of Complexity, 19. Reading, Mass.: Addison-Wesley.

Cowan, Ruth Schwartz. 1983. *More Work for Mother: The Ironies of Household Technology from the Open Hearth to the Microwave*. New York: Basic Books.

Crafts, N.F.R. 1985. *British Economic Growth during the Industrial Revolution* Oxford: Clarendon Press.

Crafts, N.F.R., and C. K. Harley. 1992. "Output Growth and the British Industrial Revolution." *Economic History Review* 45: 703–30.

Crafts, N.F.R., S. J. Leybourne, and T. C. Mills. 1992. "Britain," 109–52. In Sylla and Toniolo 1992.

Crombie, A. C. 1959. "Commentary," 66–78. In M. Clagett 1959.

Cunningham, Andrew, and Perry Williams 1993. "De-centering the 'Big Picture': *The Origins of Modern Science* and the Modern Origins of Science." *British Journal of the History of Science* 26: 407–32.

Cutler, Robert S., ed. 1991. "Engineering in Japan: Education, Practice and Future Outlook." Science in Japan Symposium, 17 February, at American Association for the Advancement of Science, Washington, D.C.

————, ed. 1993. "Technology Management in Japan: R & D Policy, Industrial Strategies, and Current Practice." Science in Japan Symposium, 13 February, at American Association for the Advancement of Science, Washington, D.C.

Danhof, Clarence H. 1969. *Change in Agriculture: The Northern United States, 1820–1870*. Cambridge: Harvard University Press.

David, Paul A. 1971. "The Mechanization of Reaping in the Ante-bellum Midwest," 214–27. In Fogel and Engerman 1971.

————. 1991. "The Hero and the Herd in Technological History: Reflections on Thomas Edison and the System Dynamics of Technological Change," 72–119. In Higonnet et al. 1991.

————. 1993. "Knowledge, Property, and the System Dynamics of Technological Change." *Proceedings of the World Bank Annual Conference on Development Economics* 1992: 215–48.

Deane, Phyllis. 1965. *The First Industrial Revolution*. Cambridge: Cambridge University Press.

Department of Agriculture. 1960. *Power to Produce. Yearbook of Agriculture*. Washington, D.C.: U.S. Government Printing Office.

————. 1970. *Contours of Change. Yearbook of Agriculture*. Washington, D.C.: U.S. Government Printing Office.

de Vries, Jan. 1993. "Between Purchasing Power and the World Goods: Understanding the Household Economy in Early Modern Europe," 85–132. In Brewer and Porter 1993.

Dieffenbach, E. M., and R. B. Gray. 1960. "The Development of the Tractor," 25–45. In Department of Agriculture 1960.

DiMaggio, Paul. 1990. "Cultural Aspects of Economic Action and Organization," 113–36. In Friedland and Robertson 1990.

Dosi, Giovanni, Renato Giannetti, and Pier Angelo Toninelli. 1992. "Introduction," 1–26. In Dosi et al. 1992.

———, eds. 1992. *Technology and Enterprise in a Historical Perspective*. Oxford: Clarendon Press.

Douglas, Mary. 1985. *Risk Acceptability According to the Social Sciences*. New York: Russell Sage Foundation.

———. 1990. "Risk as a Forensic Resource." *Daedalus* 119: 4, 1–16.

Dupree, A. Hunter. [1957] 1986. *Science in the Federal Government: A History of Policies and Activities*. Baltimore: Johns Hopkins University Press.

Durost, Donald, and Warren Bailey. 1970. "What's Happened to Farming," 2–10. In Department of Agriculture 1970.

Edwards, Paul N. 1994. "From 'Impact' to Social Process: Computers in Society and Culture," 257–85. In Jasanoff et al. 1994.

Elbaum, Bernard, and William Lazonick, eds. 1986. *The Decline of the British Economy*. Oxford: Clarendon Press.

Eldredge, N., and S. J. Gould. 1972. "Punctuated Equilibria: An Alternative to Phyletic Gradualism," 305–32. In Schopf 1992.

Elster, Jon. 1983. *Explaining Technical Change: A Case Study in the Philosophy of Science*. Cambridge: Cambridge University Press.

Engels, Friedrich. [1845] 1968. *The Condition of the Working Class in England*. Stanford: Stanford University Press.

Feller, Irwin. 1962. "Inventive Activity in Agriculture, 1837–1890." *Journal of Economic History* 22: 560–77.

Ferguson, Charles H., and Charles R. Morris. 1993. *Computer Wars: How the West Can Win in a Post-IBM World*. New York: New York Times Books.

Ferguson, Eugene S. 1981. "History and Historiography," 1–23. In Mayr and Post 1981.

Ferleger, Lou, ed. 1990. *Agriculture and National Development: Views on the Nineteenth Century*. Ames: Iowa State University Press.

Field, Alexander J. 1985. "On the Unimportance of Machinery." *Explorations in Economic History* 22: 378–401.

Fine, Ben; and Ellen Leopold. 1990. "Consumerism and the Industrial Revolution." *Social History* 15: 151–79.

Finley, Moses I. 1965. "Technical Innovation and Economic Progress in the Ancient World." *Economic History Review* 17: 29–45.

———. 1973. *The Ancient Economy*. Berkeley: University of California Press.

Fisher, F. J. 1935. "The Development of the London Food Market, 1540–1640." *Economic History Review* 5: 46–64.

Fishlow, Albert. 1966. "Levels of Nineteenth-century American Investment in Education." *Journal of Economic History* 26: 418–36.

Fitzgerald, Deborah. 1991. "Beyond Tractors: The History of Technology in American Agriculture." *Technology and Culture* 32: 114–26.

Flamm, Kenneth. 1988. *Creating the Computer: Government, Industry, and High Technology.* Washington, D.C.: Brookings Institution.

Floud, Roderick, and Donald McCloskey, eds. 1981. *The Economic History of Britain since 1700, vol. 1. 1700–1860.* Cambridge: Cambridge University Press.

Fogel, Robert W. [1964] 1971. "Railroads and American Economic Growth," 187–206. In Fogel and Engerman 1971.

Fogel, Robert W., and Stanley L. Engerman, eds. 1971. *The Reinterpretation of American Economic History.* New York: Harper and Row.

Fox, Mary F. 1994. "Women and Scientific Careers," 205–28. In Jasanoff et al. 1994.

Frangipane, M. et al., eds. 1993. *Between the Rivers and over the Mountains. Archaeologica Anatolica et Mesopotamica Alba Palmieri Dedicata.* Rome: Gruppo Editoriale Internazionale.

Frankfort, Henri. 1939. *Cylinder Seals.* London: Macmillan.

Freeman, Christopher, ed. 1982. 2d ed. *The Economics of Industrial Innovation.* Cambridge: MIT Press.

Freeman, Christopher, J. Clark, and L. Soete. 1982. *Unemployment and Technical Innovation: A Study of Long Waves and Economic Development.* Westport, Conn.: Greenwood.

Freeman, Richard B., ed. 1994. *Working under Different Rules.* New York: Russell Sage Foundation.

Friedland, Roger, and A. F. Robertson, eds. 1990. *Beyond the Market Place: Rethinking Economy and Society.* New York: Aldine.

Galison, Peter. 1988. "Physics between War and Peace," 47–86. In Mendelsohn et al. 1988, vol. 1.

Gallman, Robert E., ed. 1977. *Recent Developments in the Study of Business and Economic History: Essays in Memory of Herman E. Kroos.* Research in Economic History, Supplement 1. Greenwich, Conn.: JAI Press.

Gallman, Robert E., and Edward S. Howle. 1971. "Trends in the Structure of the American Economy since 1840," 25–37. In Fogel and Engerman 1971.

Gamota, George. 1992. "Technology Assessment in the U.S.-Japan Context," 21–41. In Arrison et al. 1992.

Geiger, Roger L. 1990. "The American University and Research," 15–35. In Government-University-Industry Research Roundtable 1990.

Gillispie, Charles C. 1957. "The Natural History of Invention." *Isis* 48: 398–407.

———, ed. 1970. *Dictionary of Scientific Biography.* New York: Charles Scribner's Sons.

Gimpel, Jean. 1988. *The Medieval Machine: The Industrial Revolution of the Middle Ages.* 2d ed. Aldershot, Hampshire: Wildwood House.

Girifalco, Louis A. 1991. *Dynamics of Technological Change.* New York: Van Nostrand Reinhold.

Gomory, Ralph E. 1983. "Technology Development." *Science* 220: 576–80.

———. 1988. "Reduction to Practice: The Development and Manufacturing Cycle," 11–17. In Academy Industry Program 1988.

Gordon, Robert B., and Patrick M. Malone. 1994. *The Texture of Industry: An Archaeological View of the Industrialization of North America*. New York: Oxford University Press.

Gould, Leroy C. et al. 1988. *Perceptions of Technological Risks and Benefits*. New York: Russell Sage Foundation.

Gould, Stephen Jay, and Niles Eldredge. 1993. "Punctuated Equilibrium Comes of Age." *Nature* 366: 223–27.

Government-University-Industry Research Roundtable. 1990. *The Academic Research Enterprise within the Industrialized Nations, Comparative Prospects: Report of a Symposium*. Washington, D.C.: National Academy Press.

Grabowski, Henry. 1991. "The Changing Economics of Pharmaceutical Research and Development," 35–52. In Committee on Technological Innovation in Medicine 1991.

Granovetter, Mark. 1993. "The Nature of Economic Relationships," 3–41. In Swedberg 1993.

Graubard, Stephen R., ed. 1980. *Modern Technology: Problem or Opportunity?* Daedalus 109, 1.

Greene, John C. 1984. *American Science in the Age of Jefferson*. Ames: Iowa State University Press.

Griffiths, Trevor, Phillip A. Hunt, and Patrick K. O'Brien. 1992. "Inventive Activity in the British Textile Industry, 1700–1800." *Journal of Economic History* 52: 881–906.

Griliches, Zvi. 1957. "Hybrid Corn: An Exploration in the Economics of Technological Change." *Econometrica* 25: 501–22.

———. 1971. "Hybrid Corn and the Economics of Innovation," 207–13. In Fogel and Engerman 1971.

———, ed. 1984. *R & D, Patents, and Productivity*. National Bureau of Economic Research Conference Report. Chicago: University of Chicago Press.

Group of Lisbon. 1995. *Limits to Competition*. Cambridge: MIT Press.

Habakkuk, H. J. 1962. *American and British Technology in the Nineteenth Century: The Search for Labour-saving Inventions*. Cambridge: Cambridge University Press.

———. 1970. "The Economic Effects of Labour Scarcity," 23–76. In Saul 1970.

Habakkuk, H. J., and M. Postan, eds. 1966. *The Industrial Revolution and after: Incomes, Population and Technological Change*. The Cambridge Economic History of Europe, vol. 6. Cambridge: Cambridge University Press.

Hall, Peter, and Praschal Preston. 1988. *The Carrier Wave: New Information Technology and the Geography of Information, 1866–2003*. London: Unwin-Hyman.

Hall, Rupert. 1959. "The Scholar and the Craftsman in the Scientific Revolution," 3–23. In Clagett 1959.

———. 1978. "On Knowing, and Knowing How to . . . ," 92–96. In *History of Technology, Third Annual Volume*. London: Mansell.

Halm, Ethan A., and Annetine C. Gelijns. 1991. "An Introduction to the Changing Economics of Technological Innovation in Medicine," 1–20. In Committee on Technological Innovation in Medicine 1991.

Hansen, Michael, Lawrence Busch, Jeffrey Burkhardt, William B. Lacy, and Laura R. Lacy. 1986. "Plant Breeding and Biotechnology: New Technologies Raise Important Social Questions." *BioScience* 36: 1, 29–39.

Harley, C. Knick 1993. "Reassessing the Industrial Revolution: A Macro View," 171–226. In Mokyr 1993.

Harris, John R. 1988. *The British Iron Industry 1700–1850.* Houndmills, Hampshire: Macmillan Education Ltd.

———. 1992a. "Introduction," 1–17. In Harris 1992.

———. 1992b. "Skills, Coal and British Industry in the Eighteenth Century," chapter 1. In Harris 1992.

———. ed. 1992c. *Essays in Industry and Technology in the Eighteenth Century: England and France.* Great Yarmouth: Variorum.

Harrison, Bennett. 1994. "The Dark Side of Flexible Production," *Technology Review* (May–June): 39–45.

Hatsopoulos, George N., Paul R. Krugman, and Lawrence H. Summers. 1988. "U.S. Competitiveness: Beyond the Trade Deficit." *Science* 241: 299–307.

Hattaway, Herman, and Archer Jones. 1983. *How the North Won: A Military History of the Civil War.* Urbana: University of Illinois Press.

Hayami, Yujiro, and Vernon W. Ruttan. 1985. Rev. ed. *Agricultural Development: An International Perspective.* Baltimore: Johns Hopkins University Press.

Herman, Robert; Siamak A. Ardekani; and Jesse H. Ausubel. 1989. "Dematerialization," 50–69. In Ausubel and Sladovich 1989.

Higonnet, Patrice, David S. Landes, and Henry Rosovsky, eds. 1991. *Favorites of Fortune: Technology, Growth, and Economic Development since the Industrial Revolution.* Cambridge: Harvard University Press.

Hill, Donald. 1984. *A History of Engineering in Classical and Medieval Times.* Beckenham, Kent: Croom, Helm.

Hills, Richard L. 1989. *Power from Steam: A History of the Stationary Steam Engine.* Cambridge: Cambridge University Press.

Hilton, Rodney H., and P. H. Sawyer. 1963. "Technical Determinism: The Stirrup and the Plough." *Past and Present* 24: 90–100.

Hindle, Brooke. 1981. *Emulation and Invention.* New York: New York University Press.

Hindle, Brooke, and Steven Lubar. 1986. *Engines of Change: The American Industrial Revolution.* Washington, D.C.: Smithsonian Institution Press.

Hohenberg, Paul M., and Lynn Hollen Lees. 1985. *The Making of Urban Europe 1000–1950.* Cambridge: Harvard University Press.

Hoke, Donald R. 1990. *Ingenious Yankees.* New York: Columbia University Press.

Holmes, Bob. 1993. "A New Study Finds There's Life Left in the Green Revolution." *Science* 261: 1517.

Hoover, Herbert Clark, and Lou Henry Hoover, eds. [1912] 1950. *Georgius Agricola: De re Metallica, translated from the first Latin Edition of 1556.* New York: Dover Publications.

Hopper, R. J. 1979. *Trade and Industry in Classical Greece.* London: Thames and Hudson.

Houghton, W. E., Jr. 1941. "The History of Trades: Its Relation to Seventeenth-century Thought." *Journal of the History of Ideas* 2: 33–60.

Hounshell, David A. 1981. "The *System*: Theory and Practice," 127–52. In Mayr and Post 1981.

———. 1984. *From the American System to Mass Production, 1800–1932.* Baltimore: Johns Hopkins University Press.

Huck, Paul. 1995. "Infant Mortality and Living Standards of English Workers during the Industrial Revolution." *Journal of Economic History* 55: 528–50.

Hudson, Pat. 1989a. "The Regional Perspective," 5–38. In Hudson 1989b.

———, ed. 1989b. *Regions and Industries: A Perspective on the Industrial Revolution in Britain.* Cambridge: Cambridge University Press.

———. 1992. *The Industrial Revolution.* London: Edward Arnold.

Hughes, Thomas P. 1983. *Networks of Power: Electrification in Western Society, 1880–1930.* Baltimore: Johns Hopkins University Press.

———. 1987. "The Evolution of Large Technological Systems," 51–82. In Bijker et al. 1987.

———. 1989. *American Genesis: A Century of Invention and Technological Enthusiasm.* New York: Viking.

———. 1994. "Technological Momentum," 102–13. In Smith and Marx 1994.

Hugill, Peter J. 1993. *World Trade since 1431: Geography, Technology and Capitalism.* Baltimore: Johns Hopkins University Press.

Hyde, Charles K. 1977. *Technological Change and the British Iron Industry, 1700–1870.* Princeton, N.J.: Princeton University Press.

Imai, Ken-Ichi, and Akiko Yamazaki. 1994. "Dynamics of the Japanese Industrial System from a Schumpeterian Perspective," 217–51. In Shionoya and Perlman 1994.

Inkster, Ian. 1991. *Science and Technology in History: An Approach to Industrial Development.* New Brunswick, N.J.: Rutgers University Press.

International Institute for Applied Systems Analysis (IIASA). 1992. *Science and Sustainability: Selected Papers on IIASA's 20th Anniversary.* Vienna: IIASA.

Jackman, William T. 1962. 2d ed. *The Development of Transportation in Modern England.* London: Frank Cass.

Jackson, R. V. 1992. "Rates of Industrial Growth during the Industrial Revolution." *Economic History Review* 45: 1–23.

Jacob, Margaret C. 1988. *The Cultural Meaning of the Scientific Revolution.* Philadelphia, Pa.: Temple University Press.

Janich, Peter. 1978. "Physics—Natural Science or Technology," 3–27. In Krohn 1978.

Jasanoff, Sheila. 1990. "American Exceptionalism and the Political Acknowledgment of Risk." *Daedalus* 119: 4, 61–81.

———. 1992. "Pluralism and Convergence in International Science Policy," 157–80. In IIASA 1992.

Jasanoff, Sheila, Gerald E. Markle, James C. Petersen, and Trevor Pinch, eds. 1994. *Handbook of Science and Technology Studies.* Thousand Oaks, Calif.: Sage Publications.

Jeremy, David J. 1973. "Innovation in American Textile Technology during the early 19th Century." *Technology and Culture* 14: 40–76.

———. 1977. "Damming the Flood: British Government Efforts to Check the Outflow of Technicians and Machinery, 1780–1843." *Business History Review* 51: 1–34.

———. 1981. *Transatlantic Industrial Revolution: The Diffusion of Textile Technologies between Britain and America, 1790–1830s.* Cambridge: MIT Press.

————. 1990. *Technology and Power in the Early American Cotton Industry*. Philadelphia, Pa.: American Philosophical Society.

————, ed. 1991. *International Technology Transfer: Europe, Japan and the U.S.A., 1700–1914*. Aldershot, Hampshire: Edward Elgar.

Jeremy, David J., and Darwin H. Stapleton. 1991. "Transfers between Culturally-related Nations: The Movement of Textile and Railroad Technologies between Britain and the United States, 1780–1840," 31–48. In Jeremy 1991.

Jones, Eric L. 1968. "Agricultural Origins of Industry." *Past and Present* 40: 58–71.

————, ed. 1974. *Agriculture and the Industrial Revolution*. Oxford: Oxford University Press.

————. 1981. "Agriculture, 1700–1800," 66–86. In Floud and McCloskey 1981.

Jordanova, Ludmilla. 1993. "Gender and the Historiography of Science." *British Journal of the History of Science* 26: 469–83.

Joyce, Patrick. 1991. *Visions of the People: Industrial England and the Question of Class, c. 1848–1914*. Cambridge: Cambridge University Press.

Kahn, Alan. 1991. "The Dynamics of Medical Device Innovation: An Innovator's Perspective," 89–95. In Committee on Technological Innovation in Medicine 1991.

Kanefsky, John, and John Robey. 1980. "Steam Engines in the Eighteenth Century: A Quantitative Assessment." *Technology and Culture* 21: 161–86.

Kasson, John F. 1976. *Civilizing the Machine: Technology and Republican Values in America 1776–1900*. New York: Grossman.

Kates, Robert W. 1986. "Managing Technological Hazards: Success, Strain, and Surprise," 206–20. In National Academy of Engineering 1986.

Katznelson, Ira. 1993. *Marxism and the City*. Oxford: Clarendon Press.

Kaufmann, William J. III, and Larry J. Smarr. 1993. *Supercomputing and the Transformation of Science*. Scientific American Library, 43. New York: W. H. Freeman.

Kelley, Maryellen R., and Todd A. Watkins. 1995. "In from the Cold: Prospects for the Conversion of the Defense Industrial Base." *Science* 268: 525–32.

Kelly, Patrick, and Melvin Kranzberg, eds. 1978. *Technological Innovation: A Critical Review of Current Knowledge*. Atlanta: Georgia Institute of Technology Press.

Keniston, Kenneth 1990. "Defining 'Technology'." *STS News* (March): 1–3. Transcript of Comments, Dibner Workshop on Technological Determinism.

Kevles, Daniel J. 1978. *The Physicists: The History of a Scientific Community in Modern America*. New York: Alfred A. Knopf.

————, 1979. "The Physics, Mathematics, and Chemistry Communities: A Comparative Analysis," 139–72. In Oleson and Voss 1979.

————. 1988. "An Analytical Look at R & D and the Arms Race," 465–80. In Mendelsohn et al. 1988b, vol. 2.

King, Gregory [1696] 1936. "Natural and Political Observations and Conclusions upon the State and Condition of England," 13–56. In Barnett, ed. 1936.

Kingery, W. David, ed. 1991. *Japanese/American Technological Innovation: The Influence of Cultural Differences on Japanese and American Innovation in Advanced Materials*. Amsterdam: Elsevier.

Klein, Randolph S., ed. 1986. *Science and Society in Early America: Essays in*

Honor of Whitfield J. Bell, Jr. Philadephia, Pa.: American Philosophical Society, Memoir 166.

Kline, Stephen J., and Nathan Rosenberg. 1986. "An Overview of Innovation," 275–305. In Landau and Rosenberg 1986.

Krohn, Wolfgang, ed. 1978. *The Dynamics of Science and Technology.* Dordrecht: D. Reidel.

Kuhn, Thomas S. 1969. "Comment." *Comparative Studies in Society and History* 11: 426–30.

———. 1970. 2d ed. *The Structure of Scientific Revolutions.* Chicago: University of Chicago Press.

———. 1976. "Mathematical vs. Experimental Traditions in the Development of Physical Science." *Journal of Interdisciplinary History* 7: 1–31.

Kuznets, Simon. 1971. "Notes on the Pattern of U.S. Economic Growth," 17–24. In Fogel and Engerman 1971.

———. 1978. "Technological Innovations and Economic Growth," 335–56 In Kelly and Kranzberg 1978.

———. 1979. *Growth, Population, and Income Distribution: Selected Essays.* New York: W.W. Norton.

Landau, Rachel, ed. 1984. *The Nature of Technological Knowledge: Are Models of Scientific Change Relevant?* Dordrecht: D. Reidel.

Landau, Ralph, and Nathan Rosenberg, eds. 1986. *The Positive Sum Strategy: Harnessing Technology for Economic Growth.* Washington, D.C.: National Academy Press.

———. 1992. "Successful Commercialization in the Chemical Process Industries," 73–119. In Rosenberg et al. 1992.

Landels, J. G. 1978. *Engineering in the Ancient World.* Berkeley: University of California Press.

Landes, David S. 1966. "Technological Change and Development in Western Europe, 1750–1914," 274–601. In Habakkuk and Postan 1966.

———. 1969. *The Unbound Prometheus: Technological Change and Industrial Development in Western Europe from 1750 to the Present.* Cambridge: Cambridge University Press.

———. 1983. *Revolution in Time: Clocks and the Making of the Modern World.* Cambridge: Harvard University Press.

———. 1991. "Introduction: On Technology and Growth," 1–71. In Higonnet et al. 1991.

———. 1992. "Homo Faber, Homo Sapiens: Knowledge, Technology, Growth, and Development." *Contention* 3: 81–107.

———. 1993. "The Fable of the Dead Horse; or, the Industrial Revolution Revisited," 132–70. In Mokyr 1993b.

———. 1994. "What Room for Accidents in History?: Explaining Big Changes by Small Events." *Economic History Review* 47: 637–56.

Larsen, Mogens T. 1987. "Commercial Networks in the Ancient Near East," 47–56. In Rowlands et al. 1987.

Layton, Edwin T., Jr. 1971. "Mirror-image Twins: The Communities of Science and Technology in 19th-century America." *Technology and Culture* 12: 562–80.

———. 1974. "Technology as Knowledge." *Technology and Culture* 15: 31–41.

————. 1976. "American Ideologies of Science and Engineering." *Technology and Culture* 17: 688–701.

————. 1986. *The Revolt of the Engineers: Social Responsibility and the American Engineering Profession*. Baltimore: Johns Hopkins University Press.

————. 1987a. "Through the Looking Glass, or News from Lake Mirror Image." *Technology and Culture* 28: 594–607.

————. 1987b. "European Origins of the American Engineering Style of the Nineteenth Century," 151–66. In Reingold and Rothenberg 1987.

————. 1988. "Science as a Form of Action: The Role of the Engineering Sciences." *Technology and Culture* 29: 82–97.

Lazonick, William. 1986. "The Cotton Industry," 18–50. In Elbaum and Lazonick 1986.

————. 1991. *Business Organization and the Myth of the Market Economy*. Cambridge: Cambridge University Press.

————. 1992. "Business Organization and Competitive Advantage: Capitalist Transformations in the Twentieth Century." 119–63. In Dosi et al. 1992.

Lee, Susan P., and Peter Passell 1979. *A New Economic View of American History*. New York: W.W. Norton.

Lee, Thomas H. 1986. "Dynamics of Technology," 28–31. In International Institute for Applied Systems Analysis 1992.

Lemon, James T. 1987. "Colonial America in the Eighteenth Century," 121–46. In Mitchell and Groves 1987.

Letwin, William. 1989. "American Economic Policy, 1865–1939," 641–90. In Mathias and Pollard 1989.

Levine, David. 1987. *Reproducing Families: The Political Economy of English Population History*. Cambridge: Cambridge University Press.

Lieberman, Victor. 1993. Abu-Lughod's Egalitarian World-order: A Review Article." *Comparative Studies in Society and History* 35: 544–50.

Lilley, Samuel. 1973. "Technological Progress and the Industrial Revolution 1700–1914," 187–254. In Cipolla 1973.

Lindberg, David C. 1992. *The Beginnings of Western Science: The European Scientific Tradition in Philosophical, Religious, and Institutional Context, 600* B.C. to A.D. 1450. Chicago: University of Chicago Press.

Lindert, Peter H., and Keith Trace. 1971. "Yardsticks for Victorian Entrepreneurs," 239–74. In McCloskey 1971.

Lindert, Peter H., and Jeffrey G. Williamson. 1983. "English Workers' Living Standards during the Industrial Revolution: A New Look." *Economic History Review* 36: 1–25.

Logan, John R., and Harvey L. Molotch. 1987. *Urban Fortunes: The Political Economy of Place*. Berkeley: University of California Press.

Lynch, Lisa M. 1994. "Payoffs to Alternative Training Strategies," 63–95. In Freeman 1994.

Macdonald, Anne L. 1992. *Feminine Ingenuity: Women and Invention in America*. New York: Ballentine Books.

MacKenzie, Donald, and Judy Wajcman. 1985a. "Introductory Essay," 2–25. In MacKenzie and Wajcman 1985b.

————, eds. 1985b. *The Social Shaping of Technology*. Milton Keynes: Open University Press.

MacLeod, Christine. 1988. *Inventing the Industrial Revolution: The English Patent System, 1660–1800*. Cambridge: Cambridge University Press.

Manning, Thomas G. 1988. *U.S. Coast Survey vs. Naval Hydrographic Office: A 19th Century Rivalry in Science and Politics*. Tuscaloosa: University of Alabama Press.

Marx, Leo. 1964. *The Machine in the Garden: Technology and the Pastoral Ideal in America*. New York: Oxford University Press.

———. 1992a. Letter to the Editor. *Technology and Culture* 33: 407.

———. 1992b. "Environmental Degradation and the Ambiguous Social Role of Science and Technology." *Journal of the History of Biology* 25: 449–68.

Marx, Leo, and Merritt Roe Smith. 1994. "Introduction," ix–xv. In Smith and Marx 1994.

Mathias, Peter. 1972a. "Who Unbound Prometheus? Science and Technical Change, 1600–1800," 54–80. In Mathias 1972b.

———. 1972b. *Science and Society 1600–1900*. Cambridge: Cambridge University Press.

Mathias, Peter. 1991. "Resources and Technology," 18–42. In Mathias and Davis 1991b.

Mathias, Peter, and John A Davis. 1991a. "Editors' Introduction," 1–5. In Mathias and Davis 1991b.

———, eds. 1991b. *Innovation and Technology in Europe, from the Eighteenth Century to the Present Day*. Oxford: Basil Blackwell.

Mathias, Peter, and Sidney Pollard, eds. 1989. *The Industrial Economies: The Development of Economic and Social Policies*. Cambridge Economic History of Europe, vol. 8. Cambridge: Cambridge University Press.

Mathias, Peter, and M. M. Postan, eds. 1978. *The Industrial Economies: Capital, Labour, and Enterprise. Part I, Britain, France, Germany, and Scandinavia; Part II, The United States, Japan, and Russia*. Cambridge Economic History of Europe, vol. 7. Cambridge: Cambridge University Press.

Maurice, Klaus, and Otto Mayr. 1980. *The Clockwork Universe: German Clocks and Automata 1550–1650*. Washington, D.C.: Smithsonian Institution Press.

Mauskopf, Seymour, ed. 1993. *Chemical Sciences in the Modern World*. Philadelphia: University of Pennsylvania Press.

Mayntz, Renata, and Thomas P. Hughes, eds. 1988. *The Development of Large Technical Systems*. Frankfurt am Main: Campus.

Mayr, Ernst. 1988a. "Speciational Evolution through Punctuated Equilibria," 457–88. In Mayr 1988b.

———. ed. 1988b. *Toward a New Philosophy of Biology: Observations of an Evolutionist*. Cambridge: Belknap Press of Harvard University Press.

Mayr, Otto. 1976. "The Science-Technology Relationship as a Historiographic Problem." *Technology and Culture* 17: 663–73.

———. 1986. *Authority, Liberty and Automatic Machinery in Early Modern Europe*. Baltimore: Johns Hopkins University Press.

Mayr, Otto, and Robert C. Post. 1981a. "Introduction," xi–xx. In Mayr and Post 1981b.

———, eds. 1981b. *Yankee Enterprise: The Rise of the American System of Manufacture*. Washington, D.C.: Smithsonian Institution Press.

McCloskey, Donald N., ed. 1971. *Essays in a Mature Economy: Britain after 1840.* London: Methuen.

———. 1981. "The Industrial Revolution 1780–1860: An Overview," 103–27. In Floud and McCloskey 1981.

———. 1985. "The Industrial Revolution 1780–1860: A Survey," 53–74. In Mokyr 1985.

McGuire, Patrick, Mark Granovetter, and Michael Schwartz. 1993. "Thomas Edison and the Social Construction of the Early Electricity Industry in America," 213–46. In Swedberg 1993.

McKelvey, John. 1985. "Science and Technology: The Driven and the Driver." *Technology Review* 88, 1: 38–47.

McKendrick, Neil. 1973. "The Role of Science in the Industrial Revolution: A Study of Josiah Wedgwood as a Scientist and Industrial Chemist," 274–319. In Teich and Young 1973.

———. 1982. "The Commercial Revolution in Eighteenth-century England," 9–33. In McKendrick et al. 1982.

McKendrick, Neil, John Brewer, and J. J. Plumb, eds. 1982. *The Birth of a Consumer Society: The Commercialization of Eighteenth-Century England.* Bloomington: University of Indiana Press.

McPherson, James M. 1988. *Battle Cry of Freedom: The Civil War Era.* New York: Oxford University Press.

Mendelsohn, Everett, Merritt Roe Smith, and Peter Weingart. 1988a. "Introductory Essay: Science and the Military: Setting the Problem," 1, xi–xxix. In Mendelsohn et al. 1988b.

———. 1988b. *Science, Technology and the Military.* 2 vols. Sociology of Sciences Yearbook, 12/1. Dordrecht: Kluwer Academic Publishers.

Mensch, Gehard. 1979. *Stalemate in Technology.* Cambridge: Ballinger Publishing.

Merrill, Robert. 1968. "The Study of Technology," 576–89. In *International Encyclopedia of the Social Sciences*, 15. New York: Macmillan and the Free Press.

Merton, Robert. 1937. "Some Economic Factors in Seventeenth-Century English Science." *Scientia* 29: 142–52.

———. [1938] 1970. *Science, Technology, and Society in Seventeenth-Century England.* New York: Harper and Row.

———. [1957] 1973a. "Priorities in Scientific Discovery," 286–324. In Merton 1973b.

———. ed. 1973b. *The Sociology of Science: Theoretical and Empirical Investigations.* Chicago: University of Chicago Press.

Mikulski, Barbara A. 1994. "Science in the National Interest." *Science* 264: 221–22.

Miller, Lillian B. 1972. *The Lazzaroni: Science and Scientists in Mid-nineteenth Century America.* Washington, D.C.: Smithsonian Institution Press.

Mitchell, Robert A., and Paul A. Groves, eds. 1987. *North America: The Historical Geography of a Changing Continent.* New York: Rowman and Littlefield.

Mokyr, Joel, ed. 1985. *The Economics of the Industrial Revolution.* New York: Rowman and Littlefield.

———. 1989. Review of Hudson 1989b. *Journal of Interdisciplinary History* 22: 306–309.

Mokyr, Joel, ed. 1990. *The Lever of Riches: Technological Creativity and Economic Progress*. New York: Oxford University Press.

———. 1993a. "Editor's Introduction: The New Economic History and the Industrial Revolution," 1–131. In Mokyr 1993b.

———, ed. 1993b. *The British Industrial Revolution: An Economic Perspective*. Boulder: Westview Press.

Molella, Arthur P. 1976. "The Electric Motor, the Telegraph,and Joseph Henry's Theory of Technological Progress." *Proceedings of the Institute of Electrical and Electronics Engineers* 64: 1273–76.

Molella, Arthur P., and Nathan Reingold. 1973. "Theorists and Ingenious Mechanics: Joseph Henry Defines Science." *Science Studies* 3: 323–35.

Mowery, David C. 1986. "Industrial Research, 1900–1950," 189–222. In Elbaum and Lazonick 1986.

Mowery, David C., and Nathan Rosenberg. 1989. *Technology and the Pursuit of Economic Growth*. Cambridge: Cambridge University Press.

———. 1993. "The U.S. National Innovation System," 29–75. In Nelson 1993.

Mukaibo, Takashi. 1991. "Engineering in Japan," 1–12. In Cutler 1991.

Muldrew, Craig. 1993. "Interpreting the Market: The Ethics of Credit and Community Relations in Early Modern England." *Social History* 18: 163–83.

Musson, A. E. 1981. "British Origins," 25–48. In Mayr and Post 1981.

Musson, A. E., and E. H. Robinson. 1969. *Science and Technology in the Industrial Revolution*. Toronto: University of Toronto Press.

Nader, John. 1994. "The Rise of an Inventive Profession: Learning Effects in the Midwestern Harvester Industry." *Journal of Economic History* 54: 397–408.

Nakicenovic, Nebojsa, and Arnulf Gruebler, eds. 1991. *Diffusion of Technologies and Social Behavior*. Berlin: Springer-Verlag.

National Academy of Engineering. 1986. *Hazards: Technology and Fairness*. Washington, D.C.: National Academy Press.

National Academy of Sciences, Committee on Science and Public Policy. 1969. *Technology: Processes of Assessment and Choice*. Report to the Committee on Science and Astronautics, U.S. House of Representatives. Washington, D.C.: U.S. Government Printing Office.

National Critical Technologies Panel. 1991. *Report of the National Critical Technologies Panel*. Washington, D.C.: Department of Commerce.

———. 1993. *Second Biennial Report*. Washington, D.C.: National Technical Information Service.

National Science Board. 1977. *Science Indicators, 1976*. Washington, D.C.: U.S. Government Printing Office.

———. 1979. *Science Indicators, 1978*. Washington, D.C.: U.S. Government Printing Office.

———. 1981. *Science Indicators, 1980*. Washington, D.C.: U.S. Government Printing Office.

———. 1983. *Science Indicators, 1982*. Washington, D.C.: U.S. Government Printing Office.

———. 1993. *Science and Engineering Indicators, 1992*. Washington, D.C.: U.S. Government Printing Office.

———. 1994. *Science and Engineering Degrees: 1966–1991*. National Science Foundation Publication 94–305. Arlington, Va.

Nelson, Richard R. 1962a. "Introduction," 3–16. In Nelson 1993.

———. ed. 1962b. *The Rate and Direction of Economic Activity: Economic and Social Factors*. Universities-NBER Conference Series, 13. Princeton, N.J.: Princeton University Press for the National Bureau of Economic Research.

———. 1990. "Capitalism as an Engine of Progress." *Research Policy* 19: 193–214.

———. 1992. "What is 'Commercial' and What Is 'Public' about Technology, and What Should Be?" 57–71. In Rosenberg et al. 1992.

———, ed. 1993. *National Innovation Systems: A Comparative Analysis*. New York: Oxford University Press.

Nelson, Richard R., and S. G. Winter. 1982. *An Evolutionary Theory of Economic Change*. Cambridge, Mass.: Harvard University Press.

Neumann, Hans. 1987. *Handwerk in Mesopotamien*. Schriften zur Geschichte und Kultur des Alten Orients, 19. Berlin: Akademie Verlag.

Nicholas, Stephen J., and Jacqueline M. Nicholas. 1992. "Male Literacy, 'Deskilling', and the Industrial Revolution." *Journal of Interdisciplinary History* 23: 1–18.

Niemi, Albert W., Jr. 1980. *U.S. Economic History*. 2nd ed. Chicago: Rand McNally College Publishing Co.

Nissen, Hans J. 1993. "The Early Uruk Period—A Sketch," 124–31. In Frangipane et al. 1993.

Nissen, Peter Damerow, and Robert K. Englund. 1993. *Archaic Bookkeeping: Early Writing and Techniques of Economic Administration in the Ancient Near East*. Chicago: University of Chicago Press.

North, Douglass C. 1966. "Industrialization in the United States," 673–705. In Habakkuk and Postan 1966.

O'Brien, Patrick. 1991. "The Mainsprings of Technological Progress in Western Europe 1750–1850," 6–17. In Mathias and Davis 1991b.

———. 1993a. "Introduction: Modern Conceptions of the Industrial Revolution," 1–30. In O'Brien and Quinault 1993.

———. 1993b. "Political Preconditions for the Industrial Revolution," 124–55. In O'Brien and Quinault 1993.

O'Brien, Patrick, and Roland Quinault, eds. 1993. *The Industrial Revolution and British Society*. Cambridge: Cambridge University Press.

OECD Development Centre. 1989. *Programme of Research, 1990–92*. Paris: Organization for Economic Cooperation and Development.

Office of Japan Affairs. 1990. *Learning the R & D System: Industrial R & D in Japan and the United States*. Office of International Affairs, National Research Council. Washington, D.C.: National Academy Press.

Ohmae, Kenichi. 1985. *Triad Power: The Coming Shape of Global Competition*. New York: Free Press.

Oleson, Alexandra, and John Voss, eds. 1979. *The Organization of Knowledge in Modern America, 1860–1920*. Baltimore: Johns Hopkins University Press.

Olmstead, Alan L., and Paul W. Rhode. 1995. "Beyond the Threshold: An Analysis of the Characteristics and Behavior of Early Reaper Adopters." *Journal of Economic History* 55: 27–57.

Olson, Richard. 1990. *Science Deified and Science Defied: The Historical Significance of Science in Western Culture, 2: From the Early Modern Age through the*

Early Romantic Era ca. 1640 to ca. 1820. Berkeley: University of California Press.

Ovitt, George, Jr. 1987. *The Restoration of Perfection: Labor and Technology in Medieval Culture.* New Brunswick, N.J.: Rutgers University Press.

Pagel, Walter. 1970. "Paracelsus, Theophrastus Philippus Aureolus von Hohenheim," 10: 304–13. In Gillispie 1970.

Parker, William N. 1971. "Productivity Growth in American Grain Farming: An Analysis of Its 19th-century Sources," 175–86. In Fogel and Engerman 1971.

Payne, Peter L. 1978. "Industrial Entrepreneurship and Management in Great Britain," 180–230. In Mathias and Postan 1978.

Perrow, Charles. 1991. *Accidents in High-risk Systems.* Working Paper 19. New York: Russell Sage Foundation.

Persson, Karl Gunnar. 1988. *Pre-industrial Economic Growth: Social Organization and Technological Progress in Europe.* London: Basil Blackwell.

Pleket, H. W. 1967. "Technology and Society in the Graeco-Roman World." *Acta Historica Neerlandica* 2: 1–25.

Polanyi, Karl, Conrad Arensberg, and Harry Pearson, eds. 1957. *Trade and Market in the Early Empires.* New York: Free Press.

Pollard, Sidney. 1978. "Labour in Great Britain," 1: 97–179. In Mathias and Postan 1978.

———. 1992. "The Concept of the Industrial Revolution," 29–62. In Dosi et al. 1992.

Population Summit of the World's Scientific Academies. 1994. *A Joint Statement by Fifty-eight of the World's Scientific Academies.* Washington, D.C.: National Academy Press.

Postan, M. M. 1975. *The Medieval Economy and Society.* London: Pelican.

Postan, M. M., and John Hatcher. 1985. "Population and Class Relations in Feudal Society," 64–78. In Aston and Philpin 1985.

Price, Derek deSolla 1959. "On the Origin of Clockwork, Perpetual Motion Devices and the Compass." Contributions from the Museum of History and Technology, *Bulletin* 218: 81–112.

———. 1984. "Notes toward a Philosophy of the Science/ Technology Interaction," 105–14. In Landau 1984.

Purcell, C. W., Jr. 1969. *Early Stationary Steam Engines in America: A Study in the Migration of a Technology.* Washington, D.C.: Smithsonian Institution Press.

Pursell, Carroll. 1979. "Science Agencies in World War II: The OSRD and Its Challengers," 359–78. In Reingold 1979.

Raepsaet, Georges. 1995. "Les Prémises de la Mécanisation Agricol entre Seine et Rhin de l'Antiquité au 13ᵉ Siècle." *Annales* (July–August), 911–42.

Randsborg, Klaus. 1991. *The First Millennium A.D. in Europe and the Mediterranean.* Cambridge: Cambridge University Press.

Rasmussen, Wayne D. 1962. "The Impact of Technological Change on American Agriculture." *Journal of Economic History* 22: 578–91.

Reingold, Nathan, ed. 1979. *The Sciences in the American Context: New Perspectives.* Washington, D.C.: Smithsonian Institution Press.

Reingold, Nathan, and Marc Rothenberg, eds. 1987. *Scientific Colonialism: A Cross-cultural Comparison.* Washington, D.C.: Smithsonian Institution Press.

Reynolds, Joyce. 1988. "Cities," 15–51. In David Braund 1988.

Reynolds, Terry S. 1984. "Medieval Roots of the Industrial Revolution." *Scientific American* (July): 123–31.

Robinson, Eric H. 1974. "The Early Diffusion of Steam Power." *Journal of Economic History* 34: 91–107.

Rogers, Everett M. [1962] 1983. *Diffusion of Innovations*. New York: Macmillan.

Rose, M. E. 1981. "Social Change and the Industrial Revolution," 253–92. In Floud and McCloskey 1981.

Rosenberg, Nathan. 1963. "Technological Change in the Machine Tool Industry, 1840–1910." *Journal of Economic History* 23: 414–43.

———. 1972. "Factors Affecting the Diffusion of Technology." *Explorations in Economic History* 10: 3–33.

———. 1979. "Technological Interdependence in the American Economy." *Technology and Culture* 20: 25–50.

———. 1981. "Why In America?" 49–61. In Mayr and Post 1981b.

———1982. *Inside the Black Box: Technology and Economics*. Cambridge: Cambridge University Press.

———. 1988. "U.S. Economic Policy: Implications for R & D," 1–9. In Academy Industry Program 1988.

———. 1991. "Critical Issues in Science Policy Research." *Science and Public Policy* 18: 335–46.

———. 1992. "Science and Technology in the Twentieth Century," 63–96. In Dosi et al. 1992.

———, ed. 1994. *Exploring the Black Box: Technology, Economics, and History*. Cambridge: Cambridge Unversity Press.

Rosenberg, Nathan, and L. E. Birdzell, Jr. 1986. *How the West Grew Rich: The Economic Transformation of the Western World*. New York: Basic Books.

Rosenberg, Nathan, and Claudio Frischtak. [1984] 1994. "Technological Innovation and Long Waves," 62–84. In Rosenberg 1994.

Rosenberg, Nathan, Ralph Landau, and David C. Mowery. 1992a. "Introduction," 1–14. In Rosenberg et al. 1992.

———, eds. 1992b. *Technology and the Wealth of Nations*. Stanford: Stanford University Press.

Rosenberg, Nathan, and Richard R. Nelson. 1993. *American Universities and Technical Advance in Industry*. Center for Economic Policy Research, Discussion Paper Series, 342. Stanford: Stanford University Press.

Rosenbloom, Joshua L. 1993. "Anglo-American Technological Differences in Small Arms Manufacture." *Journal of Interdisciplinary History* 23: 683–98.

Rousseau, G. S., and R. Porter, eds. 1980. *The Ferment of Knowledge: Studies in the Historiography of Eighteenth-century Science*. Cambridge: Cambridge University Press.

Rowlands, Marie B. 1989. "Continuity and Change in an Industrializing Society: The Case of the West Midlands Industries," 103–31. In Hudson 1989b.

Rowlands, Michael, Mogens T. Larsen, and Kristian Kristiansen, eds. 1987. *Centre and Periphery in the Ancient World*. Cambridge: Cambridge University Press.

Ruttan, Vernon W. 1994a. "Challenges to Agricultural Research in the 21st Century," 343–57. In Ruttan 1994b.

Ruttan, Vernon W., ed. 1994b. *Agriculture, Environment, and Health: Sustainable Development in the 21st Century.* Minneapolis: University of Minnesota Press.

Sagan, Scott D. 1993. *The Limits of Safety: Organization, Accidents, and Nuclear Weapons.* Princeton, N.J.: Princeton University Press.

Ste. Croix, G.E.M. de. 1981. *The Class Struggle in the Ancient Greek World, from the Archaic Age to the Arab Conquests.* Ithaca, N.Y.: Cornell University Press.

Salsbury, Stephen. 1988. "The Emergence of an Early Large-Scale Technical System: The American Railroad Network," 37–68. In Mayntz and Hughes, eds. 1988.

Sapolsky, Harvey M. 1979. "Academic Science and the Military: The Years since the Second World War," 379–99. In Reingold 1979.

Saul, S. B., ed. 1970. *Technological Change: The United States and Britain in the 19th Century.* London: Methuen.

Sawyer, John E. 1954. "The Social Basis of the American System of Manufacturing." *Journal of Economic History* 14: 361–79.

Scherer, F. M. [1965] 1984a. "Invention and Innovation in the Watt-Boulton Steam Engine Venture," 8–31. In Scherer 1984b.

———, ed. 1984b. *Innovation and Growth: Schumpeterian Perspectives.* Cambridge: MIT Press.

Schlebecker, John T. 1975. *Whereby We Thrive: A History of American Farming, 1607–1972.* Ames: Iowa State University Press.

Schmookler, Jacob. 1966. *Invention and Economic Growth.* Cambridge, Mass.: Harvard University Press.

Schopf, T.J.M., ed. 1972. *Models in Paleobiology.* San Francisco: Freeman, Cooper and Co.

Schumpeter, Joseph. 1934. *The Theory of Economic Development.* Cambridge, Mass.: Harvard University Press.

———. 1939. *Business Cycles: A Theoretical, Historical and Statistical Analysis of the Capitalist Process.* 2 vols. London: McGraw-Hill University Press.

———. 1950. *Capitalism, Socialism and Democracy.* 3d ed. New York: Harper.

Scranton, Philip. 1991. "Diversity in Diversity: Flexible Production and American Industrialization, 1880–1930." *Business History Review* 65: 27–90.

———. 1994a. "Determinism and Indeterminacy in the History of Technology," 143–68. In Smith and Marx 1994.

———. 1994b. "Manufacturing Diversity: Production Systems, Markets, and an American Consumer Society, 1870–1930." *Technology and Culture* 35: 476–505.

Searle, Eleanor. 1974. *Lordship and Community.* Toronto: University of Toronto Press.

Shammas, Carole. 1984. "The Eighteenth-Century English Diet and Economic Change." *Explorations in Economic History* 21: 254–69.

———. 1990. *The Pre-industrial Consumer in England and America.* New York: Oxford University Press.

———. 1993. "Changes in English and Anglo-American Consumption from 1550 to 1800," 177–205. In Brewer and Porter 1993.

Shapin, Steven. 1989. "The Invisible Technician." *American Scientist* 77: 554–63.

Shionoya, Yuichi, and Mark Perlman, eds. 1994. *Innovation in Technology, Industries, and Institutions: Studies in Schumpeterian Perspectives*. Ann Arbor: University of Michigan Press.

Silverman, Sydel, ed. 1991. *Inquiry and Debate in the Human Sciences: Contributions from Current Anthropology 1960–90*. Chicago: University of Chicago Press.

Singer, Charles, E. J. Holmyard, A. R. Hall, and Trevor I. Williams, eds. 1954–58. *A History of Technology*. 5 vols. Oxford: Clarendon Press.

Singer, Maxine. 1993. "President's Commentary." In *Year Book 91: July 1991–June 1992*. Washington, D.C.: Carnegie Institution of Washington.

Sladovich, Hedy E., ed. 1991. *Engineering as a Social Enterprise*. Washington, D.C.: National Academy Press.

Slovic, Paul. 1987. "Perception of Risk." *Science* 236: 280–85.

Smith, Adam. [1776] 1976. *An Inquiry into the Nature and Causes of the Wealth of Nations*. Chicago: University of Chicago Press.

Smith, Cyril S. 1972. "Art, Technology, and Science: Notes on their Historical Interaction." *Technology and Culture* 11: 493–549.

Smith, John Kenley. 1993a. "Thinking about Technological Change: Linear and Evolutionary Models," 65–79. In Thomson 1993.

———. 1993b. "The Evolution of the Chemical Industry: A Technological Perspective," 137–57. In Mauskopf 1993.

Smith, Merritt Roe. 1985a. "Introduction," 1–37. In Smith 1985.

———, 1985b. "Army Ordnance and the 'American System' of Manufacture," 39–86. In Smith 1985c.

———, ed. 1985c. *Military Enterprise and Technological Change: Perspectives on the American Experience*. Cambridge: MIT Press.

Smith, Merritt Roe, and Leo Marx, eds. 1994. *Does Technology Drive History? The Dilemma of Technological Determinism*. Cambridge: MIT Press.

Sokoloff, Kenneth L. 1988. "Inventive Activity in Early Industrial America: Evidence from Patent Records, 1790–1846." *Journal of Economic History* 48: 813–48.

Sokoloff, Kenneth L., and B. Zorina Khan. 1990. "The Democratization of Invention during Early Industrialization: Evidence from the United States, 1790–1846." *Journal of Economic History* 50: 363–78.

Solow, Robert M. 1957. "Technical Change and the Aggregate Production Function." *Review of Economics and Statistics* 39: 312–20.

———. 1992. *An Almost Practical Step toward Sustainability: An Invited Lecture on the Occasion of the Fortieth Anniversary of Resources for the Future*. Washington, D.C.: Resources for the Future.

Solow, Robert M., and Peter Temin. 1978. "The Inputs for Growth," 1: 1–27. In Mathias and Postan 1978.

Stanley, Autumn. 1993. *Mothers and Daughters of Invention: Notes for a Revised History of Technology*. Metuchen, N.J.: Scarecrow Press.

Stapleton, Darwin H. 1987. *The Transfer of Early Industrial Technologies to America*. American Philosophical Society, Memoir 177.

Staudenmaier, John M. 1994. "Rationality versus Contingency in the History of Technololgy," 260–73. In Smith and Marx 1994.

Steward, Julian H. 1955. *Theory of Culture Change.* Urbana: University of Illinois Press.

Stewart, Larry. 1986. "The Selling of Newton: Science and Technology in Early Eighteenth-Century England." *Journal of British Studies* 25: 178–92.

Stone, Lawrence, and Jeanne C. F. Stone. 1986. *An Open Elite? England 1540–1880.* Abridged ed. Oxford: Oxford University Press.

Sullivan, Richard J. 1989. "England's 'Age of Invention': The Acceleration of Patents and Patentable Invention during the English Industrial Revolution." *Explorations in Economic History* 26: 424–52.

———. 1990. "The Revolution of Ideas: Widespread Patenting and Invention during the English Industrial Revolution." *Journal of Economic History* 50: 349–62.

Swedberg, Richard, ed. 1993. *Explorations in Economic Sociology.* New York: Russell Sage Foundation.

Sylla, Richard, and Gianni Toniolo. 1991. "Introduction," 1–26. In Sylla and Toniolo 1991.

———, eds. 1991. *Patterns of European Industrialization: The Nineteenth Century.* London: Routledge.

Teich, Miklas, and R. M. Young, eds. 1973. *Changing Perspectives in the History of Science: Essays in Honor of Joseph Needham.* London: Heineman.

Temin, Peter. 1964. *Iron and Steel in Nineteenth-Century America: An Economic Inquiry.* Cambridge: MIT Press.

———. 1966. "Steam and Waterpower in the Early Nineteenth Century." *Journal of Economic History* 26: 187–205.

Teuber, Andreas. 1990. "Justifying Risk." *Daedalus* 119, 4: 235–54.

Thackray, Arnold. 1974. "Natural Knowledge in Cultural Context: The Manchester Model." *American Historical Review* 79: 672–709.

Thaler, Richard H., ed. 1993. *Advances in Behavioral Finance.* New York: Russell Sage Foundation.

Thirtle, Colin G., and Vernon W. Ruttan. 1987. *The Role of Demand and Supply in the Generation and Diffusion of Technical Change.* Fundamentals of Pure and Applied Economics, 21. Chur, Switzerland: Harwood Academic Publishers.

Thomas, Brinley. 1993. *The Industrial Revolution and the Atlantic Economy.* London: Routledge.

Thompson, E. P. 1963. *The Making of the English Working Class.* New York: Vintage.

———. 1967. "Time, Work-discipline, and Industrial Capitalism." *Past and Present* 38: 56–97.

Thompson, F.M.L., ed. 1993. *The Cambridge Social History of Britain, 1750–1850.* 3 vols. Cambridge: Cambridge University Press.

Thomson, Ross, ed. 1993. *Learning about Technological Change.* New York: St. Martin's Press.

Travis, Anthony S. 1993. *The Rainbow Makers: The Origins of the Synthetic Dyestuffs Industry in western Europe.* Cranbury, N.J.: Lehigh University Press.

Tschinkel, Victoria J. 1989. "The Rise and Fall of Environmental Expertise," 159–66. In Ausubel and Sladovich 1989.

Tyson, Laura D'Andrae. 1992. *Who's Bashing Whom? Trade Conflict in High-technology Industries.* Washington, D.C.: Institute for International Economics.

U.S. Department of Commerce. 1991. *Report of the National Critical Technologies Panel.* Washington, D.C.: National Technical Information Service.

———. 1993. *National Critical Technologies Panel: Second Biennial Report.* Washington, D.C.: National Technical Information Service.

Uselding, Paul. 1977. "Studies of Technology in Economic History," 159–219. In Gallman 1977.

———. 1981. "Measuring Techniques and Manufacturing Practice," 103–26. In Mayr and Post 1981.

Vance, James E., Jr. 1990. *Capturing the Horizon: The Historical Geography of Transportation since the Sixteenth Century.* Baltimore: Johns Hopkins University Press.

van Helden, Albert. 1983. "The Birth of the Modern Scientific Instrument, 1550–1700," 49–84. In Burke 1983.

Veblen, Thorstein [1915] 1939. *Imperial Germany and the Industrial Revolution.* 2d ed. New York: Viking Press.

Vernant, Jean-Pierre. 1980. *Myth and Society in Ancient Greece.* Brighton, Sussex: Harvest Press.

Vicenti, Walter G. 1979. "The Air-propellor Tests of W. F. Durand and E. P. Lesley: A Case Study in Technological Methodology." *Technology and Culture* 20: 712–51.

———. 1990. *What Engineers Know and How They Know It: Analytical Studies from Aeronautical History.* Baltimore: Johns Hopkins University Press.

Vickers, Brian. 1984a. "Introduction," 1–55. In Vickers 1984b.

———, ed. 1984b. *Occult and Scientific Mentalities in the Renaissance.* Cambridge: Cambridge University Press.

von Hayek, F. A., ed. 1954. *Capitalism and the Historians.* Chicago: University of Chicago Press.

von Hippel, Eric. 1988. *The Sources of Innovation.* New York: Oxford University Press.

von Tunzelmann, G. N. 1978. *Steam Power and British Industrialization to 1860.* Oxford: Clarendon Press.

———. 1981. "Technical Progress in the Industrial Revolution," 143–63. In Floud and McCloskey 1981.

Waggoner, Paul E. 1994. *How Much Land Can Ten Billion People Spare for Nature?"* Council for Agricultural Science and Technology, Task Force Report 121. Ames, Iowa.

Wajcman, Judy. 1994. "Feminist Theories of Technology," 189–204. In Jasanoff et al. 1994.

Waldbaum, Jane. 1978. *From Bronze to Iron: The Transition from the Bronze Age to the Iron Age in the Eastern Mediterranean.* Atlantic Highlands, N.J.: Humanities Press.

Waldrop, M. Mitchell. 1992. *Complexity: The Emerging Science at the Edge of Order and Chaos.* New York: Simon and Schuster.

Walker, D. P. 1975. *Spiritual and Demonic Magic: From Ficino to Campanella.* South Bend: University of Notre Dame Press.

Wallace, Anthony F. C. 1982. *The Social Context of Innovation.* Princeton, N.J.: Princeton University Press.

Walton, John K. 1989. "Proto-industrialisation and the First Industrial Revolution: The Case of Lancashire" 41–68. In Hudson 1989b.

Ward, J. R. 1994. "The Industrial Revolution and British Imperialism." *Economic History Review* 47: 44–65.

Warner, Deborah Jean. 1990. "What Is a Scientific Instrument, When Did It Become One, and Why?" *British Journal of the History of Science* 23: 83–93.

Weart, Spencer R. 1979. "The Physics Business in America, 1919–1940: A Statistical Reconnaissance," 295–358. In Reingold 1979.

Weatherill, Lorna. 1988. *Consumer Behaviour and Material Culture in Britain, 1660–1760*. New York: Routledge.

Webster, Charles. 1975. *The Great Instauration: Science, Medicine and Reform 1626–1660*. New York: Holmes and Meier.

Wertime, Theodore A., and Steven F. Wertime, eds. 1982. *Early Pyrotechnology: The Evolution of the First Fire-using Industries*. Washington, D.C.: Smithsonian Institution Press.

Westfall, Richard. 1983. "Robert Hooke, Mechanical Technology, and Scientific Investigation," 85–110. In Burke 1983.

———. 1984. "Newton and Alchemy," 315–35. In Vickers 1984b.

White, K. D. 1984. *Greek and Roman Technology*. Ithaca, N.Y.: Cornell University Press.

White, Lynn, Jr. 1972. "The Expansion of Technology 500–1500," 143–74. In Cipolla 1972.

———. [1962] 1978. *Medieval Religion and Technology: Collected Essays*. Berkeley: University of California Press.

White, Robert M. 1991. "National Technology Policy in an Age of Global Technology." *Naval Engineers Journal* July, 51–54.

———. 1993. "What Is at the End of the Technological Rainbow?" *The Bridge* 23, 4: 3–8.

Whitney, Elspeth. 1990. *Paradise Restored: The Mechanical Arts from Antiquity through the Thirteenth Century*. American Philosophical Society, *Transactions* 80, 1.

Wikander, Örjan. 1984. *Exploitation of Water-power or Technological Stagnation? A Reappraisal of the Productive Forces in the Roman Empire*. Scripta Minora Regiae Societatis Humaniorum Litterarum Lundensis 1983–84. Lund: CWK Gleerup.

———. 1986. "Archaeological Evidence for the Early Water-mill: An Interim Report." *History of Technology. Tenth Annual Volume, 1985*, 151–79. London: Mansell.

Willey, Gordon R. 1988. *Portraits in American Archaeology*. Albuquerque: University of New Mexico Press.

Winter, John C. 1960. "Railroads, Trucks, and Ships," 297–307. In Department of Agriculture 1960.

Womack, James, Daniel Jones, and Daniel Roos. 1990. *The Machine that Changed the World*. New York: Macmillan.

Wright, Gavin. 1990. "The Origins of American Industrial Success, 1879–1940." *American Economic Review* 80: 651–58.

Wrigley, E. A. 1985. "Urban Growth and Agricultural Change: England and the Continent in the Early Modern Period." *Journal of Interdisciplinary History* 15: 683–728.

————. 1987. *People, Cities and Wealth: The Transformation of Traditional Society*. Oxford: Basil Blackwell.

————. 1988. *Continuity, Chance and Change: The Character of the Industrial Revolution in England*. Cambridge: Cambridge University Press.

Index

Throughout, page references to note numbers indicate the page of the *text* that corresponds to the note listed. The endnotes themselves are not indexed.

sive roles of, 70, 173, 199, 206, 259–65; military requirements of, spinoffs from, 69–70, 150, 174–76, 217–26, 261, 263–64; and trade restrictions, 70, 139, 143–44, 201, 250–51, 261; and unemployment and plant-closing concerns, 239–40, 266–67

steam engines, 43, 64–67, 95, 97–99, 109–10, 113, 115, 144–46, 178

Steinmetz, C. P., 198

Stephenson, G., 115

Stevens, J., 156

Steward, J. H., 27

sustainability: technology's role in, 269–71

Taylor, F. W., 188

technology: acceleration of, 12–13, 21, 40, 92, 253; advances in, generalized, 16, 18, 147, 250, 258–62; as artifact, 11; complexities in, and study of, 8–10, 17, 36–37, 41; definitions of, 11–12, 17, 213–14; diffusion, transfer, and adoption of, 27–29, 138, 142–45, 156; and economic incentives, 4, 30; environmental impacts of, 135, 267–76; Graeco-Roman, 42–46, 226, 260; "high technology," 213–16, 225, 232, 234–39, 241, 246; and human agents, 4; indeterminacy, unpredictability, and limited autonomy of, 4–5, 29–35, 40, 69, 71, 107, 187, 258; and information, 49, 213, 216, 236; and instrumentation and instrument makers, 33, 59–63; medieval, 47–53, 260; Mesopotamian, 38–42, 258; modernization, its role in, 1, 56; and path dependency, 4, 25, 28, 33–34; priorities and frontiers of, 231–38, 267; pyro-technology, 42, 258; as residual in economic growth, 31; and "reverse salients," 26; and risk and uncertainty, 159, 257, 265–66, 271–76; societal context of, 50, 69–71, 254–55; and socio-technical systems, 12, 23–27, 258, 265; supply "push" and demand "pull" of, 31–32, 80, 85–87, 90–91. *See also* engineers; technology-science

technology-science, 4; contrasts of, 7, 30, 32, 55–59, 88, 169–70, 192, 194–95, 213; convergences and interrelations of, 19, 32, 64, 66, 194–96, 199, 209, 213–14, 238, 255; and direction of stimuli, 100–102, 255–56; federal support for,

211, 214, 217, 227–31; and religion, role of, 44, 46, 51, 53, 55, 72, 94, 105, 130; and research and development (R&D), 17, 192–99, 208, 210, 212, 215, 217–18, 221–25, 239–45, 247–49, 252, 262–63. *See also* science; technology

telegraph, 22–23, 153, 170, 175

telephone, 34

telescope, 60

Temin, P., 146, 179n.18, 180

textile industry: cotton, Britain, 86–87, 92, 109, 110–12, 114, 129, 134; cotton, U.S., 142–44, 178; Mesopotamian, 38, 41; wool, Britain, 67–68

Thackray, A., 122

Third Dynasty of Ur, 41

Thompson, E. P., 105–6, 123–29, 124nn.54 and 57

Thompson, F. M. L., 124n.56

Thomson, E., 191

Toynbee, A., 106

Trevithick, R., 115, 146

Tull, J., 101

Tyson, L. D'A., 250

United States: Civil War in, 112, 173–76; Coast Survey, 168, 170, 171; Constitution, 138, 166, 170, 173, 175; Department of Agriculture, 174, 196, 199, 206, 224, 233, 235; Department of Commerce, 267; Department of Defense, 224–25, 232–33, 235; federal-state issues in, 165–66, 168, 171–74; and frontier, effects of, 138–39, 141, 161; during great depression, 179, 189, 209; Military Academy, 166–67, 193; natural setting and resources of, 139–41, 177–78; Revolution, American, 138–39; in World War I, 191, 206–9, 262; in World War II, 216–19, 262. *See also under* coal; demographic factors; industrial growth in U.S.; textile industries; universities

universities: British, 72, 137; U.S. (*see also* science: basic), 168–69, 196, 209, 223, 226–28

urban revolution, Mesopotamian, 38, 254, 259

Uselding, P., 149

Vance, J. E., 185

van Helden, A., 60

About the Author

ROBERT McC. ADAMS is a former Provost of the University of Chicago and Secretary Emeritus of the Smithsonian Institution. Now an Adjunct Professsor of Anthropology at the University of California, San Diego, he is currently a Fellow (1995–96) at the Institute for Advanced Study in Berlin. His many books and articles, mostly taking a very long-term perspective similar to this study, have previously concentrated on urban and agricultural development over the past six millennia in the Near East.

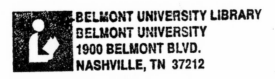